# Mining Technology and Metallurgy

# Mining Technology and Metallurgy

Edited by Beth Thorpe

SYRAWOOD
PUBLISHING HOUSE

New York

Published by Syrawood Publishing House,
750 Third Avenue, 9th Floor,
New York, NY 10017, USA
www.syrawoodpublishinghouse.com

**Mining Technology and Metallurgy**
Edited by Beth Thorpe

© 2017 Syrawood Publishing House

International Standard Book Number: 978-1-68286-460-9 (Hardback)

**Cataloging-in-publication Data**

Mining technology and metallurgy / edited by Beth Thorpe.
    p. cm.
Includes bibliographical references and index.
ISBN 978-1-68286-460-9
1. Mining engineering. 2. Mining geology. 3. Metallurgy. 4. Mineral industries. I. Thorpe, Beth.
TN153 .M56 2017
622--dc23

Printed in the United States of America.

# Contents

Preface..................................................................................................................................IX

Chapter 1  **A Finite Element Approach of Stability Analysis of Internal Dump
Slope in Wardha Valley Coal Field, India, Maharashtra**............................................1
Dhananjai Verma, Ashutosh Kainthola, S S Gupte, T N Singh

Chapter 2  **Optimisation of Dump Slope Geometry Vis-à-vis Flyash Utilisation using
Numerical Simulation**.....................................................................................................7
S.P. Pradhan, V. Vishal, T. N. Singh, V.K. Singh

Chapter 3  **Setting, Hardening and Mechanical Properties of Some Cement / Agrowaste
Composites**......................................................................................................................14
H. H. M. Darweesh, M. R. Abo El-Suoud

Chapter 4  **Seismic Effect Prognosis for Objects with Different Geometric Configuration
of Fundament in Close Blasting**....................................................................................23
Viktor Boiko

Chapter 5  **Pulp White Liquor Waste as a Cement Admixture**.......................................................27
H. H. M. Darweesh, M. G. El-Meligy

Chapter 6  **Economic non Metallic Mineral Resources in Quaternary Sediments of
Tehran and its Environmental Effects**.........................................................................33
Kaveh Khaksar

Chapter 7  **Geological Analysis of Zakiganj Upazila and Feasibility Study of Available
Geo Resources**................................................................................................................39
Mohammad Masudul Alam, Mir Raisul Islam, Md. Ashraful Islam Khan

Chapter 8  **Precambrian Stratigraphy of Central Iran and its Metallogenic**..............................44
Kaveh Khaksar, Keyvan Khaksar, Saeid Haghighi

Chapter 9  **Ventilation Air Methane of Coal Mines as the Sustainable Energy Source**.............49
Junjie Chen, Deguang Xu

Chapter 10  **Operational Dependence of Galvanized Steel Corrosion Rate on its Structural
Weight Loss and Immersion-Point pH in Sea Water Environment**.........................57
C. I. Nwoye, E. C. Chinwuko, I. E. Nwosu, W. C. Onyia, N. I. Amalu, P. C. Nwosu

Chapter 11  **Study of Corrosion and Corrosion Protection of Stainless Steel in Phosphate
Fertilizer Industry**.........................................................................................................64
Rajesh Kumar Singh, Rajeev Kumar

Chapter 12    **Groundwater Quality and Hydrogeochemistry of Toungo Area, Adamawa State, North Eastern Nigeria**................................................................69
J.M. Ishaku, B.A. Ankidawa, A.M. Abbo

Chapter 13    **Assessment of Fire Risk of Indian Coals using Artificial Neural Network Techniques**............................................................................................................ 80
Devidas S. Nimaje, Debi P. Tripathy

Chapter 14    **Magnetic Basement Depth Re-Evaluation of Naraguta and Environs North Central Nigeria, Using 3-D Euler Deconvolution**.............................91
Opara A.I., Emberga T.T., Oparaku O.I., Essien A.G., Onyewuchi R.A.,
Echetama H.N., Muze N.E., Onwe R.M

Chapter 15    **Cut-off Grade and Hauling Cost Varying with Benches in Open Pit Mining**....................... 105
Siwei He, Xianli Xiang, Gun Huang

Chapter 16    **Knowledge-Based Intellectual DSS of Steel Deoxidation in BOF Production Process**...........................................................................................109
Zheldak T.A., Slesarev V.V., Volovenko D.O.

Chapter 17    **Stream Sediment Geochemical Survey of Gouap-Nkollo Prospect, Southern Cameroon: Implications for Gold and LREE Exploration**.........................113
Soh Tamehe Landry, Ganno Sylvestre, Kouankap Nono Gus Djibril,
Ngnotue Timoleon, Kankeu Boniface, Nzenti Jean Paul

Chapter 18    **Economic Feasibility Study of Hard rock Extraction using Quarry Mining Method at Companiganj Upazila in Sylhet District, Bangladesh**.........................122
Mohammad Kashem Hossen Chowdhury, Md. Ashraful Islam Khan,
Mir Raisul Islam

Chapter 19    **Empirical Evaluation of Slag Cement Minimum Setting Time (SCMST) by Optimization of Gypsum Addition to Foundry Slag during Production**...........................126
C. I. Nwoye, I. Obuekwe, C. N. Mbah, S. E. Ede, C. C. Nwangwu, D. D. Abubakar

Chapter 20    **A Comparative Study of Various Empirical Methods to Estimate the Factor of Safety of Coal Pillars**........................................................................ 132
A. K. Verma

Chapter 21    **Assessment of the Geotechnical Properties of Lateritic Soils in Minna, North Central Nigeria for Road design and Construction**.............................138
Amadi A.N., Akande W. G., Okunlola I. A., Jimoh M.O., Francis Deborah G.

Chapter 22    **Simulative Analysis of Emitted Carbon during Gas Flaring Based on Quantified Magnitudes of Produced and Flared Gases**...................................... 144
C. I. Nwoye, I. E. Nwosu, N. I. Amalu, S. O. Nwakpa, M. A. Allen, W. C. Onyia

Chapter 23    **Pd-based Catalysts for Ethanol Oxidation in Alkaline Electrolyte**...........................150
A. M. Sheikh, E. L. Silva, L. Moares, L. M. Antonini,
Mohammed Y. Abellah, C. F. Malfatti

Chapter 24    **Reliability Level of Al-Mn Alloy Corrosion Rate Dependence on its As-Cast Manganese Content and Pre-Installed Weight in Sea Water Environment**...................... 156
C. I. Nwoye, P. C. Nwosu, E. C. Chinwuko, S. O. Nwakpa,
I. E. Nwosu, N. E. Idenyi

Chapter 25    **Statistical Assessment of Groundwater Quality in Ogbomosho,
Southwest Nigeria**..........................................................................................................................163
Olasehinde P. I., Amadi A. N., Dan-Hassan M. A., Jimoh M. O., Okunlola I. A

**Permissions**

**List of Contributors**

**Index**

# PREFACE

Metallurgy as a branch of materials science and engineering refers to the study of metallic elements, intermetallic compounds and alloys. The most important technique used to extract metals is mining. Therefore, metallurgy and mining are important to various industries. This book will talk about the various metals and their mining techniques and their importance in the industrial sector. The various advancements in mining and metallurgy are glanced at and their applications as well as ramifications are looked at in detail in this text. This book elucidates the concepts and innovative models around prospective developments with respect to these fields. While understanding the long terms perspective of the topics, the book makes an effort in highlighting their impact as a modern tool for the growth in mining and metallurgy. Students, researchers, experts and all associated with this subject will benefit alike from the book.

The information contained in this book is the result of intensive hard work done by researchers in this field. All due efforts have been made to make this book serve as a complete guiding source for students and researchers. The topics in this book have been comprehensively explained to help readers understand the growing trends in the field.

I would like to thank the entire group of writers who made sincere efforts in this book and my family who supported me in my efforts of working on this book. I take this opportunity to thank all those who have been a guiding force throughout my life.

**Editor**

# A Finite Element Approach of Stability Analysis of Internal Dump Slope in Wardha Valley Coal Field, India, Maharashtra

Dhananjai Verma, Ashutosh Kainthola[*], S S Gupte, T N Singh

Department of Earth Sciences, Indian Institute of Technology Bombay, Mumbai, India
*Corresponding author: ashuddn@live.com

**Abstract** Designing of a stable overburden disposal slope is vital in large opencast coal mines. Spoil generated during extraction of coal which is dumped externally requires larger land to remain stable and also poses problems to surrounding environment due to limited land availability. This has lead to the preference of internal dumping in which the waste is dumped in de-coaled region which is beneficial during extraction and reclamation of mine. Internal dumping is also the most economical and environment friendly method of waste disposal and is being adopted everywhere. It has certain limitations and inherent dangers of failures posing operational and safety threats. In this paper, a numerical study for stability of 80 m high internal dump slopes from an opencast coal mine of Wardha Valley Coal Field, Maharashtra, India has been carried out using Finite Element Method (FEM). Different scenarios as per the dump heights have been accounted and simulated using Plaxis2D-8 to understand the failure mechanism and the changes in factor of safety with variation in bench height and the number of benches.

*Keywords:* *internal dump, slope stability, numerical modeling, Coal Field*

## 1. Introduction

Surface mining activity in India is increasing at a rapid rate to bridge the gap between demand for and supply of coal to the energy sector. During the process of surface mining operations, huge amount of waste material is generated, hauled and then loosely dumped on the ground surface or used to fill unused open pits. Overburden/waste generated during extraction of coal are being dumped both internal as well as on external dump. Internal dumping is the most economical and environment friendly waste management being adopted widely. But it has certain limitations and inherent danger of dump slides posing operational and safety threats [1,2]. The presence of water reduces the frictional strength of the slope material, and the geo-mechanical properties are reduced further due to the presence of pore water pressure. The migration of water may augment the seepage, leading to the formation of tension cracks parallel to the internal dump slopes. Tension cracks are also generated due to shocks and vibration caused by poor blasting in slopes [3,4,5].

With the increase in size and the stripping ratio of the opencast mines, the amount of overburden generate has also increase considerably. Coal India Limited (CIL) is the chief coal producing organization in India. CIL has removed 21, 160, 462 and 695 million cubic metre of overburden during 1976, 1986-87, 1999-2000 and 2009-2010 respectively. The paucity in the land area availability for dumping the waste rock has compelled the mine managers to think ways for the safe, stable and economic disposal of the dump material. Overburden dumps can be external dumps created at a site away from the coal bearing area or it can be internal- dumps created by in-pit dumping (IPD) concurrent to the creation of voids by extraction of coal. Practice of dumping overburden in the external dumps have some serious problems [6] foremost amongst them are requirement of additional land, involves very high transport and rehandling cost which will increase the cost of coal production substantially, stability and reclamation at the site. It is not possible to eliminate the external dumps concept completely, even if we adopt internal dumping practice.

Waste material dump stability is essential to ensure the safety of haul trucks during placement and long term safety. Many research articles had been published since the publication of the first method of analysis by [7] that were either related to slope stability or involved slope stability analysis subjects. Among the available analysis methods, ordinary method of slices [7], Bishop's modified method [8], Janbu's generalized procedure of slices [9] and Spencer's method [10].

The demerit with all the equilibrium methods is that they are based on the assumptions that the failing soil mass can be divided into slices, which necessiates further assumptions relating to side force directions between slices, with consequent implications for equilibrium.

Because of the certain advantages of finite element method like no assumptions needs to be made in advance about the shape or location of failure surface, it has been used widely used for slope stability analysis over traditional equilibrium methods. [11,12,13,14] have used the finite element method for slope stability analysis for further confidence in the method for dum material as well as rock slopes.

In this paper, a study of numerical analysis for stability of internal dump slope (Figure 1) has been carried out using FEM with Strength Reduction approach. A case from an opencast coal mine has been considered in which in-pit dumping is taking place, currently at 80 m height formed of two benches of 60m and 20m height. The dump has been executed at an angle of $40^0$, composed of weak sandstone and shale rocks. As the amount of dump material is increasing day by day, efficient and stable disposal of the dump material will be a key factor for sustainable production. Hence, in the present study the effect of increase in height and inclusion of benches in the dump slope have been studied in a finite element code to a reach conclusive methodology for proper dumping.

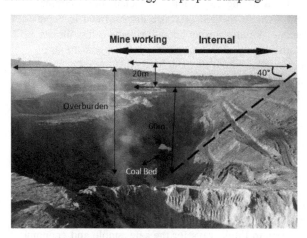

**Figure 1.** An opencast mine with internal dumping in progress

Finite element method has been increasingly used in slope stability analysis. The method can be applied with complex slope configurations and dump deposits in two or three dimensions to model virtually all types of mechanisms. General debris material models that include Mohr-Coulomb and numerous other tools can be employed. The equilibrium stresses, strains, and the associated shear strength in the dump mass can be computed very precisely and accurately. The critical failure mechanism developed can be extremely general and need not be simple circular or logarithmic spiral arcs.

## 2. Stability Features of Internal Dump

In the case of open pit mines, the underground minerals are accessed by removing the overburden material which is placed in the de-coaled area. Many of these waste dumps possess environmental or extraction problems, slope stability concern and undesirable aesthetic attributes. There is an increased need for basic information, understanding of construction, characterization and its stability status as unplanned dumping can be a threat to life and property [15, 16, 17, 18]. The Internal dumps are affected by the particle size of the waste material,

geometry, unit weight, shear strength, pore pressure, and the foundation of the dump material [1,19,20]. Therefore it is necessary to study important feature in respect of the construction of dump, factor influencing shear strength of dump, characterization of dump as well as stability of dump.

## 3. Stability Assessment Using FEM

There are three major aspects involve in slope stability analysis. The first is about the material properties of the slope forming material. The second is the calculation of factor of safety and third is the definition of the slope failure [21].

### 3.1. Model Material Properties

The Mohr-Coulomb constitutive model has been used to describe the dump material properties. The criterion of Mohr-Coulomb model relates the shear strength of the material to cohesion, normal stress and angle of internal friction.

### 3.2. Factor of Safety (*FOS*) and Strength Reduction Factor (*SRF*)

Slope fails because its material shear strength on the sliding surface is insufficient to resist the actual shear stresses. Factor of safety is a value that is used to examine the stability of slopes. If *FOS* is greater than 1, it means the slope is stable, while values lower than 1 indicates unstable slope.

### 3.3. Slope Collapse

Non-convergence within a specified number of iterations in finite element program can be taken as a suitable indicator for slope failure, which means that no stress distribution can be achieved to satisfy both the Mohr-Coulomb criterion and global equilibrium. Slope failure and numerical non-convergence take place at the same time and are joined by an increase in the displacements. Usually, value of the maximum nodal displacement just after slope failure has a sharp rise as compared to the one before failure.

For the present work detailed systematic sampling has been carried out. The dump material was tested in the laboratory for the assessment of their strength properties as per standards [22,23,24]. The samples were tested in dry as well as saturated condition when pores were fully charged. The dump material mainly consists of sandstone, shale and carbonaceous shale. The geo-mechanical properties of material are listed in Table 1. There are six type of model prepared with different height and angle using the geo-mechanical properties listed below in Table 1.

**Table 1. Input Parameters used in simulation**

| Parameters | Symbol | Values |
|---|---|---|
| Saturation weight (kN/m³) | $\gamma_{saturated}$ | 1.9 |
| Unsaturation weight (kN/m³) | $\gamma_{unsaturated}$ | 1.7 |
| Young's Modulus (GPa) | E | 57 |
| Poisson's Ratio | $\nu$ | 0.28 |
| Angle of Internal Friction($^0$) | $\phi$ | 22.4 |
| Cohesion (KPa) | C | 91 |
| Dilatancy Angle(Degrees) | $\psi$ | 0 |

## 4. Results and Discussion

**Case 1**: In this case an internal dump slope with the height of 60m and slope angle of 40 degrees is considered. Results obtained from numerical analysis of this slope using Table 1 parameter are shown in Figure 2.

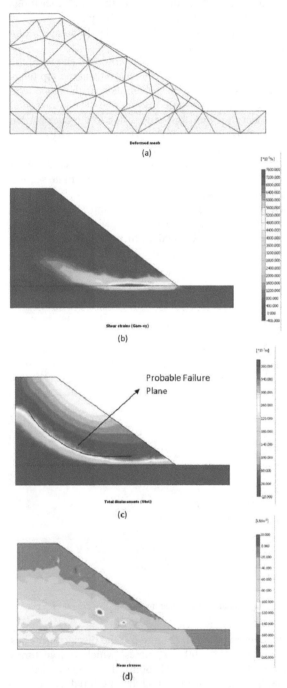

**Figure 2.** (a) Deformed mesh after solving, (b) Shear strain in x-y plane, (c) Zones of distribution of Total displacements and (d) Zones of Mean stresses

It is seen that mesh has been deformed (Figure 2a) which represents consolidation and subsequent subsidence by 7m in the dump material due to the effect of gravity dump slope is stable with factor of safety 1.784. Shear strains are higher in bottom and toe region of the dump slope (Figure 2b) due to weight of OB and corresponding normal force from the platform below. The possible

failure plane can be seen from Figure 2c, the total displacement of the dump region is shown along with their displacement scaling. The mean stresses (Figure 2d) the amount of stresses at the bottom of the dump is more due to the overburden weight. A possible toe failure is predicted with the FEM analysis of the dump slope in the current state.

**Case 2**: In this case, materials are dumped as a bench of height 20m and slope angle is 40 degrees above existing internal dump slope of 60m depth. Results obtained from numerical analysis of this model using Table 1 parameter are shown in Figure 3.

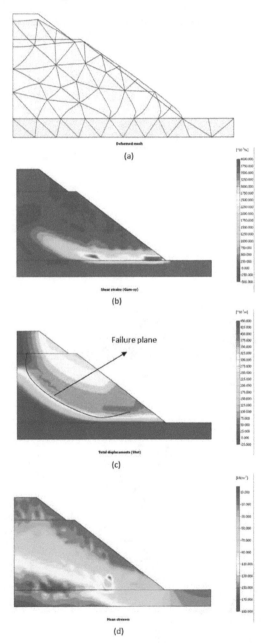

**Figure 3.** (a) Model of deformed mesh, (b) Model of Shear strain in x-y plane, (c) Zones of distribution of Total displacements and (d) Zones of Mean stresses

Mesh has been deformed more (Figure 3a) with compared to (Figure 2a) which represents consolidation of material due to addition of one bench, yet slope is stable with factor of safety 1.378. The displacement of the materials has been scaled up by $20 \times 10^{-6}$ times. Shear strains are more densely packed at near bottom of the dump and the toe region; it has also extended towards the center due to a additional weight of the 20m bench (Figure 3b). The probable failure plane can be seen from (Figure 3c), the total displacement of the dump region is shown along with their displacement scaling. Displacement zone is more at the top of the bench and has more probability of failure in this region. The mean stress distribution is shown in the (Figure 3d). The stress concentration is more near the bottom of the dump, the value of which is -180 $KN/ft^2$, the negative value indicates that stress is compressive in nature.

**Case 3**: In this case one more bench of height 20m is added keeping same slope angle. Results obtained from numerical analysis of this model using Table 1 parameter are shown in Figure 4.

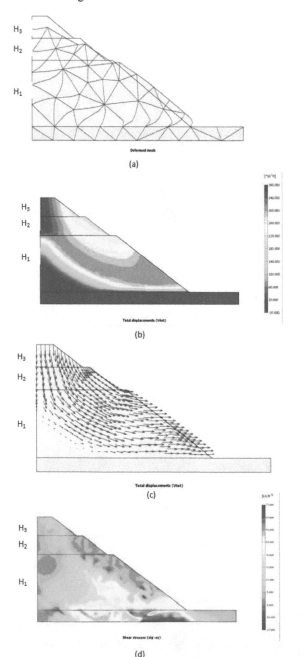

**Figure 4.** (a) Model of deformed mesh, (b) Zones of distribution of Total displacements along the slope, (c) Components of Total displacements along the slope and (d) Zones of Mean stresses along the slope

Deformed mesh is shown in the Figure 4a, it can be seen that material is falling down from the top of the dump but still the whole model has become critical with a factor of safety 1.161. The displacement of the materials has been scaled up by $50\times10^{-6}$ times, which is twice when compared to previous case (Figure 3a).

Total displacement of the dump in the form of shadings are shown in Figure 4b, the maximum displacement of the materials will occur at the left top of the dump and along the slopes it comparatively lesser.

The total displacement of the materials are shown in the form of arrows in Figure 4c, the materials at the top tends to move down, where as the materials near the slope tend to move towards right side due to free face. Horizontal displacement near the slope side is due to the over burden and vertical displacement is due to force of gravity.

The shear stress shown in Figure 4d points out that the stress is more near the bottom of the dump, it is comparatively lesser near the top and at the middle section of the lower most benches.

**Case 4**: In this case an internal dump slope with the depth of working 80m and slope angle of 40 degrees is considered. Results obtained from numerical analysis of this slope using Table 1 parameter are shown in Figure 5.

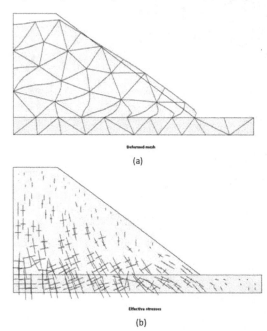

**Figure 5.** (a) Model of deformed mesh and (b) Zones of Effective stresses

In the deformed mesh (Figure 5a) the displacement has been scaled up by $100\times10^{-6}$ times, which is five times more than that of case 1 (H1=60m), this suggests that the material has displaced more and also FOS reduced due to the increase in height.

The effective stresses are shown in Figure 5b, the development of stress is higher and it is confined at the toe of bench due to the overburden weight from the top and also increases in height of the dump.

**Case 5**: In this case materials are dumped as a bench of height 20m and slope angle is 40 degrees above existing internal dump slope of 80m depth. Results obtained from numerical analysis of this model using Table 1 parameter are shown in Figure 6.

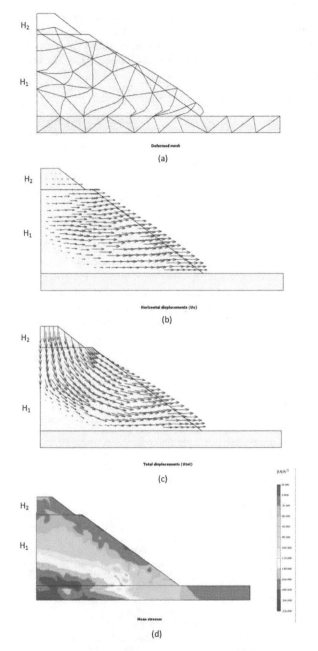

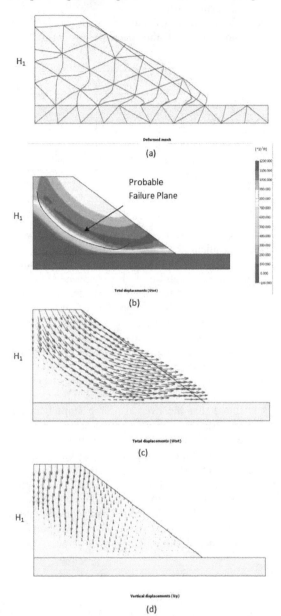

increase in dump height (H1) and also addition of one more bench (H2).

**Case 6**: In this case an internal dump slope with the depth of working 100m and slope angle of 40 degrees is considered. Results obtained from numerical analysis of this slope using Table 1 parameter are shown in Figure 7.

**Figure 6.** (a) Model of deformed mesh when height, (b) Components of horizontal displacement, (c) Components of Total displacements and (d) Zones of Mean stresses along the slope

The deformed mesh is shown in the Figure 6a. In this case, mesh is getting loosened and node points are expanding as compared to the previous case, the displacement of the materials has scaled up $50 \times 10^{-6}$ times. The materials fall along the slope reducing the height, slope angle also the stability of the dump.

The horizontal displacement of the materials (Figure 6b) at H2 bench is very less as we can see very few resultant vectors but the horizontal displacement near the slope side of H1 bench is large due to free face of the slope.

Total displacement (Figure 6c) of the material is large at the left top of the dump and hence the material falls towards the slope side making dump slope unstable. Since the FOS is 1.130 the model is in a critical state and may collapse with high displacement in the material.

Mean stresses of the dump (Figure 6d) shows that the stresses are more near the bottom of the bench due to the

**Figure 7.** (a) Model of deformed mesh, (b) Zones of distribution of Total displacements, (c) Components of distribution of Total displacements and (d) Components of vertical displacements

The deformed mesh shown in the Figure 7a indicates the displacement of the material, which is scaled up by $20 \times 10^{-6}$ times, the mesh is widened hence the material is more loosely packed, large materials overflows along the slope resulting in the change of dump shape. FoS, in this case, is 1.098 which shows that slope is very much likely to fail. The triangles near the bottom of the dump has distorted in their shape which shows high probability of failure in that region.

The total displacement of the dump in the Figure 7b shows the probable failure plane. More displacement is occurring near slope edge, which has its effects in the center also, this suggests that even a small load from the top may lead to collapsing of the dump, which can also be further confirmed by its FoS, which is 1.098; small disturbance in the present model may collapse the structure.

Total displacement can also be seen in Figure 7c, which is shown in the form of arrows.

Vertical displacement of the materials (Figure 7d) shows the movement of the material in the vertical direction, materials at the top of the bench moves in the vertical direction due to force of gravity and increase in dump bench height (H1). Materials along the slope side displace very little in the vertical direction due to free face availability.

Table 2 shows the factor of safety obtained and corresponding analysis of stability for all dump slope models.

**Table 2. Factor of safety calculated for each dump slopes**

| Case | Height-H (m) | | | Slope angle ($^0$) | Factor of safety (FOS) | Analysis result |
|---|---|---|---|---|---|---|
| | H1 | H2 | H3 | | | |
| 1 | 60 | 0 | 0 | 40 | 1.784 | Stable |
| 2 | 60 | 20 | 0 | 40 | 1.378 | Stable |
| 3 | 60 | 20 | 20 | 40 | 1.161 | Critical |
| 4 | 80 | 0 | 0 | 40 | 1.335 | Stable |
| 5 | 80 | 20 | 0 | 40 | 1.130 | Critical |
| 6 | 100 | 0 | 0 | 40 | 1.098 | Most Critical |

# 5. Conclusion

In this paper, number of numerical models were generated to simulate the Internal dump in an opencast coal mine of Wardha Valley Coal Field, Maharashtra, India, to analyse their stability with varying height of the bench slopes. It is found that the FOS reduces drastically with increase in height of the dump slope. FOS were also depend on the nature of material and its geomechanical properties,but here the dump material is manily consisting of sandstone, shale and carbonaceous shale. The height of working is 60m the slope was found stable. As the mine goes deeper the waste materials generated also increases, which can be dumped over the existing Internal dump slope. It is found that the over all dump slope becomes unstable after addition of two increased overburden benches on existing Internal dump slope. In case of 80m high Internal dump slope FoS decreased to less than one after addition of one bench. When the depth of working increases to 100m, the stability of Internal dump slope becomes critical and addition of overburden results in dump slope failure. It can be concluded from the study that an internal dump is more stable when it is composed of a number of benches rather than a single slope.

# References

[1]    Gomez, P., Diaz, M., and Lorig, L., 2003, Stability analysis of waste dumps at Chuquicamata Mine, Chile: Gluckauf-Forschungshefte, v. 64, no. 3, 93-99.

[2]    Quine, R.L., 1993, Stability and deformation of mine waste dumps in north-central Nevada:M.S.thesis, University of Nevada, Reno

[3]    Singh, T.N. and Singh, D.P. (1992) Assessing Stability of voids in a Multi-seam Opencast Mining Block, Int. J. of Colliery Guardian, 240 (4),159-164.

[4]    Singh, T.N. Goyal, M. and Singh, D.P. (1994) Blast Casting Technique for Over Burden Removal, Ind. Min & Eng. J. 35(9), 21-26.

[5]    Khandelwal, M. and Singh, T.N. (2009) Prediction of blast-induced ground vibration using artificial neural network, Int. J. of Rock Mech. & Min Sci.,46(7), 1214-1222.

[6]    Upadhyay, O.P., Sharma, D.K., & Singh, D.P., 1990, Factors affecting stability of waste dumps in mines: International Journal of Surface Mining and Reclamation (4): 95-99.

[7]    Fellienius, W.(1927). Erdstatische Nerechnungen mit Reibung and Kohansion. Ernst, Berlin, 40pp.

[8]    Bishop, A.W., (1955). The use of the slip circle in the stability analysis of slopes. Geotechnique 5(1): 7-17.

[9]    Jambu, N.(1968).Slope stability computations. Soil mechanics and foundation engineering report, Technical University of Norway, Trondheim.

[10]   Spencer, E (1967). A methos of analysis of the stability of embankments assuming parallel interslices forces. Geotechnique 17(1): 11-26.

[11]   Matsui, T., and San, K. C., 1992, "Finite Element Slope Stability Analysis by Shear Strength Reduction Technique," Soils and Foundations, 32 (1); 59-70.

[12]   Duncan, J. M., 1996, "State of the Art: Limit Equilibrium and Finite Element Analysis of Slopes," Journal of Geotechnical Engineering, ASCE, 122 (7); 577-596.

[13]   Griffiths, D. V., and Lane, P. A., 1999, "Slope Stability Analysis by Finite Elements," Geotechnique, 49(3) ;387-403.

[14]   Kainthola, A., Verma, D., Gupte, S.S. and Singh, T.N., (2011a). A Coal mine dump stability analysis – a case study, Int. Journal of Geomaterial. 1, 1-13.

[15]   Kainthola, A., Verma, D., Gupte, S.S. and Singh, T.N., (2011b). Analysis of failed dump slope using limit equilibrium approach, Mining Engineers Journal, .12(12) 28-32.

[16]   Kainthola, A., Verma, D., T. N. Singh, (2011c), Computational Analysis for the Stability of Black Cotton Soil Bench in an Open Cast Coal Mine in Wardha Valley Coal Field, Maharashtra, India, Int. J. Econ. Env.Geol. 2(1):11-18.

[17]   Monjezi M. and Singh T.N., 2000. Slope Instability in an Opencast Mine, Coal International, 45-147.

[18]   Kripamoy Sarkar, Sazid M., Khandelwal M., and Singh T. N., 2009. Stability analysis of soil slope in Luhri area, Himachal Pradesh, Mining Engineers Journal, 10(6), 21-27.

[19]   Singh T.N. and Monjezi M., 2000. Slope Stability Study in Jointed Rockmass - A Numerical Approach, Mining Engineering Jl., 1(10), 12-13.

[20]   Goodman R.E. and John C. St., 1977. Finite element analysis for discontinuous rock. In: C.S. Desai and J.T. Christian, Editors, Numerical Methods in Geotechnical Engineering, Mc Graw Hill, New York. 148-175.

[21]   Rocscience Inc., 2004. Application of the Finite Element Method to Slope Stability, Toronto, 2-6.

[22]   ISRM, (1972). Suggested methods for determining water content, porosity, density, absorption and related properties and swelling and slake durability index properties. Int. J. Rock Mech. Min. Sci. Geomech.Abst.1-12.

[23]   ISRM, (1977). Suggested methods for determining the strength of rock materials in triaxial compression.Int. J. Rock Mech. Min. Sci. Geomech..15, 47-51.

[24]   ISRM, (1981). Rock characterization testing and monitoring, ISRM suggested method. Int. Soc. Rock Mech. 211.

# Optimisation of Dump Slope Geometry Vis-à-vis Flyash Utilisation Using Numerical Simulation

**S.P. Pradhan[1,*], V. Vishal[2], T. N. Singh[1], V.K. Singh[2]**

[1]Department of Earth Sciences, Indian Institute of Technology Bombay, Mumbai, India
[2]Mine Fire Division, CSIR-Central Institute of Mine and Fuel Research, Dhanbad, India
*Corresponding author: saradaiitb@gmail.com

**Abstract**   Stability of waste dump is now gaining importance due to increasing depth and size of mine. Management of dump nearby mining areas is one of the most critical and crucial task for mine management due to limited land and other governing laws related to environment and forest conservation. In this paper, a study was conducted to establish the effect of slope angle on the stability of waste dump for accommodation of flyash is carried out. Based on numerical simulation, it was found that the dump slope of 60 m height with 36° slope can be critically stable with 20% flyash randomly mixed with overburden materials whereas flatter slopes provide higher factor of safety. Keeping other parameters constant, the optimum slope of 32° is the best possible to accommodate the mine dump for its long term stability. These findings were further supported by study of maximum velocity vectors and shear strain rates in every case and the extent of damage zone due to tensile pull. It is hoped that this technical note will find utility wherever a design of dump of chosen material type is being planned where the wastes can be managed alongside ulitisation of flyash.

*Keywords: dump, dump slope, factor of safety (FoS), finite difference method (FDM), FLAC/Slope*

## 1. Introduction

Mining industries are one of the oldest industries on this earth, next to agriculture and are the basic examples of harnessing the mother Earth for the benefit, development and growth of mankind. Demand for raw materials to meet the growing needs has led to exploitation of earth resources since centuries of years. With decreasing near surface deposits, the demands are being met by gradual deepening of the crust to obtain the ores. This in turn leads to a large scale removal of overburden materials. Large dump accumulations are therefore formed in and around the mining areas. Studies on management of mine waste and dumps have increased in the last decade since it has been established that the waste do play a significant role in mine functioning and mining economics [25]. Several approaches have been utilised and case studies discussed for development of better dumps and landfills [12,13,26].

In India, a large number of surface mines are functional where economic deposits are extracted by several operations viz. drilling, blasting, and excavation, loading and dumping called as mining cycle. All these parameters play significant role in the economics of excavation [15]. The overburden has to be removed and transferred to an internal or external spoil heap. The dumping of overburden is itself a tedious task and formation of spoil heap and its stability is crucial for the progress of the mine. The dumping has to be done judiciously to ensure optimum use of ground and to reduce chances of any slide back or to avoid dangerous accidents in future. There are several parameters that affect the stability of dump slope like dump geometry, strength of the underlying rocks, mining methods, hydrological conditions, etc [16,22,23]. Once the location for dump accumulation is determined, it is mainly the design parameters that can be modified as per the existing foundation conditions. Dumps with low height and flatter slopes are ideal from the stability point of view but these not only occupy lot of ground space but also prove to be expensive due to transport and other handling costs involved. Hence, it is necessary to obtain a benchmark for an allowable dump height, slope angle and number of benches for a long term stable dump accumulation of the material type and should also be most economical. This not only provides with safe and stable conditions of mining but also minimizes the ill effects that a slope failure may have on the surrounding ecosystem [17,19,20,21,24].

Coal is a major source of energy and is believed to remain so for decades to come particularly in India. This is because nearly 64% of power generation in India is mainly based on coal production. Deeper and deeper coal deposits are being looked for and exploited to meet the needs of India's growing economy, industrialisation and population. The large amounts of fly ash generated from the capture power plants are a major environmental concern now a day [16]. Although methods of reutilizing these are being implemented, yet, large quantities need to be kept in the waste systems to achieve the goal of zero waste discharge. In this work, the stability of flyash wastes has been studied to understand and minimise the chances of devastating slope failures which is quite possible due to low cohesion of such materials. A fixed percentage of flyash is mixed with coal mine dump materials for long term safe containment of both of them.

Waste dump management is of prime concern today mainly as a consequence of many cases when slope failures have hampered the regular process of mine production which directly affects the mine economics. The consequence of slope failure can be very devastating when men and heavy earth moving machines work or come closer to the unstable zone [1,7,20]. Moreover, such failures cause severe damage to the niche of the surrounding ecosystem. Hence, prior testing of waste dump and flyash generated must be carried out to obtain their strength characteristics and accordingly a safe, stable and cost effective solution in the form of optimised dump slope design must be proposed and implemented with regular monitoring. The studies for the dump slopes with varying dump height and its effect on change in safety considerations have been carried out as a part of the detail dump slope analysis by Vishal et. al (2010) [26]. Apart from other parameters such as foundation strength, ground conditions, water conditions and dynamic forces, dump slope geometry needs to be critically analysed for elimination of chances of failures.

There are various methods of assessment of dump slope stability like limit equilibrium methods, stress analysis, kinematic analysis, physical modelling, numerical modelling etc. In this paper, we present the numerical modelling of combined slope (waste + flyash) to obtain the most appropriated and optimum solution with respect to dump geometry which can accommodate more and more dump material and yet remain stable. A two dimensional Finite Difference model was used to simulate various situations to see the behaviour of dump with increasing the dump slope for a most economical and stable solution of the problem in question.

# 2. Numerical Simulation of Dump

Numerical modelling for the purpose of dump slope stability has several advantages over the other methods. They not only have a shorter running time but also get quite detailed solutions with simple assumptions. They enable more number of trials in the design and other parameters. One major advantage of numerical simulation is that it provides information well in advance to take care of dump at a particular point during actual working. Keeping in view, a two dimensional dump slope analysis was performed in different geo-mining conditions to understand not only failure but also what best can be done to prevent it and protect the accumulation for long term stability. Variations in constituent size and types of material are not distributed uniformly. As a result of which the assessment of the internal strength of the waste dumps is difficult and often challenging due to inadequate engineering design for such type of dumps [28]. Hence, for the purpose of simulation, a homogenous material

property equivalent to the average of those composing the dump accumulation has been calculated from laboratory tests and assigned to the material type used [8,9]. The rock properties play a significant and an important role in any kind of research involving geomaterials [14,18,22,27].

## 2.1. Slope Design

To calculate the individual bench stability for a dump slope can be expressed as Girard and McHugh (2000):

$$\tan A = 1 / \left[ (W/H) + (1 / \tan B) \right] \qquad (1)$$

where;
A = overall (average) slope angle;
B = bench face angle;
H = vertical height of bench and
W = horizontal width of bench.

## 2.2. Factor of Safety (FOS) and Strength Reduction Factor (SRF)

There are several methods to analyze slope which fails because its material shear strength on the sliding surface is insufficient to resist the actual shear stresses. Factor of safety is a value that is used to examine the stability state of slopes. In other words, it is the ratio of collapse load to working load. A better definition for FOS will be the ratio of maximum available shear strength to the shear strength needed for equilibrium. For FOS values greater than unity means the slope is stable, whereas, values lower than unity means unstable conditions. In accordance to the shear failure, the factor of safety against slope failure is simply calculated as:

$$FOS = \frac{\tau}{\tau_f} \qquad (2)$$

Where $\tau$ is the shear strength of the slope material, which is calculated through Mohr-Coulomb criterion as:

$$\tau = C + \sigma_n \tan\Phi \qquad (3)$$

C stands for material cohesion where as $\Phi$ is the angle of internal friction.

Where $\tau_f$ is the shear stress along the sliding surface. It can be calculated as:

$$\tau_f = C_{f} + \sigma_n \tan\Phi_f \qquad (4)$$

where the factored shear strength parameters $C_f$ and $\Phi_f$ are:

$$C_f = \frac{C}{SRF} \qquad (5)$$

$$\phi_F = \tan^{-1}[\frac{\tan\phi}{SRF}] \qquad (6)$$

where SRF is strength reduction factor. This method has been referred to as the 'shear strength reduction technique'. To achieve the correct SRF, it is essential to trace the value of FOS that will just cause the slope to fail.

This technique is employed where the state of effective stresses in slope is calculated [2,4]. By the use of this method, factor of safety calculations are done by progressively reducing the shear strength of the material to bring the slope to a state of limiting equilibrium.

A series of simulations are made using trial values of factor $F^{trial}$ to reduce the cohesion, C, and friction angle, Φ, until slope failure occurs. In the used numerical tool, namely FLAC/Slope, a bracketing approach similar to that proposed by Dawson et al (1999) is used [4].

## 2.3. FLAC/Slope

FLAC/Slope is a mini-version of FLAC (Fast Langrangian Analysis of Continua) that is designed specifically to perform factor-of-safety calculation for slope-stability analysis. This version is operated entirely from FLAC's graphical interface (the GIIC) which provides for rapid creation of models for soil and/or rock slopes and solution of their stability condition [10]. It provides an alternative to traditional "limit equilibrium" programs to determine factor of safety. Limit equilibrium codes use an approximate scheme — typically based on the method of slices — in which a number of assumptions are made [11]. Several assumed failure surfaces are tested, and the one giving the lowest factor of safety is chosen. Equilibrium is only satisfied on an idealized set of surfaces.

The procedure in FLAC/Slope is as follows:

First, the code finds a "characteristic response time", which is a representative number of steps (denoted by Nr) that characterizes the response time of the system. Nr is found by setting the cohesion and tensile strength to large values, making a large change to the internal stresses, and finding how many steps are necessary for the system to return to equilibrium then, for a given factor of safety, F, Nr steps are executed.

If the unbalanced force ratio is less than $10^{-3}$, then the system is in equilibrium. The factor-of-safety solution stops when the difference between the upper and lower bracket values becomes smaller than 0.005. The strength reduction technique gives FOS with respect to geo-material shear strength [5,6]. It gives the advantage of automatic critical failure mechanism for dump.

## 3. Modelling of Dump Using FLAC/Slope

In the present study, the explicit two-dimensional finite difference program FLAC/SLOPE version 4.00 has been used for the analysis by simulating the similar geometrical condition [3]. Assuming FDM techniques to be the most rigorous method, it has been chosen here to investigate the stability of dump slope under gravity loading. The geometrical and geotechnical parameters used were determined in the field as well as laboratory as per standards [8,9].

Figure 1 shows the general geometry of a model of dump slope having a height of 60m and overall slope angle equal to 28°. The dump slopes were evaluated at this constant dump height, base width and bench width, while the slope angle was varied as 30°, 32°, 34°, 35°, 36°, 37°, 38°, 39°, 40°. The shear strain rates and the velocity vector distribution of the dump material were evaluated at each inclination. The factor of safety is found to decrease with the increase in dump slope inclination. Although this fact is known like a thumb rule but it is important to quantify the rate at which FoS decreases and particularly to help in deciding the critical slope angle along with the FoS. In this article, the results of dump having dump slope

angle of 28°, 30°, 32°, 34° and 36° are discussed. Cases with yet higher angles of slopes have not been shown as under normal conditions the dump with same material types and dump height, increase in dump slope only leads to further reduction in FOS close to 1 or even less. This implies an unstable condition and the dump slopes would fail unless the material cohesion is increased using various methods such as biostablisation which itself needs yet another optimisation study as carried out by Pradhan et al (2013) [16]. The samples collected from the field site are investigated for the following geomechanical properties which are used as input parameters for simulating a real environment model using. These include porosity, permeability, density, elastic modulus, tensile strength, cohesion.

**Figure 1.** General Geometry of dump slope with Height, H = 60m

Dump slopes should be designed in accordance with the dump material characteristics, the availability of moisture as well as other operational features which may hamper its stability without giving much time for rectification. With progressive mining and exploitation of near surface ore deposits, as the mining advances deeper, more and more waste materials are taken out. These conditions of increasing stripping ratio results in generation of large quantities of dump which need to be managed with a scientific approach.

In first case, as shown in Figure 2, a 60 m high dump with 28° slope angle was simulated using the geomechanical properties of flyash and dump material to understand the failure behaviour of slope. The Factor of Safety obtained is very high i.e. 4.47 with a maximum shear strain rate equal to 8.00E-06 $s^{-1}$ and maximum velocity vector equal to 1.247E-05 m/s. It can be seen that the shear strain rate and velocity values are quite low indicating imperceptibly low displacement of material particles. The distribution of velocity vectors is along most parts of the slope showing that the slopes although stable are not a static entity and hence monitoring at every stage is important. This is mainly because the materials are weak with low cohesion values unlike the rock cut or open pit slopes with intact rock masses. Yet the slope angle and height are optimised in this case and most coherent in the standards for maintaining the dump slopes which leads to a stable accumulation of dump materials as indicated by the FoS values. The concentration of higher shear strain rate is along the toe of the dump slope. Few tension cracks are developed at the top of the dump. Due to gravity loading, there are plastic deformations in the rear end of the dump indicates settlement of dump material.

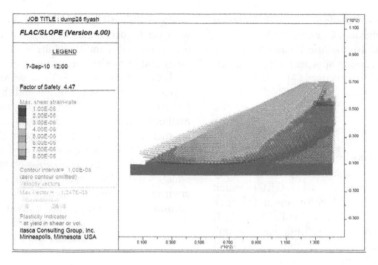

**Figure 2.** Simulated model for dump slope with H = 60 m and slope angle = 28°

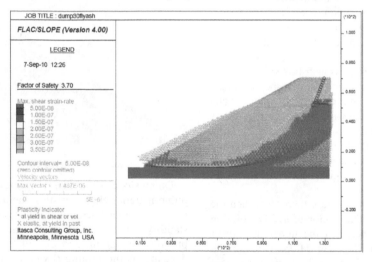

**Figure 3.** Simulated model for dump slope with H = 60m and slope angle = 30°

In the second case, the slope angle is increased by two degrees, equal to 30° while the slope height remains 60 m (Figure 3). This case clearly indicates the drop in factor of safety values from the previous case. Yet, the model is quite similar in outlook. This is because the dump material itself is weak unconsolidated heap of crushed rock grains. Hence, some movements are noticed in most of the parts of the dump. Yet it is noticeable that though the maximum velocity vector is less as compared to first case, the numerical value being 1.487E-06 m/s, the shear strain rate has increased almost ten times and the maximum shear strain rate in the model equals to 3.50E-07 $s^{-1}$. The tensile domain developed at the top rear end of the slope is due to the increased rate of shearing and the material is slowly being pulled out through the free slope surface. The slope at this angle is also quite stable.

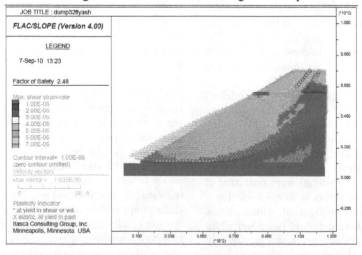

**Figure 4.** Simulated model for dump slope with H = 60m and slope angle = 32°

The slope angle is further increased up to 32° while keeping the base and dump height constant, the factor of safety is found to decrease up to 2.48, the value dipping by almost 45% from the 28° dump slope. The maximum velocity vector is found to be higher than the previous cases 1.635E-05 m/s while the maximum shear strain rate equals to 7.00E-06 s$^{-1}$ (Figure 4). The tension cut off points form tension zones parallel to the slope at the top rear of the slope. The distribution of velocity vectors indicate a typical circular failure pattern according to which the health of slope with passage of time is evaluated. In this case, it is also worthy to notice that the disposition of shear strain rate also predicts a development of weak plane parallel to the face of the slope which may lead to a combination type of failure comprising of a planar slide along with circular failure. This situation if adopted may remain stable unless very heavy and sudden torrential rainfall suddenly weakens the frontal portion of the slope and reduction in FoS may occur.

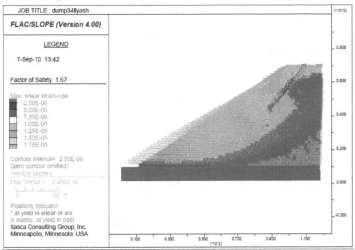

**Figure 5.** Simulated model for dump slope with H = 60m and slope angle = 34°

In fourth case, the flyash mixed dump slope was yet steepened to 34°. Here, Factor of Safety substantially reduces to 1.67 (Figure 5). The concentration of shear strain rates all along the dump has increased significantly and several weak zones are developed. The magnitude of maximum velocity vector and maximum shear strain rate are 2.438E-05 m/s and 1.75E-05 s$^{-1}$ respectively. It is also noticed that some significant velocity vectors of constituent particles in this simulated model have developed on the rear end of the slope indicating the increased vulnerability of the dump slope where weakening is encroaching on the rear side as well. The velocity vectors have although been moving out with high magnitudes along the lower part of the slopes in each case but in with the increase in slope angle to 34°, the velocity vectors have become significantly higher in the top frontal portions as well. This demonstrates that the slope is slowly weakening and once these vectors are of sufficient magnitude as well as quantity in the major portions of the dump, more and more failure zones may develop and eventually the slope may fail due to contributions from every part of the slope.

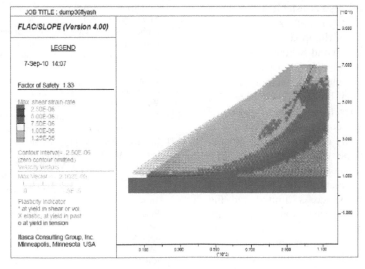

**Figure 6.** Simulated model for dump slope with H = 60m and slope angle = 36°

The fifth case studied is of a dump slope with 20% flyash mixing estimated at a slope angle equal to 36° for the same material types. In this case the factor of safety has decreased to a nearly critical state, the value equal to 1.33 (Figure 6). The quantity of velocity vectors with sizeable magnitude has increased significantly. The maximum velocity vector is equal to 2.102E-05 m/s while the maximum shear strain rate is equal to 1.25E-05 s$^{-1}$.

There as more distinct zones of increased shear strain rates in various parts of the slope and high abundance of elastic and tension points. The velocity vectors of particles are more spread out affecting larger domains including the rear end of the slope as well. This is the most vulnerable condition of maintaining the dump slope stable at 36° slope angle as the already-not-so-high FoS value may decrease fast in case of increase in moisture content of the dump material during rainfall and with time.

The results show that there is a drastic change in the safety conditions as the slope of waste accumulations in a mine site is increased keeping other parameters like slope height and material types constant. The changes in FOS as a function of dump slope angle for mine wastes generated are plotted in Figure 7.

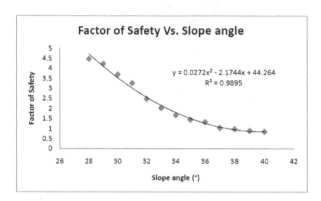

**Figure 7.** Variation of FOS values as a change in dump slope angle

The FoS vs. slope angle curve clearly indicates a strong correlation between the dump slope angle and Factor of Safety (Figure 7). The relation obtained has a high second order correlation. However, this relationship must be used with due care as site specific studies are always preferred and suggested for safe disposal of mine wastes.

## 4. Conclusions

The FLAC/Slope simulated results establish that for a dump of specific material type and keeping other parameters constant, the angle of dump slope can only be increased to a certain level beyond which it becomes critically stable and can lead to a major failure and hamper the surrounding establishments, men and machinery.

From this study, it is clear that high angle slopes are not recommended and dump slopes for any given choice of dump materials (as in this case coal mines) should not be made steeper than 36° under any circumstance to prevent large scale failure of the slope which will hamper the regular operations of waste disposal from the mine. The gravity of the situations may increase in case of in pit mining conditions where the space is limited and the process of coal extraction and waste management are interdependent and operate hand in hand. The quantification helps in considerations for optimum design of dump geometry. The cases where the ground conditions are weak, having low bearing capacity or more moisture saturated, the climate is moist and temperature variations high, the dump slope angle should be maintained well within the stable limits and kept low, preferable close to 32°. Even without external driving forces like blast induced vibrations or rainfall, dump with steeper slopes

may fail under its own weight without giving much prior signal. Hence, the standards must be maintained to achieve safety and stability of mine and mine waste containment systems for regular and progressive mining and uninterrupted productivity of the mine. Careful monitoring and safe slope design by qualified geotechnical engineers at mine sites is important. Additionally, proper bench design, safe blasting patterns should be followed to minimize the adverse effects of nearby dump slopes.

## Reference

[1]   Bishop, A W, The stability of tips and soil heaps. Quarterly Journal of Engineering Geology & Hydrogeology, Vol. 6, Nos. 3 & 4, 1973, pp. 335-376.

[2]   Chang, Y L and Huang, T K, Slope stability analysis using strength reduction technique. Journal of the Chinese Institute of Engineers, Vol. 28, No. 2, 2005, pp. 231-240.

[3]   Cundall, P, Explicit finite difference methods in geomechanics. In Numerical Methods in Engineering, Proceedings of the International Conference on Numerical Methods in Geomechanics, Blacksburg, Vol. 1, 1976, pp. 132-150.

[4]   Dawson, E M, Roth, W H and Drescher, A, Slope stability analysis by strength reduction. Géotechnique, Vol. 49, No. 6, 1999, pp. 835-840.

[5]   Girard, J M and McHugh, E, Detecting problems with mine slope stability. In: 31st Annual Institute on Mining Health, Safety, and Research, Roanoke, VA, 2000.

[6]   Han, J and Leshchinsky, D, Limit equilibrium and continuum mechanics based numerical methods for analyzing stability of MSE walls. Proceedings of the 17th Engineering Mechanics Conference, ASCE, 2004.

[7]   Hebil, K E, Spoil pile stabilization at the Paintearth mine, Forestburg, Alberta. International Symposium on Geotechnical Stability in surface mining, Calgary, Canada, Vol. 181, 1986, pp. 8.

[8]   ISRM, Commission on Standardization of Laboratory and Field Tests, Suggested methods for the quantitative description of discontinuities in rock masses. International Journal of Rock Mechanics and Mining Science & Geomechanical Abstracts, Vol. 15, 1978, pp. 319-368.

[9]   ISRM, Rock Characterization Testing and Monitoring. ISRM Suggested Methods, International Society for Rock Mechanics, Pergamon Press, 1981, pp. 211.

[10]  Itasca, FLAC/SLOPE Users' Guide. Itasca Consulting Group, Command Reference, FISH and Theory and Background, Minneapolis, 1995.

[11]  Janbu, N, Discussion of dimensionless parameters for homogeneous Earth. Journal of the Soil Mechanics and Foundations Division, ASCE, Vol. 93, No. SM6, 1967, pp. 367-374.

[12]  Kuraz, V, Soil properties and water regime of reclaimed surface dumps in the North Bohemian brown-coal region - a field study. Waste Management, Vol. 21, 2001, pp. 147-151.

[13]  Lakshmikantha, H, Report on waste dump sites around Bangalore. Waste Management, Vol. 26, 2006, pp. 640–650.

[14]  Monjezi, M and Singh, T N, Slope Instability in an Opencast Mine. Coal International, 2000, pp. 145-147.

[15]  Pradhan, S P, Vishal, V and Singh, T N, Stability of slope in an open cast mine in Jharia coalfield, India -A Slope Mass Rating approach. Mining Engineers' Journal, Vol. 12 No. 10, 2011, pp. 36-40.

[16]  Pradhan, S P, Vishal, V and Singh, T N, Influence of bio-stabilisation on dump slopes-A discrete element modeling approach. 47th US Rock Mechanics / Geomechanics Symposium, San Francisco, USA. 2013, ARMA-736.

[17]  Rai, R and Singh, T N, Cost benefit and its environmental impact in mining. Journal of Industrial Pollution Control, 20, No. 1, 2004, pp. 17-24.

[18]  Sarkar, K, Vishal, V and Singh, T N, 2012. An empirical correlation of index geomechanical parameters with the compressional wave velocity. Geotechnical and Geological Engineering.

[19] Singh, T N and Chaulya, S K, External dumping of overburden in open cast mine. Indian Journal of Engineers, Vol. 22, Nos. 1 & 2, 1992, pp. 65-73.

[20] Singh, T N and Monjezi, M, Slope Stability Study in Jointed Rockmass - A Numerical Approach. Mining Engineering Journal, Vol. 1, No. 10, 2000, pp. 12-13.

[21] Singh, T N, Chaulya, S K and Singh, J, Effect of mine waste disposal on environment and its protection. Eurock 93, Lisbon, Portugal, 1993, pp. 283-391.

[22] Singh, T N, Pradhan, S P and Vishal, V, Stability of slopes in a fire-prone mine in Jharia Coalfield, India. Arabian Journal of Geosciences, Vol. 6, No. 2, 2013, pp. 419-427.

[23] Swati, M and Joseph, K, Settlement analysis of fresh and partially stabilised municipal solid waste in simulated controlled dumps and bioreactor landfills. Waste Management, Vol. 28, 2008, pp. 1355-1363.

[24] Turer, D and Turer, A, A simplified approach for slope stability analysis of uncontrolled waste dumps. Waste Management and Research, Vol. 29, 2011, pp. 146-156.

[25] Vishal, V, Pradhan, S P and Singh, T N, Instability assessment of mine slope-A Finite Element Approach. International Journal of Earth Sciences and Engineering, Vol. 3, 2010, pp. 11-23.

[26] Vishal, V, Pradhan, S P and Singh, T N, Mine sustainable development vis-a-vis dump stability for a large open cast mine. Proceedings of International Conference on Earth Sciences and Engineering, 2010, pp. 7-14.

[27] Vishal, V, Pradhan, S P and Singh, T N, Tensile strength of rock under elevated temperature. Geotechnical and Geological Engineering Vol. 29, 2011, pp. 1127-1133.

[28] Watters, R J, Influence of internal water and material properties on mine dump stability. Geological Society of America Abstracts with Programs, Vol. 37, No. 7, 2005, pp. 394.

# Setting, Hardening and Mechanical Properties of Some Cement / Agrowaste Composites

**H. H. M. Darweesh[1,*], M. R. Abo El-Suoud[2]**

[1]Refractories, Ceramics and Building Materials Department, Egypt
[2]Botany Department, National Research Centre, Egypt
Corresponding author: hassandarweesh2000@Yahoo.com

**Abstract** The main objective of this study is to reutilize the barely and rice husks (BH & RH) after its conversion to ashes by firing at 600°C (BHA & RHA) as replacing materials of Ordinary Portland cement (OPC) to prevent or at least to reduce the problems of air pollution and energy consumption. The results showed that the water of consistency and setting times of fresh cement pastes increased gradually by the addition of either BHA or RHA. Generally, all the studied properties are improved and enhanced with curing time up to 90 days. The combined water content, bulk density, flexural and compressive strengths decreased with curing time up to 3 days, whereas the apparent porosity increased. During the later ages from 3 up to 28 days and then decreased onward, these properties increased, while the apparent porosity decreased. These characters improved and increased only with 16 wt.% BHA or RHA content and then decreased. The free lime content of the OPC pastes increased as the curing time proceeded up to 90 days, while those containing either BHA or RHA increased only up to 3 days and then decreased up to 90 days and then reincreased onward. It was concluded that the higher contents of BHA and RHA (20 wt.%) must be avoided due to its outstanding cementing properties. The FT-IR spectra and SEM images appeared a slight improve in the crystals and microstructures of the newly formed phases.

*Keywords:* OPC, BHA, RHA, hydration, setting, combined water, free lime, porosity, density, strength, FT-IR, SEM

## 1. Introduction

In Egypt, the rice production is concentrated in a limited agricultural area around the Nile Delta, where nearly about 8 million tons of rice is produced annually. This is creating a large volume of rice byproducts. Often, an uncontrolled burning of rice husk was done for its disposal. Random burning of either rice or even barely husks creates what is known as "Black Cloud" which is very dangerous to the environment. The seasonal and highly localized massive burning usually generated an excessive air pollution that lowers air quality in the surrounding megacity of Cairo. This has become a serious healthy concern for citizens and authorities [1,2,3,4].

Since 2-3 decades, the pozzolanic cements are widely used all over the world. Such cements are employed for their economical, ecological and technological importance, i.e. reduction of energy consumption and $CO_2$ emission [4,5,6]. These cements reduce the resulting lime during the hydration process and replace it with pore-filling cement hydrates, which are known to improve the ultimate strength, impermeability and durability to the aggressive attack of chemical environments around cement structures. Several pozzolanic materials are used such as natural pozzolans, low and high calcium fly ashes, silica fume (SF), Perlite and Granulated blast furnace slag (GbfS) and also crystalline materials which generally known as fillers. The pozzolanic activity of these materials is mainly associated with their vitreous and/or amorphous structure [5,6,7,8]. Among of these additives is the ash resulted from the controlled combustion of some agricultural wastes such as rice and barley husks [1,2,3]. Burning these husks under a controlled temperature, a highly reactive material-like ash is obtained.

About 20% of the rice and barley paddy are husks, the majority of these husks are either burnt or dumped as a waste [4-8]. RHA and BHA contain high amounts of $SiO_2$, and their reactivity related to lime depends on a combination of two factors, namely the noncrystalline silica content and their specific surface. The ashes are very light and easily carried by wind and water in its dry state. Prior to 1970 the rice husk ash was usually produced by uncontrolled combustion, and the ash so produced was generally crystalline and had poor pozzolanic properties [9,10,11,12]. In 1973, Mehta had shown that burning rice husks at 600°C produces an ash with an optimum composition for pozzolanic materials [11].

The main advantage of using RHA as a mineral admixture in a concrete is the significant reduction in the permeability of the concrete. Controlled combustion influences the surface area of RHA, since that time, temperature and environment are considered to produce ash of maximum reactivity [13]. Some of the advantages include improved workability, reduced permeability, increased ultimate strength, reduced bleeding, a better surface finish and reduced heat of hydration [14,15]. Mineral admixtures or pozzolans are used to improve strength, durability and workability in concretes [14,15,16,17].

Cement notation: C: CaO, S: $SiO_2$, A: $Al_2O_3$, F: $Fe_2O_3$, CS: $CaSO_4$, H: $H_2O$, CH: $Ca(OH)_2$, CSH: Calcium silicate hydrate, CAH: Calcium Aluminate hydrate.

Smoother mixtures are typically produced if the mineral admixture is substituted for sand rather than cement, but highly reactive or cementitious pozzolans can cause loss of workability through early hydration. Very finely divided mineral admixtures, such as silica fume, can have a very strong negative effect on water demand and hence workability, unless high-range water-reducing admixtures are used. The main objectives of this work are the production of the barley and rice husk ashes and its utilization as additives for Ordinary and Slag Portland cements to study their effect on the physicochemical and mechanical properties of its hardened pastes.

## 2. Experimental

### 2.1. Raw Materials

The raw materials are Ordinary Portland cement (OPC-1, 32.5 R) with a Blaine surface area of 3.350 $m^2/g$ was delivered from National Cement Company, El-Tibbin, Egypt as well as the barley (BH) and rice husks (RH). The chemical analysis of the OPC cement as well as RHA and BHA ashes using the X-ray fluorescence (XRF) technique is shown in Table 1, while the mineralogical phase composition of OPC as calculated from Bogue equations [17,18] is given in Table 2.

Table 1. The chemical composition of the raw materials, wt.%

| Oxides | L.O.I | $SiO_2$ | $Al_2O_3$ | $Fe_2O_3$ | CaO | MgO | $Na_2O$ | $K_2O$ | $TiO_2$ | $SO_3$ |
|---|---|---|---|---|---|---|---|---|---|---|
| Material | | | | | | | | | | |
| OPC | 2.64 | 20.12 | 5.25 | 1.29 | 63.13 | 1.53 | 0.55 | 0.3 | 0.23 | 2.54 |
| RHA | 0.84 | 77.17 | 8.83 | 1.29 | 1.26 | 0.11 | 0.03 | 1.06 | 0.11 | 0.02 |
| BHA | 0.73 | 78.30 | 9.78 | 1.42 | 2.84 | 0.08 | 0.05 | 0.85 | 0.13 | 0.03 |

Table 2. Mineralogical composition of the OPC sample, mass%

| Phases | $C_3S$ | $\beta$-$C_2S$ | $C_3A$ | $C_4AF$ |
|---|---|---|---|---|
| Material | | | | |
| OPC | 46.81 | 28.43 | 5.90 | 12.56 |

### 2.2. Physical Properties

The physical properties (19,20) of raw materials are calculated from the following relations:

$$Kb = \left(CaO + MgO / SiO_2 + Al_2O_3\right) \quad (1)$$

$$Hm = \left(CaO / SiO_2 + Al_2O_3 + Fe_2O_3\right) \quad (2)$$

$$Sm = \left(SiO_2 / Al_2O_3 + Fe_2O_3\right) \quad (3)$$

$$Am = \left(Al_2O_3 / Fe_2O_3\right) \quad (4)$$

$$Lm(100xCaO / 2.8\ SiO_2 + 1.1\ Al_2O_3 + 0.7\ Fe_2O_3) \quad (5)$$

where, Kb, Hm, Sm, Am and Lm are the basicity coefficient, hydration modulus, silicate modulus, aluminate modulus and lime modulus, respectively.

### 2.3. Preparation of Cement Pastes

At first, the husks of rice (RH) and barely (BH) were processed and washed separately with running water for few minutes and then washed with distilled water, well dried under sun, burned at 600°C for 8 hrs and then screened to pass through 200 mesh sieve (63 μm). The OPC cement was mixed with 0, 4, 8, 12, 16 and 20 wt.% of RHA and BHA ashes as shown in Table 3. The mixes are taken the symbols of R0, R1, R2, R3, R4, R5 for RHA and B0, S1, B2, B3, B4, B5 for BHA, respectively. The pastes were moulded into one inch cubic stainless steel moulds (2.5 x 2.5 x 2.5 $cm^3$), vibrated manually for two minutes and on a mechanical vibrator for another two minutes. The surfaces of pastes were smoothed with a spatula and then were kept inside a humidity cabinet for 24 hrs at 23 ±1°C and 100% R.H, demoulded and soon cured under water till the time of testing for bulk density, apparent porosity and compressive strength after 1, 3, 7, 28 and 90 days.

Table 3. Cement mixes containing BHA and RHA, wt.%

| Group | Raw materials | OPC | BHA | RHA |
|---|---|---|---|---|
| | B0 | 100 | --- | --- |
| | B1 | 96 | 4 | --- |
| | B2 | 92 | 8 | --- |
| BHA | B3 | 88 | 12 | --- |
| | B4 | 84 | 16 | --- |
| | B5 | 80 | 20 | --- |
| | R0 | 100 | --- | --- |
| | R1 | 96 | --- | 4 |
| | R2 | 92 | --- | 8 |
| RHA | R3 | 88 | --- | 12 |
| | R4 | 84 | --- | 16 |
| | R5 | 80 | --- | 20 |

### 2.4. Methods of Investigation

The standard water of consistency (or mixing water) as well as setting times (initial and final) of the prepared cement pastes were directly determined by Vicat Apparatus [21,22]. The water of consistency was determined from the following relation:

$$WC, \% = H / W \times 100 \quad (6)$$

Where, WC is the water of consistency, H is the amount of water taken to produce a suitable paste and W is the weight of the cement sample (300 g). The initial setting time is the time taken to reach the initial set while the final setting time is the time taken to reach the final set of the paste.

The bulk density and apparent porosity [17] of the hardened cement pastes were calculated from the following equations:

$$B.\ D,\left(g\,/\,cm^3\right) = W_1\,/\left(W_1\text{-}W_2\right)\times 1 \qquad (7)$$

$$A.\ P,\% = \left(W_1 - W_3\right)/\left(W_1 - W_2\right)\times 100 \qquad (8)$$

Where, B.D, A.P, $W_1$, $W_2$ and $W_3$ are the bulk density, apparent porosity, saturated, suspended and dry weights, respectively.

The compressive strength [23] was measured by using a hydraulic testing machine of the Type LPM 600 M1 SEIDNER (Germany) having a full capacity of 600 KN and the loading was applied perpendicular to the direction of the upper surface of the cubes as follows:

$$Cs = L\,(KN)\,/\,Sa\,\left(cm^2\right)KN\,/\,m^2$$
$$x\ 102\left(Kg\,/\,cm^2\right)/\,10.2\,(MPa) \qquad (9)$$

Where, Cs: Compressive strength (MPa), L: load (KN), Sa: surface area ($cm^2$).

The flexural or bending strength [24] was carried out using three point adjustments system (Figure 1), WAM-VEB THÜRIHGER INDUSTRIEWERK, testing machine - model RAUENSTEIN WPM, Berlin. The beam load was applied perpendicular to the axis of the sample. The flexural strength was determined from the following equation:

$$\delta = 3\,/\,2\left(F.S\,/\,W.T^2\right)KN\,/\,cm^2$$
$$x102\ Kg\,/\,cm^2\,/\,10.2MPa \qquad (10)$$

Where, $\delta$ is flexural strength (MPa), F is the loading force (KN), S is the span (cm), W and T are width and thickness of the sample (cm).

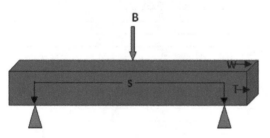

**Figure 1.** Schematic diagram of the bending strength, B: beam, S: span, W: width and T: thickness

After measuring the mechanical strengths, the hydration of the cement pastes at each interval must be stopped using a mixture of methyl alcohol and diethylether. Then, the chemically-combined water content at each hydration age was determined on the basis of ignition loss [17,21] as follows:

$$Wn,\ \% \ = W1 - W2\,/\,W2\ x100 \qquad (11)$$

Where, Wn, W1 and W2 are combined water content, weight of sample before and after ignition, respectively.

The free lime content of the hydrated samples pre-dried at 105°C for 24 hours was also determined [25,26,27]. 0.5 g sample + 40 ml ethylene glycol → heating to about 20 min. (without boiling). Add 1-2 drops of ph. ph. indicator to the filtrate and then titrate against freshly prepared 0.1 N HCl till the pink colour disappears. The 0.1 N HCl was prepared using the following equation:

$$V1 = N\ x\ V2\ x\ W\ x\ 100\ /\ D\ x\ P\ x\ 1000 \qquad (12)$$

where, V1 is the volume of conc. HCl, V2 is the volume required, N is the normality required, W is the equivalent weight, D is the density of Conc. HCl and P is the purity%. Repeat heating and titration several times till the pink colour does not appear on heating. Calculate the free lime content from the following relation: CaO% = (V x 0.0033/1) x100, where V is the volume of 0.1 N HCl taken on titration.

The phase compositions of some selected samples were investigated using infrared spectroscopy (IR) and scanning electron microscopy (SEM). The IR spectra were performed by Pye-Unicum SP-1100 in the range of 4000-400 $cm^{-1}$. The SEM images of the fractured surfaces, coated with a thin layer of gold, were obtained by JEOL-JXA-840 electron analyzer at accelerating voltage of 30 KV.

# 3. Results and Discussion

## 3.1. Composition of Raw Materials

The chemical composition of the starting raw materials, OPC, BHA and RHA is shown in Table 1. It shows that the cement is composed mainly of CaO (64.63%) and $SiO_2$ (20.12%) with reasonable amounts of $Fe_2O_3$ (1.29%), little ratios $Al_2O_3$ (5.25%), MgO (1. 53%), SO3 (2.54%), Na2O + K2O (0. 85%), free lime (1.09%) and ignition loss (2.64). Due to ASTM-Standards [28], the used cement is the Type I-Ordinary Portland cement (OPC),. The table also illustrates that the two ash samples BHA and RHA are composed mainly of silica ($SiO_2$); 77.17 and 78.30%, respectively. They possess higher amounts of alumina ($Al_2O_3$); 8.83 and 9.78, but little amounts of $Fe_2O_3$; 1.29 and 1.42. There are another constituents in variable amounts e.g. MgO (0.11, 0.08), Na2O (0.03, 0.05), K2O (1.06, 0.), Ti2O (0.11, 0.13) and $SO_3$ (0.02, 0.03), respectively. The ignition loss of the two ash samples is less than 2%. When the sum of $SiO_2$ + $Al_2O_3$ + $Fe_2O_3$ is $\geq$ 70%, the material could be considered as a pozzolanic material. On this basis, the sum of $SiO_2$, $Al_2O_3$ and $Fe_2O_3$ of BHA and RHA is 89.5 and 87.29%, respectively, i.e. > 70%. So, the used two ash samples are pozzolanic in nature and the pozzolanicity of BHA is slightly higher than that of RHA. So, they can be used successfully as a partial substitution for Ordinary Portland cement (OPC) to produce blended cement.

## 3.2. Physical Properties

The physical properties of the OPC, BHA (B1-B5) and RHA (R1-R5) are listed in Table 4, while the relationship between the blaine surface area or fineness of the various cement batches and its densities is plotted in Figure 2. As it is clear, the OPC has a higher Kb and Hm whereas Lm is much higher than those of BHA or RHA, but Sm and Am of BHA or RHA are much higher than those of the OPC. Figure 2 showed that as the fineness of starting raw batches increased, the density decreased. This indicates that BHA or RHA are siliceous and have no hydraulic properties in nature. The data shown in Table 5 indicate that either BHA or RHA are conformed the specifications of ASTM Standards to be used as mineral admixtures for cement pastes, mortars or even concretes [28,29,30,31].

**Table 4. The physical properties of raw materials, wt.%**

| Property | Kb | Hm | Sm | Am | Lm | S. Area, cm²/g | density, g/cm³ |
|---|---|---|---|---|---|---|---|
| Materials | | | | | | | |
| CEM-I | 2.55 | 2.196 | 2.331 | 1.55 | 97.91 | 3350 | 2.8897 |
| BHA | 0.033 | 0.032 | 6.991 | 6.89 | 1.23 | 1100 | 2.4866 |
| RHA | 0.018 | 0.014 | 7.625 | 6.84 | 0.558 | 1000 | 2.3781 |

**Table 5. The Conformity of the BHA and RHA to ASTM-C618-01**

| | Of control | | |
|---|---|---|---|
| | Character | | |
| 1- (SiO₂+Al₂O₃+Fe₂O₃) | 89.50 | 87.29 | Min. 70% |
| Content,% | | | |
| 2- SO₃ content,% | 0.03 | 0.02 | Max. 4% |
| 3- L.O.I | 0.73 | 0.84 | Max. 10% |
| 4- Fineness, retained on | ≈24 | ≈25 | Max. 34% |
| 45 µm sieve,% | | | |

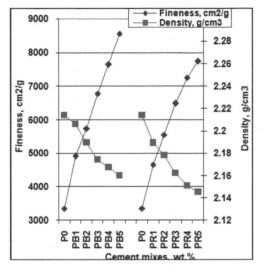

**Figure 2.** Relationship between the blaine surface area and density of the OPC, BHA (PB0-PB5) and OPC/RHA (PR0-PR%) cement mixes

## 3.3. Water of Consistency and Setting Times

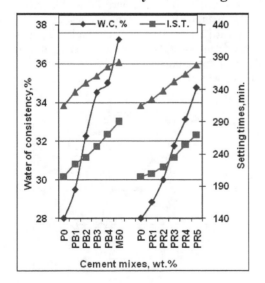

**Figure 3.** Water of consistency and setting times of cement pastes with BHA and RHA

The water of consistency as well as setting times are represented as a function of BHA (B1-B5) and RHA (R1-R5) content in Figure 3 and Figure 4. It is obvious that the water of consistency of OPC pastes is increased with the increase of BHA and RHA contents. This is mainly due to

the higher blaine surface area of the two ash specimens (10 and 11 m²/g) than the OPC (3100 cm²/g) in addition to the hygroscopic nature of the two ash samples which consumes relatively higher amounts of mixing water. These observations are limited to that observed with other several researchers [30,31,32].

The setting times (initial and final) of the OPC pastes (B0 or R0) are also increased gradually as the BHA (B1-B5) or RHA (R1-R5) contents increased. These values are in a good agreement with several studies [4,14,17,33,34,35,36].

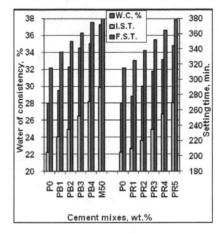

**Figure 4.** Water of consistency and setting times of cement pastes with BHA and RHA

## 3.4. Combined Water Content

The combined water content (CWn) of the OPC (P0) cement paste mixed with 4, 8, 12, 16 and 20 wt.% of BHA (B1-B5) and RHA (R1-R5) are plotted as a function of curing time up to 90 days in Figure 5. Generally, the CWn of all cement mixes increased continuously with curing time up to 90 days. The Wn of the control mix (P0) are gradually increased sharply with curing time up to 28 days due to the rapid normal hydration process and then slightly increased up to 90 days due to the fact that about 75% of the hydration was completed at the first 28 days and the rest 25% behaved as a slight increase up to 90 days. The cement phases $C_3S$, $C_3A$ and $C_4AF$ are responsible for the early hydration up to 28 days while $\beta$–$C_2S$ for the older hydration from 28 days onwards [17,18].

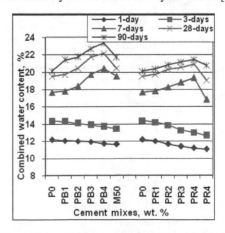

**Figure 5.** Combined water content of cement pastes with BHA and RHA cured up to 90 days

On the other side, the CWn of cement pastes containing either BHA (B1-B5) or RHA (R1-R5) are decreased up to 3 days compared with those of the blank, i.e. the values of CWn are lower than those of the blank, but then started to increase up to 28 days and then redecreased again (Figure 5), i.e. after the first 3 days of hydration, the CWn enhanced with the increase of BHA or RHA contents only up to 16 wt.%, i.e. the optimum content of both agrowastes is 16 wt,%. This is essentially due to the pozzolanic reactivity of the two used agrowastes are the maximum [4,25,34], and then decreased due to the fact that the higher amounts of the agrowastes prevent or hinder the hydration of cement phases to some extent, and so it affected adversely on it [27,35,36,37]. Moreover, the values of CWn are slightly higher with BHA than with RHA. As a result, it can be concluded that the higher amounts of either agrowastes must be avoided.

## 3.5. Free Lime Content

The free lime content (FLn) of the OPC pastes (P0) mixed with 4, 8, 12, 16 and 20 wt.% of BHA (B1-B5) and RHA (R1-R5) are plotted as a function of curing time up to 90 days in Figure 6. It is generally clear that the (FLn) of the OPC pastes (P0) increased with curing time up to 28 days and then seemed to be nearly constant, whilst those of the other cement blends (B1-B5) and (R1-R5) increased only up to 3 days and then decreased gradually up to 28 days and then started to reincrease again up to 90 days. The increase of the (FLn) of P0 is mainly due to the rapid hydration of $C_3S$ at early ages of hydration and the hydration of $\beta$-$C_2S$ at older ages [4,10,11,17,18,19]. The decrease of (FLn) of B1-B5 or R1-R5 is essentially due to the pozzolanic reactions that can take place between the constituents of the added BHA or RHA with the released Ca (OH)$_2$ from the hydration of $C_3S$ and $\beta$–$C_2S$ phases of OPC [4,17,18].

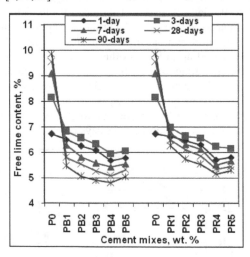

**Figure 6.** Free lime content of cement pastes with BHA and RHA cured up to 90 days

The (FLn) of B1-B5 and /or R1-R5 during the first 3 days are lower than those of the P0 at all curing ages of hydration. This is attributed to the deficiency of the main hydration material [4,8,11]. On the other side, the cement pastes containing BHA recorded the lowest values of (FLn) than those containing RHA at all curing ages. This is mainly attributed to the higher pozzolanicity rate of BHA than RHA [38,39,40]. Cement pastes of B4 containing 16

wt.% BHA shows the lowest values of (FLn) than OPC mix due to its highest rate of pozzolanity. This indicates that BHA has a higher pozzolanic reactivity than RHA. This is evidently in a good agreement with those of combined water.

## 3.6. Bulk Density and Apparent Porosity

Figure 7 and Figure 8 show the bulk density (BD) and apparent porosity (AP) of the OPC cement pastes (P0) blended with 4, 8, 12, 16 and 20 wt.% of BHA (B1-B5) and RHA (R1-R5) are plotted as a function of curing time up to 90 days, respectively. Generally, the (BD) increased as the curing time proceeded up to 90 days, while the (AP) decreased. This is mainly attributed to the normal hydration of cement phases and so the continual formation of hydration products which in turn deposited into the pore structure of the cement pastes. Hence, the (AP) gradually decreased and the compactness of samples improved. This means that the (BD) increased [17,36,41]. As the amount of BHA and/or RHA increased, the (BD) slightly decreased only up to 3 days and the (AP) increased. This is attributed to the continuous deficiency of cement which is responsible for the hydration process [17,18] and also, to the relatively lower density of both BHA and RHA compared with OPC. The (BD) slightly increased from 3 to 28 days and the (AP) decreased. This is essentially due to the pozzolanic reactivity of BHA or RHA with the releasing Ca (OH)$_2$ from the hydration of $C_3S$ and $\beta$-$C_2S$ phases of the cement [42]. From 28 to 90 days, the (BD) tended to redecrease. The same trend was achieved by all cement blends, i.e. the hydration and the pozzolanic reactions of BHA or RHA may be ceased or nearly completed at 28 days because the change in the (BD) or even the (AP) in the period from 28 to 90 days is intangible [1,2,3,4]. Furthermore, the (BD) of either PB4 or PR4 containing 16 wt.% of both BHA and RHA exhibited the highest and lowest values of (BD) and (AP). The higher amounts of these agrowastes hindered and reduced the hydration of cement phases and affected adversely on it [25,26,34,37]. So, the higher amounts of these agrowasted must be avoided. Hence, it can be concluded that the obtained results of (BD) and (AP) are in accordance with those of combined water and free lime contents.

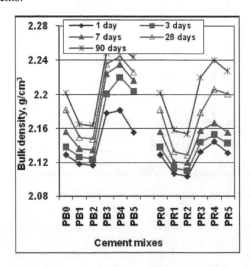

**Figure 7.** Bulk density of cement pastes with BHA and RHA cured up to 90 days

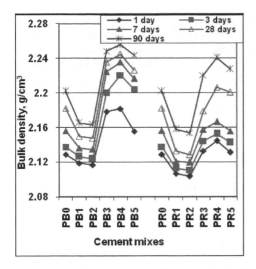

**Figure 8.** Bulk density of cement pastes with BHA and RHA Cured up to 90 days

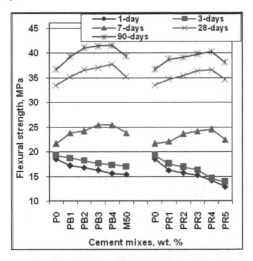

**Figure 9.** Flexural strength of cement pastes with BHA and RHA cured up to 90 days

## 3.7. Compressive Strength

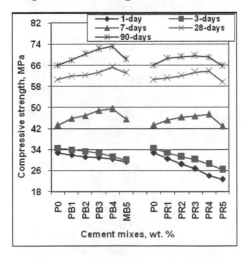

**Figure 10.** Compressive strength of cement pastes with BHA and RHA cured up to 90 days

The compressive strength (CS) of the OPC pastes (P0) blended with 4, 8, 12, 16 and 20 wt.% of BHA (B1-B5) and RHA (R1-R5) are graphically represented as a function of curing time up to 90 days in Figure 9. In a general sense, as the curing time progressed, the (CS) gradually improved and increased up to 90 days. This is mainly attributed to the continual deposition of the formed hydration products filling the pore structure. Hence, the total porosity decreased. This evidently reflected positively on the BD and often followed by an increase in the (CS) [4,16-18]. The increase of (CS) from 28 to 90 days is less significant. The (CS) of cement blends containing either BHA or RHA decreased up to 3 days and then increased up to 28 days which then began to redecrease again but still slightly higher than those of the blank. Moreover, the (CS) increased with the increase of BHA or RHA content only up to 16 wt.% and then decreased. This is principally due to the high rate of pozzolanic activity of both BHA and RHA that could consume the resulting $Ca(OH)_2$ from the normal hydration process, where it is higher with BHA than with RHA [1,3,11,17]. So, the optimum BHA or RHA content is 16 wt.%. Furthermore, the (CS) values of cement pastes containing BHA are slightly higher than those containing RHA, i.e. the pozzolanic activity of BHA is more than that of RHA. At all, the results of (CS) are in a well accordance with those of previous tests.

The (CS) values were improved and enhanced with the addition of BHA or RHA after 7 days of hydration. This is essentially attributed to the pozzolonic reactivity of these materials which started after the first 3 days of hydtration. The constituents of BHA or RHA could be reacted with $Ca(OH)_2$ evolved during the hydration of the OPC phases ($C_3S$ and $\beta-C_2S$), to create new phases which are responsible to improve the specific properties of the hardened cement pastes [31,41,42]. The increase of (CS) is also due to a filler effect beside the higher pozzolanic character of ashes because these ashes are very fine [31,33,43,44]. This means that the higher strength is primarily due to the finer particles of ashes than those of the OPC which causes the segmentation of large pores and increases the nucleation sites for the precipitation of pozzolanic reaction products in cement paste, which in turn increases the pozzolanic reaction and refines the pore structure of the paste. The increase of hydration rate leads to the reduction of $Ca(OH)2$ in the paste. The incorporation of pozzolan such as BHA or RHA reduces the average size and results in impermeable pastes [41-44]. The decrease of (CS) with the increase of ash content at early ages of hydration up to 3 days may be due to the higher deficiency of the OPC, whereas at older ages of hydration is due to the complete consumption of the $Ca(OH)2$ by the pozzolanic reactions with BHA or RHA. Accordingly, the excess contents of ash (>16 wt.%) are not involved in the pozzolanic process as a result of the marked decrease of OPC. Hence, the $Ca(OH)2$ resulting from its hydration was re-increased.

## 3.8. Flexural Strength

The flexural Strength (FS) of the OPC cement pastes (P0) blended with 4, 8, 12, 16 and 20 wt.% of BHA (B1-B5) and RHA (R1-R5) are graphically plotted as a function of curing time up to 90 days in Figure 10. The (FS) of all mixes displayed the same trend as in CS at all curing ages of hydration but with lower values. The (FS) values of all cement pastes are lower than those of the control cement (P0) during the early ages of hydration up to 3 days, but became slightly higher during the later ages (28 and 90 days). The addition of > 16 mass% BHA or RHA to the OPC decreases the (FS) at all curing stages, but still higher than that of the blank. At all, the (FS) values of BHA mixes are slightly more than those of RHA. This is due to that the higher water absorption capacity of RHA than BHA which creates more pore structure and worsens the workability of the pastes. This negatively reflected on the mechanical properties [29-31]. The cement blend containing 16 mass% BHA or RHA (B4 or R4) achieved the highest results of (FS) at all curing periods, but >16 mass% (B5 or R5), the cement blends recorded the lowest values but still higher than that of the blank. So, the high amounts of either BHA or RHA must be avoided when added to OPC.

## 3.9. Mechanism of Hydration and Pozzolanic Reactions

It is well known that the major phases of the OPC are $C_3S$, $\beta$-$C_2S$, $C_3A$ and $C_4AF$ which can be hydrated [17,18] as follows:

$$3C_3S + H_6 \rightarrow C_3S_2H_3 + 3CH \qquad (13)$$

$$2C_2S + H_4 \rightarrow C_3S_2H_3 + CH \qquad (14)$$

$$3C_3A + CS^-.H_2 + H_{10} \rightarrow C_3A.3CS^-.H_{32} \qquad (15)$$

$$C_4AF + 2CH + H_{10} \rightarrow 6CAFH_{12} \qquad (16)$$

The constituents of BHA and RHA can react of the raw gypsum in the presence of water to produce ettringite and CSH and moreover can react with a part of the free lime $Ca(OH)_2$ resulting from the hydration of the OPC phases to form ettringite and CSH [4,14,33] as follows :

$$C_{16}S_{13}A + CS^- + H \rightarrow C_3A.3CS^-.H_{32} + 13CSO_3.H_x \qquad (17)$$

$$C_6S_5A + 2CH + H \rightarrow C_3A.3CS^-.H_{32} + 5CS^-O_3.H_x \qquad (18)$$

On the other side, the main components of both ashes are $Al_2O_3$ and $SiO_2$ which can react with a larger part of the resulting $Ca(OH)_2$ from hydration to form cubic crystals of hydrogarnet ($C_3A. S_2.Hn$) [14,33] as follows:

$$A + S + CH + H \rightarrow C_3A. S_2. H_n \qquad (19)$$

The formation of these new phases is responsible for the strength development along with the whole reaction.

## 3.10. The FT-IR Spectra

The FT-IR spectra of the OPC (P0) and those blended with 16 wt.% BHA (PB4) and RHA (PR4) hydrated up to 90 days, are shown in Figure 11. The sharp absorption band at 3646-3640 cm$^{-1}$ is related to the free OH- group that coordinated to $Ca^{+2}$ (Ca (OH)$_2$ or free lime). The intensity of the broad absorption band at 3454-3428 cm$^{-1}$, which was ascribed to the OH$^-$ group associated to H$^+$ bond that related to the symmetrical stretching frequency of water, increased in presence of BHA or RHA. The two absorption bands at 1646-1636 cm$^{-1}$ and 1435-1425 cm$^{-1}$ are related to the main silicate band involving Si-O stretching vibration bands of CSH, while the band at 1122-1114 cm$^{-1}$ may be due to CAH. The intensity of the two absorption bands at 998-984 cm$^{-1}$ and 876 cm$^{-1}$ characterizing $CO_3^{2-}$ and $SO_4^{2-}$ is irregular due to the rate

of carbonation or sulphonation of CSH and /or CAH, where the vibrations of $CO_3^{2-}$ are smaller than those of $SO_4^{2-}$. Also, the intensity of the absorption bands of Si-O, CAH, $CO_3^{2-}$ and $SO_4^{2-}$ are slightly higher with BHA cement mixes. The intensities of the main characteristic peaks were slightly improved with OPC than with BHA or RHAcement pastes.

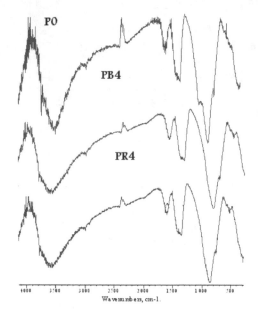

**Figure 11.** The FT-IR spectra of the OPC (P0) and those blended with BHA (PB4) and RHA (PR4) cured up to 90 days

## 3.11. The SEM Images

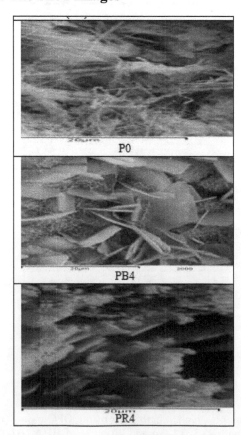

**Figure 12.** The SEM images of P0, PB4 and PR4 cement Pastes cured up to 90 days

Figure 12 illustrates the SEM micrographs of the interfacial layers of the OPC (P0) and the optimum cement pastes blended with BHA (PB4) and RHA (PR4) hydrated in water up to 90 days. It is obvious that the major hydration products are CSH in all samples. Also, the ettringite phase and portlandite or free lime, Ca (OH)$_2$ are clearly shown in the matrix (P0) as needle-like crystals and white granules which reduced or disappeared with the blended cement pastes particularly with mixes PB4 or PR4. This is mainly due to the transformation of ettringite into monosulphate and the pozzolanic reaction of BHA and RHA with the Ca (OH)$_2$. Also, the crystals either BHA or RHA are well-developed when compared with those of the OPC (P0).

# 4. Conclusion

1. The water of consistency as well as setting times of the OPC cement pastes (P0) increased gradually by the incorporation of either BHA or RHA and it is slightly higher with RHA than with RHA cement mixes.

2. The combined water content (Wn), bulk density (BD) and compressive strength (CS) of the blank increased while the apparent porosity (AP) decreased with curing times up to 90 days, but decreased with either BHA or RHA contents up to 3 days and then started to increase gradually up to 28 days and decreased onward. Moreover, this continued up to 16 wt.% BHA or RHA and then decreased.

3. The free lime content (FLn) of the OPC pastes increased as the curing time proceeded up to 90 days, while those containing either BHA or RHA increased only up to 3 days and then decreased up to 28 days of hydrationand then decreased onwards.

4. Generally, the partial substitution of BHA or RHA to OPC pastes improved the specific cementing properties in the following descending order BHA > RHA which contributed to their pozzolonic character.

5. The higher contents of BHA and RHA (>16 wt.%) must be avoided because it caused adverse cementing properties.6- The FT-IR spectra and SEM micrographs showed a slight improve in the crystals and structures of the newly formed hydration product..

# References

[1] M. Nehdi, J. Daquette and A. El-Damatty,"Performance of rice husk ash produced using a new technology as a mineral admixture in concrete", Cem. Concr. Res., 33, 2003, 1203-1210.

[2] V. Rahhal and R. Talero, "Early hydration of Portland cement with crystalline mineral additions", Cem. Concr. Res., 35, 19, 2005, 1285-1291.

[3] P. K. Mehta "Rice husk ash as a mineral admixture in concrete". In: Proceedings of the 2$^{nd}$ international seminar on durabilityof concrete: aspects of admixturesand industrial by-products, Gothenburg, Sweden; 1989. 131-136.

[4] H. H. M. Darweesh, "Effect of combination of Some Pozzolanic Wastes on the Properties of Portland cement Pastes", iiC l'italiana del Cemento, 808, 4, 2005, 298-310.

[5] D. J. M. Flower and J. G. Ganjayan "Green house gas emissions due to concrete manufacture". Int. J. LCA, 12, 5, 2007, 282-8.

[6] S. Asavapisit and N. Ruengrit "The role of RHA-blended cement in stabilizing metal-containing wastes". Cem. Concr. Compos., 27, 2005, 782-787.

[7] S. Chandrasekar, K. G. Satyanarayana and P. N. Raghavan "Processing, properties and applications of reactive silica from rice husk", J. Mat. Sci., 38, 2003, 3159-3168.

[8] B. H. Abu Bakar, P. J. Ramadhansyah, M. A. Megat and M. A. Johari "Effect of rice husk ash fineness on the chemical and physical properties of concrete", Mag. Concr. Res., 63, 2011, 313-320.

[9] C. L. Hwang and D. S. Wu "Properties of cement paste containing rice husk ash", Amer. Concr. Inst., 114, 1989, 733 765.

[10] V. P. Della, L. Kuhn and D. Hotza "Rice husk ash as an alternative source for active silica production". Mat. Letters, 57, 2002, 818-821.

[11] S. A. Memon, M. A. Shaikh and H. Akbar, "Utilization of rice husk ash as a mineral admixture", Constr. Build. Mat., 25, 13, 2011, 1044-1048.

[12] G. R.. de Sensale "Effect of rice-husk ash on durability of cementitious materials". Cem. Concr. Compos., 32, 2010, 718-725.

[13] P. Chindaprasirt, S. Rukzon and V. Sirivivatnanon "Resistance to chloride penetration of blended Portland cement mortar containing palm oil fuel ash, rice husk ash and fly ash". Const. Build. Materials, 22, 2008, 932-938.

[14] H. El-Didamony, H. M. M. Darweesh and R, A. Mostafa, "Characteristics of pozzolanic cement pastes Part II: Resiatance against some aggressive solutions" Sil. Ind. (Cer. Sci. & Techn.), (Belgium), 74, Nr. 3-4, 2009, 98-105.

[15] Q. Feng, H. Yamamichi, M, Shoya and S. Sugita, "Study on the pozzolanic properties of rice husk ash by hydrochloric acid pretreatment". Cem. Concr. Res., 34, 2004, 521-526.

[16] H. H. M. Darweesh, "Hydration, Strength Development and Sulphate Attack of Some Cement Composites", World App. Science, 2013, 52-57.

[17] P. C. Hewlett, "Lea's Chemistry of Cement and Concrete"; 4$^{th}$ Edn.; John Wiley & Sons Inc, New York, 1998.

[18] A. M.Neville "Properties of concrete". 4$^{th}$ edn. Essex (UK): Longman, 1995.

[19] Z. Kersener, H.H. M. Darweesh and L. Routil, "Mortar composites from waste materials" J. Cem. Hormigon, Spain, Vol. 2, No. 943, 2011, 12-19.

[20] Z. Kersener, H.H. M. Darweesh and L. Routi "Alkali-activated slag-rich cement pastes with sodium silicate and water glass" Cem. Hormigon, Spain, Vol. 4, No. 945, 2011, 18-26.

[21] ASTM-Standards "Standard Test Method for Normal water of Consistency of Hydraulic Cement", C187-86, 1993, 148-150.

[22] ASTM-Standards "Standard Test Method for Setting Time of Hydraulic Cement", C191-92, 1993, 866-868.

[23] ASTM-Standards "Standard Test Method for Compressive Strength of Dimension Stone", C170-90, 1993, 828-830.

[24] ASTM-Standards: Standard test method for Flexural strength of dimension stone. C674-71 (1980), 668-671,

[25] H. H. Darweesh and H. M. Awad, "Effect of the calcination temperature and calcined clay substitution on the properties of Portland cement pastes", iiC l'industria italiana del Cemento, 844, 2008, 486-401.

[26] H. El-Didamony, H. H. Darweesh and R. A. Mostafa, "Characteristics of pozzolanic cement pastes Part I: Physico-mechanical properties" Sil. Ind. (Cer. Sci. & Techn.), Belgium, 73, Nr. 11-12, 2008, 193-200.

[27] H. H. M. Darweesh "Utilization of Perlite Rock in Blended Cement-Part I: Physicomechanical Properties", J. Chemical and Materials Sciences, (DRCMS), ISSN 2354-4163, Vol. 2, No. 1, 2014, 1-12.

[28] ASTM-Standards, "Standard Specification for Coal Fly Ash and Raw or Calcined Natural Pozzolan for Use in Concrete", C 618-12 a, 1978.

[29] H. H. M. Darweesh,"Setting, hardening and strength properties of cement pastes with zeolite alone or in combination with slag" InterCeram (Intern. Cer. Review), Germany, Vol. 1, 2012, 52-57.

[30] P. K. Metha, "Proc. work shop on rice husk ash cement", Peshawar, Pakistan. Bangalore, India: Reg. l Cent. Techn. Trans., 1979, 113-122.

[31] S. Rukzon and P. Chindaprasirt, "Strength of ternary blended cement mortar containing Portland cement, rice husk ash and fly ash", J. Eng. Inst. Thailand, 17, 2006, 33-37.

[32] A. N. Givi, S. A.Rashid, A. Farah Nora, Aziz and M. A. Mohd Salleh, "Assessment the effects of RHA particle size on strength, water permeability and workability of binary blended concrete", Constr Build Mater., Vol. 24, 2010, 2145-2150.

[33] L.O. Ettu, C.A. Ajoku, K.C. Nwachukwu, C.T.G. Awodiji and U.G. Eziefula, "Strength variation of OPC-rice husk ash composites with percentage rice husk ash" Int. J. App. Sci. and Eng. Res., 2, 4, 2013, 420-424.

[34] H. H. Darweesh, "Characteristics of metakaoline blended cement pastes" Sil. Ind. (Cer. Sci. & Techn.), Belgium, Vol. 72, Nr. (1-2), 2007, 24-32.

[35] H. H. M. Darweesh and H. Youssef, "Preparation of 11Å Al-substituted Tobermorite from Egyptian Trackyte Rock and Its Effect on the Specific Properties of Portland Cement", accepted for publication in Micro structure and Mesostructure journal, 2014.

[36] K. Kartini "Rice husk ash-pozzolanic material for sustainability", Int. J. Appl. Sci.. Technol., 1, 6, 2011, 169-178.

[37] S. K. Antiohos, V. G. Papadakis and S. Tsimas, "Rice husk ash (RHA) effectiveness in cement and concrete as a function of reactive silica and fineness", Cem. Concr. Res. 2013.

[38] S. K. Antiohos, J. G. Tapali, M. Zervaki, J. Sousa-Coutinho, S. Tsimas and V. G. Papadakis, "Low embodied energy cement containing untreated RHA: A strength development and durability study", Const. Build. Materials, 49, 2013, 455-463.

[39] V. S Ramachandran and R. F. Feldman, "*Concrete Admixtures Handbook, Properties Science and Technology*", 2nd Edn.; Noyes Publications: New Jersey, 1995.

[40] P. Chindaprasirt and S. Rukzon, "Strength, porosity and corrosion resistance of ternary blend Portland cement, rice husk ash and fly ash mortar", 22, 8, 2008, 1601-1606.

[41] M. R. Karim, M F. M. Zain, M. Jamil, F. C. Lai and M. N. Islam, "Strength of Mortar and Concrete as Influenced by Rice Husk Ash: A Review", World Applied Sciences, ISSN 1818-4952, 19, 10, 2012, 1501-1513.

[42] V. Sata, C. Jaturapitakkul and K. Kiattikomol, "Influence of pozzolan from various by-product materials on mechanical properties of high-strength concrete", Con. and Build. Materials, 21, 7, 2007, 1589-1598.

[43] Viet-Thien-An Van, Christiane Rößler, Danh-Dai Bui and Horst-Michael Ludwig, "Mesoporous structure and pozzolanic reactivity of rice husk ash in cementitious system", Cons. Build. Materials, 43, 2013, 208-216.

[44] P. Chindaprasirt, S. Rukzon "Strength, porosity and corrosion resistance of ternary blend Portland cement, rice husk ash and fly ash mortar", Constr. Build. Mater., 22, 8, 2008, 1601-1606.

# Seismic Effect Prognosis for Objects with Different Geometric Configuration of Fundament in Close Blasting

**Viktor Boiko**[*]

Department of Geobuilding and Mining Technologies, Institute of Energy Saving and Energy Management, National Technical University of Ukraine "Kiev Polytechnic Institute", Kyiv, Ukraine
*Corresponding author: ptl777@ukr.net

**Abstract**  Seismic wave influence in near field from millisecond delay blast to buildings with different foundation configuration was researched. Three main type of base (square, tree-corner and elliptical shape) were described. Recommendation for charges disposition and orientation in blasting scheme were given.

*Keywords:* *blasting, building foundation, near field, seismic wave, base configuration, seismic effect*

## 1. Introduction

During blasting practice problems to forecast objects' seismic stability by building foundation's contour appears very often. The main mistake is determining boundary allowable charges in near field for guarded objects' seismic stability with different geometric configuration of foundation, when we use the same calculation as for remote zone. In this case, we keep invalid data with relation to practice. Really, if value of breakdown speed 3 cm/s is putted in Eq.(1) and coefficient K, which is depended on blast charge loading, accept as 300, Eq. (2) would be got. As rules, coefficient K value is bigger for hard rocks than for soft rock, because blast charges are more loaded.

$$v = K\rho^2 \tag{1}$$

$$Q_{max} = (0.1r)^3 \tag{2}$$

For example, from preceding follows that even on distance 10 m from building foundation it possible to blast charge with mass 0.01 kg. [1] But blasting practice shows, it could be blast charge in 30-50 times greater.

## 2. Materials and Methods

The mechanism of seismic affecting on a building in a short-range zone and in long-range are different, and it is the reason of such divergence. Local character of explosion act in a short-range zone and dominant composition of volume waves show up. Last gives oscillation with high frequency and small slowness to a subgrade. If velocity of particles displacement takes on criterion of seismic safety, it should be velocity of particles displacement in building's base by all plane. It is quantitative characteristics for energy, which transmits through foundation to whole building. This criterion does not depend on geometrical configuration of building foundation and type of construction design, but it depends on blast conditions, distance and geological conditions along way of seismic wave's distribution.

As regards remote zone, displacement particles velocity and displacement foundation velocity have small difference (1,5 times), that is why, evenly, displacement particles velocity in house footing takes on criterion of seismic safety [1].

In near field, displacement foundation velocity is much less than displacement particles velocity under foundation. It is bound up with short waves from close blasting comes to different parts of foundation with dissimilar amplitudes, inasmuch as last damps inversely proportional to distant, which, for example, independent of foundation's configuration, is bigger for back wall than to nearby wall. (Figure 1).

If square house footing is under consideration, ground keep almost motionless under separate part (in corners) of foundation. Inasmuch as all base parts and walls are connected with each other, most of construction's elements oppose to engaging of fabric in the oscillating process, which fits velocity of particles displacement in the closest zone to the blast charge. In case, when construction parts are disconnected (for example, dam), majority of elements will response independently of one another to sway, and accordingly displacement particles velocity will change in different points with certain distances from charge. It leads to substantial increase of seismic demonstration from blasting through ground base in determinate unsafe junctions, for concrete foundation configuration. That is why the task to develop method of stress field (oscillation velocity) calculation near ground base, where foundation with elements' configuration is located and seismic waves from blasting have influence, springs up. Especially it applies to near field. Concerning far zone, calculation methods for seismic safety of buildings and other guarded objects are based on determining ground oscillation velocity only in a ground base. In this case, particles oscillation velocity measuring in any spots of ground base near foundation is enough for seismic stability calculation. Seismic stability estimation, according to this methodic, does not account oscillation

pattern along all boundary of ground-building contacts and it could be used for any configuration objects. Scilicet existing design procedures of seismic stability by soil particles oscillation velocity do not allow different durability of construction elements depend on waves' distribution changing along all boundary of massif and building contacts. It is explained absence theoretical and experimental data of dynamic stress fields distribution (soil particles' oscillation velocity) through medium to construction elements of different building's configuration.

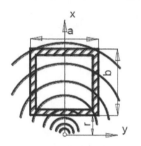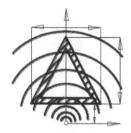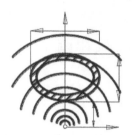

**Figure 1.** Seismic-wave motion scheme in near field through ground base to objects with square, three-angled and oval foundation

In adjacent scientific fields with geomechanics, task of loading distribution for different types elastic inclusions is has been decided [2]. Let us handle a task about plane-polarized three-dimensional wave's distribution in elastic medium, which hold non-circular cylindrical foundation (object) with dissimilar mechanical properties from medium properties. Diffraction phenomenon role in dynamic load's distribution become formed on such facility, if non- circular cylindrical coordinate system $(\rho,\theta,z)$ with axis coincide with inclusion axis will introduce. Variables $\rho,\theta,z$ are connected with Cartesian coordinates $(x,y,z)$ and circular cylindrical coordinates $(\rho,\varphi,z)$ correlation.

$$x + yi = re^{i\varphi} = R\left(\rho e^{i\theta} - \varepsilon\rho^{-N}e^{-iN\theta}\right) \quad (3)$$

Guides of cylindrical surface $\rho = 1$ by different dimensions have elliptic form or regular $(N + 1)$-gon with truncated corner.

Let polarized longitudinal wave with potential distribute in elastic medium

$$\Phi^* = A_0 \exp ik_p\left(n_{px}x + n_{py}y\right), \quad (4)$$

Where $A_o$ – amplitude of incident wave;

$n_{px} = \cos\varphi_p$, $n_{py} = \sin\varphi_p$ – guides cone of normal to waves' surface of longitudinal waves.

Solution of Helmholtz equation

$$\Delta\Phi + \alpha^2\Phi = 0, \Delta\Psi + \beta^2\Psi = 0, \quad (5)$$

(Where $\Phi$, $\Psi$ – scalar potentials accordingly longitudinal and transverse waves; $\alpha$, $\beta$ – wave numbers) for elastic medium $\rho > 1$, which characterize depictured waves in coordinates $(r, \varphi, z)$ (their potentials satisfy propagation term when $r = \infty$). Expression Eq. (5) could be written as

$$\Phi = \sum_n A_n H_n^{(1)}(\alpha r)e^{in\varphi}$$
$$\Psi = \sum_n B_n H_n^{(1)}(\beta r)e^{in\varphi} \quad (6)$$

Where $A_n$, $B_n$ – non-defined coefficients;

$H_n^{(1)}(\xi)$ – 1st rood of Hankel function.

Confines of summing from $-\infty$ till $+\infty$ are omitted here and further. Total wave field in medium defines as sum $\Phi + \Phi^*$, $\Psi$.

Potentials for inclusion $\rho < 1$ are chosen similar:

$$\Phi_f = \sum_n A_{f,n}J_n(\alpha_f r)e^{\inf} ;$$
$$\Psi_f = \sum_n B_{f,n}J_n(\beta_f r)e^{\inf}, \quad (7)$$

де $A_{f,n}$, $B_{f,n}$ – non-defined coefficients;

$\alpha_f$, $\beta_f$ – wave numbers of longitudinal and transverse waves for foundation materials;

$J_n(\xi)$ – Bessel function (index $f$ to meet an object).

Potential of incident wave Eq. (5) in circular cylindrical coordinates $(r, \varphi, z)$ could be written as

$$\Phi^* = A_0 \sum_n a_n J_n(\alpha r)e^{in\phi} \quad (8)$$

де $a_n = i^n e^{-in\phi_p}$ .

Continuity conditions of movements and tensions have to comply on surface of discontinuity $\rho = 1$.

$$U_p^f(1,\theta) = U_p(1,\theta) + U^*_p(1,\theta);$$
$$U_\theta^f(1,\theta) = U_\theta(1,\theta) + U^*_\theta(1,\theta);$$
$$\sigma_{\rho\rho}^f(1,\theta) = \sigma_{\rho\rho}(1,\theta) + \sigma^*_{\rho\rho}(1,\theta); \quad (9)$$
$$\sigma_{\rho\theta}^f(1,\theta) = \sigma_{\rho\theta}(1,\theta) + \sigma^*_{\rho\theta}(1,\theta).$$

For observance continuity conditions Eq. (9), let pass on to expressions Eq. (6)-(8) and coordinates $(\rho, z, \theta)$ using solutions of Helmholtz equation in curvilinear coordinates [3]:

$$\Phi(\rho,\theta)=\sum_n e^{in\theta}\sum_p A_{n-(N+1)p}J_p(\alpha\varepsilon\rho^{-N})H^{(1)}_{n-Np}(\alpha\rho);$$

$$\Psi(\rho,\theta)=\sum_n e^{in\theta}\sum_p B_{n-(N+1)p}J_p(\beta\varepsilon\rho^{-N})H^{(1)}_{n-Np}(\alpha\rho);$$

$$\Phi_f(\rho,\theta)=\sum_n e^{in\theta}\sum_p A_{f,n-(N+1)p}J_p\left(\alpha_f\varepsilon\rho^{-N}\right)J_{n-Np}\left(\alpha_f\rho\right);$$

$$\Psi_f(\rho,\theta)=\sum_n e^{in\theta}\sum_p B_{f,n-(N+1)p}J_p\left(\beta_f\varepsilon\rho^{-N}\right)J_{n-Np}\left(\beta_f\rho\right);$$

$$\Phi*(\rho,\theta)=\sum_n e^{in\theta}\sum_p A_0 a_{n-(N+1)p}J_p\left(\alpha\varepsilon\rho^{-N}\right)J_{n-Np}\left(\alpha\rho\right)$$

$$(10)$$

Expression Eq. (10) puts in contact conditions Eq. (9) then infinite system of algebraic equalizations be received for constants $A_n$, $B_n$, $A_{f,n}$ and $B_{f,n}$ ($n=0,\pm1,...$). This equalization system was worked with using computer.

Finite number of equalizations hold in the system for determination the nearest answers. The calculations accuracy checked through result comparison for different approximation. Maximum relative difference of results was not bigger then 5% for tensions.

When tensions analysis for massif with object was doing, computation made for most common configuration of elliptical ($N=1$; $\varepsilon=\pm0,4$; $a+b=2R$), three-cornered ($N=2$; $\varepsilon=\pm0,25$; $d=3/4R$) and square ($N=3$; $\varepsilon=+1/6$; $d=5/6R$) sections (dam, building base, etc.). These stress amplitudes $\sigma_{\rho\rho}$, $\sigma_{\rho\theta}$ and $\sigma_{\theta\theta}$ were charged to $\left|\sigma_{yy}^{(\infty)}\right|=(\lambda+2\mu)\alpha^2$.

Calculation made for points of contour $0\le\theta\le\pi$ in increments of $\Delta\theta=\pi/18$ for elliptical and three-cornered inclusion and of $\Delta\theta=\pi/12$ for square one. Poisson's ratio for massif was 0,35 and 0,2 for object when ratio of density and hardness was $\rho_f/\rho_m=2$ and $\mu_f/\mu_m=20$ respectively.

Relationship of stress amplitudes $\sigma_{\rho\rho}$, $\sigma_{\rho\theta}$, $\sigma_{\theta\theta}$ in matrix when $\rho=1$ from angle $\theta$ with fixed frequency (Value $\alpha$ is numbers near graph) are depicted on Figure. 2. Block curve meets to stress $|\sigma_{\theta\theta}|$, dashed-line curve meets to $|\sigma_{\rho\rho}|$, chain line meets to $|\sigma_{\rho\rho}|$ on the graphs.

## 3. Results and Discussion

In case of inclusions elliptical intersection, maximum tensions originate in boundaries points of the smallest crookedness radius $\theta=\pi/2$ for $\varepsilon=0,4$ (waves distribute along short axis of ellipse) and $\theta=0$, $\theta=\pi$ for $\varepsilon=-0,4$ (waves distribute lengthwise long axis of ellipse). The biggest is $\sigma_{\theta\theta}$. In area of the biggest crookedness radius points maximum tension originate $\theta=\pi$, $\theta=0$. Tension $\sigma_{pp}$ dominate in front face and shady side of contour when $\varepsilon=0,4$, as in case circular inclusion.

Maximum tension $\sigma_{pp}$ arise in corner points' areas $\theta=\pi$, $\theta=\pi/3$ for objects with three-corner section. In case of $\varepsilon=0,25$, $\theta=2/3\pi$ and $\theta=0$ for $\varepsilon=0,25$ the largest stress are $\sigma_{\theta\theta}$, as for elliptical objects. Efforts $\sigma_{pp}$ have the biggest value in black side points of contour $\theta=0$ for $\varepsilon=0,25$ and $\theta=\pi$ front side of contour for $\varepsilon=-0,25$

In case of object with square section for $\varepsilon=1/6$ stress $\sigma_{pp}$ has peak value in corner points region $\theta=3/4\pi$ and $\theta=\pi$, but tension $\sigma_{\theta\theta}$ has maximum in corner points area $\theta=\pi$, $\theta=\pi/2$ and $\theta=0$.

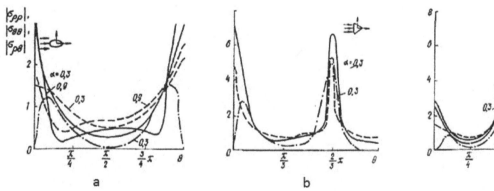

Figure 2. Plot of tension amplitude from $\theta$ angle when fixed frequencies by foundation contour of safe object with elliptical (a), three-corner (b) and square (c) configuration

It is worthwhile noting that strong stress concentration arise in the smallest radius of curvature for object with different cross-section shape, which are in three-eight times excess over tensions of incident wave. Maximum effort in incident wave $\sigma_{pp}$ is achieved in points of minimum curvature radius. Other contour points could have big value as $\sigma_{pp}$ so $\sigma_{\theta\theta}$. These theoretical calculations have found practical evidence when the largest cracks from blasting concussion have been locked in building corners with square and right-angled base (Figure 2) [4].

## 4. Conclusion

The foregoing allows affirming that take-off data of body wave's oscillation velocity, which were obtained on the corner points, should be assumed for forecasting of blasting concussion for facilities with different base configuration. Borehole charges, which blast instantly, should be oriented concurrently to flank walls (Figure 3) during drilling and blasting operations. It is not permissible when waves come from side of corner point.

In this articles just a few configuration of building base have been taken, but here are a lot building with other regular or irregular shape. It gives wide field of future research Seismic safety of engineering and nature objects could be obligatory or preventive, for example by psychological factors affecting people. Industrial implementation sphere of seismic safe methods during

special and engineering blasting will appear always.

**Figure 3.** Typical crack (place of maximum stress) formed under the impact of seismic waves from large-scale blast on mines

# References

[1] Tseitlin Y.I., Smoliy N.I. *Seismic and air-blast waves from industrial explosions*. Nedra, Moskow, 1981, 192 p.

[2] Boiko V.V., Kuzmenko A.A., Hlevnuk T.V., "Permissible seismic stability estimation for crack-weakened constructions from anthropogenic explosions", *Collection of scientific papers*, NDIBK, Kyiv, 2008.-772 c.

[3] Abdurashidov K.S., Aizenberg Y.M., Zhunusov T.Z., *Seismic stability of structure*, Nedra, Moskow, 1989, 192 p.

[4] Boiko V.V. *Seismic safety problems for blasting work in Ukrainian opencast mines,* Vudavnutstvo Stal, Kyiv, 2012, (Tseitlin Y.I., 1981) (Boiko V.V.) (Boiko V.V).

# 5

# Pulp White Liquor Waste as a Cement Admixture

**H. H. M. Darweesh[1,*], M. G. El-Meligy[2]**

[1]Refractories, Ceramics and Building Materials Dept. National Research Centre, Cairo, Egypt
[2]Cellulose and Paper Dept., National Research Centre, Cairo, Egypt
*Corresponding author: hassandarweesh2000@yahoo.com

**Abstract** The pulp white liquor waste (PWL), a byproduct from paper-making, could be applied as a cement admixture in two types of cement, namely Ordinary Portland cement (OPC) and Portland limestone cement (LPC). The results showed that the water of consistency of cement pastes premixed with 0, 1, 2 and 3 wt. % PWL was gradually decreased, while the setting times (initial and final) were increased. So, it can be used as a retarder. The compressive strength increased slightly during the early ages of hydration (1 and 3 days), but sharply increased during the later ages (28 and 90 days), specially with those premixed with PWL. The combined water content and bulk density displayed the same trend as the compressive strength, whilst the apparent porosity decreased at all curing times up to 90 days. The IR spectra of cement pastes showed that the intensities of the different peaks of cement pastes with PWL are higher than those of the pure samples. The above results proved that 2 wt. % PWL is the optimum concentration.

*Keywords:* PWL, OPC, LPC, combined water, bulk density, apparent porosity, Strength, FT-IR

## 1. Introduction

It is well known that the industrial wastes or byproducts either solid or liquid caused many environmental problems as air, soil or water pollution. So, the recent trend all over the world is to reuse these wastes in useful applications [1,2,3,4]. Although the nature and action mechanism of cement and/or concrete admixtures have not yet been explained in a satisfactory way up till now due to the huge variety of admixtures, the use of reducing water soluble polymer admixtures with dispersing, plasticizing and air-entraining effects at a considerably low polymer / cement ratios is very important because these admixtures evidently improve the workability and also modify other properties of cement pastes, mortars and/or concrete. Furthermore, they enhance the performance of concrete structures [5,6,7,8].

Black and white liquors, waste products which are coming from Paper Industry (Figure 1), have attracted our attention due to their unique advantages because it has a high solubility and low viscosity. This makes it more readily available for several applications as initiators. Black liquor is an important liquid fuel in the pulp and paper industry [9]. Darweesh et al. [10] studied and used the black liquor waste as a cement admixture. The pulp white liquor (PWL), is a strong alkaline solution used in the first stage of the kraft process in which the lignin and hemicellulose are separated from the cellulose fiber in the kraft process for the production of pulp [2].

It is called white liquor due to its white opaque color. In the pulp and paper industry's Kraft pulping process, the so-called white liquor is charged together with wood chips, into the digester, at high pressure and temperatures up to 170°C. Within the digester, lignin from the wood is removed by action of the hydroxide (OH⁻) and hydrosulfide (HS⁻) ions producing a suspension of cellulose fibers. This suspension is the pulp. The white liquor that contains polysulfide ions is known to increase the process' yield. Pure sodium hydroxide solution (NaOH) is used in a number of situations in the pulping mill [11].

Uchikawa et al. [12] reported that some organic admixtures induce physical effects, which modify the bonds between particles and can act on the chemical processes of hydration, particularly on the nucleation and crystal growth. Accordingly, the white liquor constitutes a new and promising admixture.

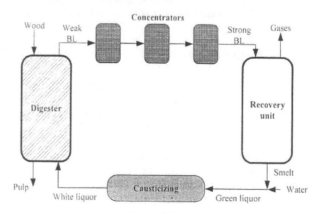

**Figure 1.** A schematic description of the pulping process

Generally, the cement admixtures are used in minor quantities as water-soluble polymers, liquid resins and monomers to confer some beneficial effects as reduction of water requirements, improving workability, control setting, accelerating hardening, improving strength, better durability, desirable appearance and volume changes [13,14,15,16]. The traditional water-soluble admixtures employed as cement modifiers are cellulose derivatives including methyl cellulose, carboxymethyl cellulose and hydroxyethyl cellulose, polyvinyl alcohol, polyethylene oxide, polyacrylamide, etc. [15] The wide achievements of cement admixtures converted our attention to look for new suitable admixtures for cement which is the main objective of the study [17,18]. So, the main objective of the current paper is to study the effect of pulp black liquor, a byproduct from the pulp production industry, on the physico-mechanical properties and microstructure of OPC and PLC cements.

## 2. Experimental

### 2.1. Raw Materials

A waste of pulp white liquor (PWL) from paper industry was provided by Paper Factory, Alexandria, Egypt. The Ordinary Portland cement (OPC) and Limestone Portland cement (LPC) with blaine surface areas of 3300 and 3100 $cm^2/g$ were delivered from Helwan and Tourah Cement Companies, Egypt, respectively. The white liquor is mainly composed of NaOH and $Na_2S$ in water. These are the active component in Kraft pulping [2]. It contains also minor amounts of sodium carbonate, $Na_2CO_3$; sodium sulfate, $Na_2SO_4$; sodium thiosulfate, $Na_2S_2O_3$; sodium chloride, NaCl, calcium carbonate, $CaCO_3$ and other salts as well as non-process elements. These additional components are considered inert in the Kraft process, except sodium carbonate that contributes to a lesser extent. The composition of typical white liquor is 3 M NaOH, 0.7 M $Na_2S$ and 0.25 M $Na_2CO_3$ as shown in Table 1, while Table 2 shows the elemental analysis of the OPC and PLC cements.

**Table 1. the composition of a typical white liquor**

| Oxides Materials | NaOH | $Na_2S$ | $Na_2CO_3$ |
|---|---|---|---|
| PWL, M | 3 | 0.7 | 0.25 |

**Table 2. The Chemical composition of the OPC and LPC cements, wt. %**

| B. S. A $cm^2/g$ | L.O.I | $SiO_2$ | $Al_2O_3$ | $Fe_2O_3$ | CaO | MgO | $Na_2O$ | $K_2O$ | $SO_3$ | Oxides Materials |
|---|---|---|---|---|---|---|---|---|---|---|
| 3300 | 2.64 | 2.64 | 5.25 | 3.38 | 63.13 | 1.53 | 0.55 | 0.3 | 2.54 | OPC |
| 3100 | 6.44 | 16.10 | 4.03 | 3.80 | 60.10 | 1.24 | 0.65 | 0.26 | 1.44 | LPC |

### 2.2. Preparation and Methods

The PWL was dissolved in the mixing water with the dosage of 0, 1, 2 and 3 wt % and then added to OPC and LPC cements. The pastes were moulded into one inch cubic stainless steel moulds (2.5 x 2.5 x 2.5 $cm^3$), vibrated manually for two minutes and on a mechanical vibrator for another two minutes. The surfaces of pastes were smoothed with a spatula and then were kept inside a humidity cabinet for 24 hrs at 23 ±1°C and 100% R.H, demoulded and soon cured under water till thehe time of testing for bulk density, apparent porosity and compressive strength after 1, 3, 7, 28 and 90 days. The samples were denoted as $P_0$, $P_1$, $P_2$, $P_3$ for OPC and L0 $L_0$, $L_1$, $L_2$, $L_3$ for LPC.

The standard water of consistency (or mixing water) as well as setting times (initial and final) of the prepared cement pastes were directly determined by Vicat Apparatus [19,20]. The bulk density and apparent porosity [1] of the hardened cement pastes were calculated from the following equations:

$$B.D, \left( g/cm^3 \right) = W_1 / \left( W_1 - W_2 \right) \times 1 \qquad (1)$$

$$A.P, \% = \left( W_1 - W_3 \right) / \left( W_1 - W_2 \right) \times 100 \qquad (2)$$

Where, B.D, A.P, $W_1$, $W_2$ and $W_3$ are the bulk density, apparent porosity, saturated, suspended and dry weights, respectively.

The compressive strength [21] was measured by using a hydraulic testing machine of the Type LPM 600 M1 SEIDNER (Germany) having a full capacity of 600 KN and the loading was applied perpendicular to the direction of the upper surface of the cubes as follows:

$$Cs = L / SaKN / m^2 x \ 102 / 10.2 \qquad (3)$$

Where, Cs: Compressive strength (MPa), L: load (KN), Sa: surface area ($cm^2$).

The chemically-combined water content at each hydration age was also determined on the basis of ignition loss [1,22] as follows:

$$Wn, \% = W1 - W2 / W2 \ x100 \qquad (4)$$

Where, Wn, W1 and W2 are combined water content, weight of sample before and after ignition, respectively.

### 2.3. Phase Composition

The phase compositions of some selected samples were investigated using infrared spectroscopy (IR) and scanning electron microscopy (SEM). The IR spectra were performed by Pye-Unicum SP-1100 in the range of 4000-400 $cm^{-1}$.

## 3. Results & Discussion

### 3.1. Water of Consistency and Setting Times

The water of consistency and setting times (initial and final) of OPC and LPC cements premixed with 0, 1, 2 and 3 wt. % PWL waste are plotted in Figure 2. Generally, the water of consistency gradually increased with PWL concentration up to 3 wt. % in both types of cements. Furthermore, the increase of water of consistency by using the same concentrations of PWL with LPC was slightly more than that with OPC. The 3 wt. % PWL waste increased the water of consistency from 28 to 31.67 wt. % with OPC and from 28.5 to 32.45 wt. % with LPC. So, the water of consistency was increased by 1.68-13.12 wt. % with OPC and by 4.11-13.85 wt. % with LPC compared with those of their blanks.

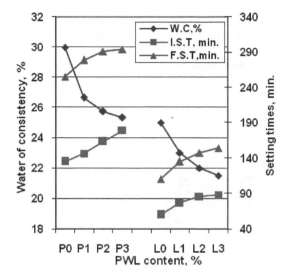

**Figure 2.** Water of consistency and setting times of the OPC and PLC cement pastes premixed with 0, 1, 2 and 3 wt. % PWL waste

On the other hand, both initial and final setting times with OPC or LPC pastes premixed with the PWL waste were gradually decreased. The 3 wt. % PWL waste decreased the initial and final setting times from 140 to 119 min. and from 255 to 231 min. with OPC, but from 148 to 128 min. and from 266 to 148 min. with LPC, respectively. Accordingly, it is clear that the setting times were faster with OPC than with LPC. Hence, it could be concluded that the PBL liquor waste can be used as a retarder for cement pastes [1,2,3,4].

## 3.2. Bulk Density and Apparent Porosity

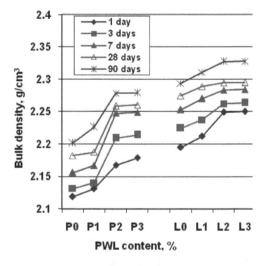

**Figure 3.** Bulk density of the OPC and PLCcement pastes premixed with 0, 1, 2 and 3 wt. % PWL waste cured up to 90 days

Figure 3 and Figure 4 show the bulk density and apparent porosity of the OPC and LPC cement pastes premixed with 0, 1, 2 and 3 wt. % PWL liquor waste, respectively. The bulk density was gradually increased with curing time, while the apparent porosity decreased. This is mainly attributed to the gradual and continual deposition of the formed hydration products in the pore system of the hardened cement pastes [1,14]. Moreover, the bulk density was further increased with PWL concentration up to 2 wt. % and be stable at 3%, whereas the apparent porosity decreased. This may be due to the

activation of cement phases by the presence of NaOH in the waste liquor and subsequently the amount of hydration products were increased when compared with those of the blanks [2]. With further increase in PWL concentration as with 3 wt. %, both bulk density and apparent porosity were almost the same. Thereby, the 3 wt. % PWL concentration is not more effective than 2 wt. % and hence it must be avoided [2,3,4,5].

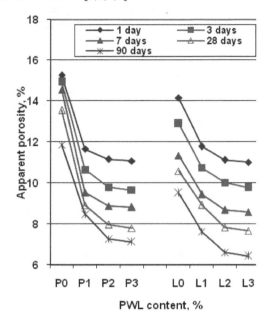

**Figure 4.** Apparent porosity of the OPC and PLC pastes premixed with 0, 1, 2 and 3 wt. % PWL waste cured up to 90 days

## 3.3. Combined Water Contents

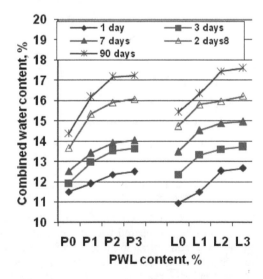

**Figure 5.** Combined water content of OPC and PLC pastes premixed with 0, 1, 2 and 3 % PWL waste cured up to 90 days

The combined water contents of the OPC and LPC cement pastes premixed with 0, 1, 2 and 3 wt. % PWL waste liquor are shown in Figure 5. The combined water contents of all cement pastes were increased with curing time up to 90 days. This was essentially attributed to the gradual and continuous formation of hydration products resulting from the hydration of the main phases of cement, particularly $C_3S$ and $\beta$-$C_2S$ [1,16]. The combined water contents increased gradually by the incorporation of PWL liquor up to 2 wt. % with both OPC and LPC and then tended to be stable with 3 wt. % in case of Bulk and apparent porosity [1,15,23]. The combined water contents of cement pastes with LPC are slightly higher than with OPC. This may be due to that the active group in PWL waste (-OH) is more effective with LPC than with OPC pastes [2]. Because the 3 wt. % PWL is not more effective than 2 wt. %, it is not desirable. Also, the PWL waste displayed the same trend with both types of cements at all curing times, but slightly more with LPC pastes [2,14].

## 3.4. Compressive Strength

The compressive strength of OPC and LPC cement pastes premixed with 0, 1, 2 and 3 wt. % of PWL waste liquor is shown in Figure 6. The compressive strength of the hardened cement pastes was generally increased with curing time up to 90 days. This is mainly attributed to the continual formation of hydration products which deposited into the pore structure of the cement pastes. So, the apparent porosity decreased gradually and the compactness of samples improved. Hence, the bulk density increased and this was positively reflected on the compressive strength [1,16,24]. The decrease of the apparent porosity and increase of the bulk density resulted from further increase of the hydration products [4,5,23].

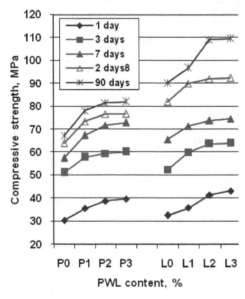

**Figure 6.** Compressive strength of OPC and PLC pastes premixed with 0, 1, 2 and 3 wt. % PWL waste cured up to 90 days

The higher compressive strength for both types of cement (OPC and LPC) compared with those of blank by increasing the PWL concentration up to 2 wt.% is mainly due to the high activation effect of the $-OH^-$ group present in the PWL particularly at later ages of hydration (28 and 90 days).

The addition of PWL waste to cement led to the formation of electrostatic repulsive forces between cement particles negatively charged by the adsorption of the liquor waste onto the cement surface which reduces the interparticle attraction between the cement particles leading to prevent the flocculation or agglomeration of cement. Accordingly a well-dispersed system is obtained [24,25,26]. Hence, the slight increase of w/c ratio and the dispersing effects (Figure 7) due to PWL waste addition helped to a large extent to improve and enhance the

compressive strength [1,26,27]. With 3 wt. % PWL waste, the compressive strength was not affected and therefore the optimum concentration is 2 wt. %. Consequently, the higher concentration is unnecessary. Figure 8 demonstrates the dissociation and orientation of the admixture particles as soon as its addition to the mixing water while Figure 9 shows the adsorption of the admixture particles on the cement grains and its rearrangement to achieve the equilibrium.

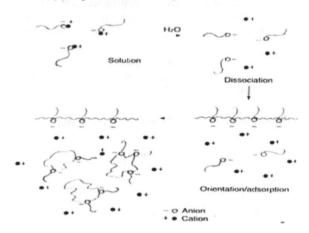

**(A)**                                    **(B)**

**Figure 7.** Dispersibility effect of admixtures, (A): In absence of admixture, (B): In presence of admixture)

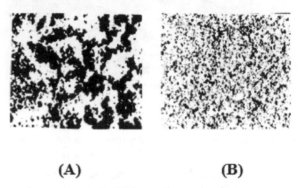

**Figure 8.** Surface reactivity and surfactant behaviour

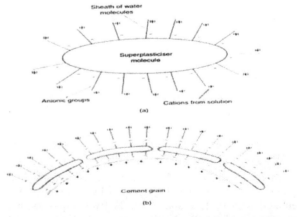

**Figure 9.** (a) Representation of an admixture molecule and (b) Mode of adsorption of admixture on cement grain

Also, the compressive strength values of OPC and LPC cement pastes are higher with the incorporation of PWL waste, where it is little more with LPC pastes than with OPC pastes. Therefore, the activation effect of the PWL liquor increased the rate of hydration which enhanced the cementing properties of the hardened cement pastes. This often has a positive action on the mechanical properties. It could then be recommended that the higher concentration of PWL waste (> 2 wt. %) is unfavorable with both types of cement. It is worth mentioning that the same trend achieved by using PWL waste was also achieved in a previous study [10] using the same concentration of black liquor waste (PBL) but with lower values, i.e. the PWL waste is more effective with both types of cements than PBL one.

### 3.5. IR Spectra

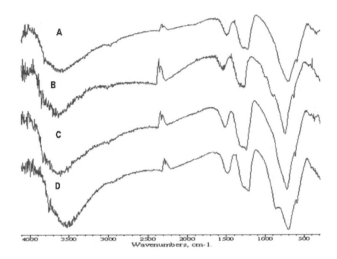

**Figure 10.** The FTIR spectra of OPC pastes cured up to 90 days (a), premixed with 1 wt. % PWL (b), 2 wt. % PWL (c) and 3 wt. % PWL (d)

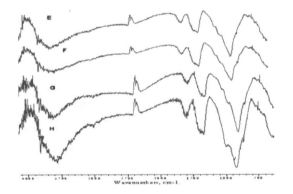

**Figure 11.** The FTIR spectra of LPC pastes cured up to 90 days (a), premixed with 1 % PWL (b), 2 wt. % PWL (c) and 3 wt. % PWL (d)

The FIR spectra of OPC (A-D) and LPC (E-H) cement pastes premixed with 0, 1, 2 and 3 wt. % of PWL waste are shown in Figure 10 and Figure 11, respectively. The sharp absorption band at 3644-3641 $cm^{-1}$ is related to the free $OH^-$ group coordinated to $Ca^{+2}$, i.e. Ca $(OH)_2$ or free lime. The intensity of the broad absorption band at 3450-3425 $cm^{-1}$, which was ascribed to the $OH^-$ group associated to $H^+$ bond that related to the symmetrical stretching frequency of water, increased in presence of PWL liquor. The two absorption bands at 2960 and 2860 $cm^{-1}$ are due to $-CH_2$ and $-CH_3$ from the residual organic mixture. The two absorption bands at 1645-1640 $cm^{-1}$ and 1430-1422 $cm^{-1}$ are related to the main silicate band involving Si-O stretching vibration bands of CSH, while the band at 1122-1112 $cm^{-1}$ may be due to CAH. The intensity of the two absorption bands at 990-980 $cm^{-1}$ and

874 $cm^{-1}$ characterizing $CO_3^{-2}$ and $SO_4^{-2}$ is irregular due to the rate of carbonation or sulphonation of CSH and /or CAH, respectively where the vibrations of $CO_3^{-2}$ are smaller than those of $SO_4^{-2}$. Also, the intensity of the absorption bands of Si-O, CAH, $CO_3^{-2}$ and $SO_4^{-2}$ are slightly higher with PWL waste cement pastes. The intensities of the main characteristic peaks were slightly improved with LPC than with OPC pastes.

## 4. Conclusion

The PWL liquor activates the cement phases and improves the rate of hydration. The incorporation of PWL waste with OPC and/or LPC pastes enhances the w/c ratio for LPC more than OPC cement pastes and increases the setting times (initial and final). So, it can be used as a retarder. The combined water contents, bulk density and apparent porosity at all curing ages of hydration are improved and gradually increased. As a result, the compressive strength was increased, particularly at later stages of hydration (28 and 90 days). No new phases are detected by IR, but only increased intensities for the formed hydrates was observed by the addition of PWL waste than those of the blanks. The optimum PWL concentration is 2 % because the higher concentration (3 wt. %) has little or no further effect on all cement properties than 2 wt. %. The IR analysis showed the highly modified crystals of hydrates in presence of PWL. Moreover, the presence of PWL eventually has preferential efficiency with LPC pastes than with OPC pastes.

## References

[1] Hewlett, P. C. "*Lea's Chemistry of Cement and Concrete*"; 4th Edn.; John Wiley & Sons Inc, New York, 1998.

[2] Johan, G. and Fogelham, C. J. "Papermaking Sciences and Technology, 6A Chemical pulping", Finland, Tappi Press, ISBN 952-5216-06-3, 2000, 41-42.

[3] Darweesh, H. H. "Effect of the combination of some pozzolanic wastes on the properties of Portland cement pastes" iiC L'industria italiana del Cemento, Italy, 808, 2005, 298-311.

[4] El-Didamony, H.; Darweesh, H. H. and Mostafa, R. A. "Characteristics of pozzolanic cement pastes Part I: Physico-mechanical properties" Sil. Ind. (Cer. Sci. & Techn.), Belgium, 73, Nr. 11-12, 2008, 193-200.

[5] Darweesh, H. M. H. "Alkali-activation of slag-rich cement pastes with sodium hydroxide", iiC l'industria italiana del Cemento, 826, 12, 2006, 992-1007.

[6] Darweesh, H. H. M. "Preparation of Ca-lignosulfonte from waste liquor and its application as cement dispersant", J. Chemistry and Materials Research (JCMR), Vol., 1, No. 2, 2014, 28 -34.

[7] Ayoub, M. M. H.; Nasr, H. E. and Darweesh, H. H. M. "Synthesis, characterization and cement application of vinyl acetate water soluble graft polymers", Polymer-Plastics Techn. and Eng. USA, 44, 2, 2005, 305-319.

[8] Ayoub, M. M. H.; Darweesh, H. H. M. and Negm, S. M. "Utilization of hydrophilic copolymers as superplastisizers for cement pastes", Cemento Hormigon, 919, 2007, 4-15.

[9] Adams, T. N. and Fredrick, W. J. "Kraft Recovery Boiler Physical and Chemical Processes", Amer. Paper Institute (1988).

[10] Darweesh, H. H.; Abdel-Kader, A.H. and El-Meligy, M.G. "Utilization of pulp black liquor waste as a cement admixture", Intern. Journ. Basic and Applied Sciences, 2, 3, 2013, 230-238.

[11] Behm, M. and Simonsson, D. "Electrochemical production of polysulfide and sodium hydroxide from white liquor Part I: Experiments with rotating disc and ring-disc electrodes", Applied Electrochemistry, 27 (1997) 507-518.

[12] Uchikawa, H.; Hanchara, S. and Sawaki, D. "The role of steric repulsive force in the dispersion of cement particles of fresh paste prepared with organic admixture", Cem. Concr. Res., Vol. 27, No. 1, 1997, 37–50.

[13] Ayoub, M. M. H.; Nasr, H. E. and Darweesh, H. H. M. "Characterization and Utilization of Polystyrene and Polyacrylamide-Graft-Methoxypolyethylene as Cement Admixtures", Polymer-Plastics Technology and Engineering, 45, 2006, 1307-1315.

[14] Amin, A;. Darweesh, H. H.; Morsi, S. M. and H. Ayoub, M. M. "Employing of some Maleic anhydride based hyperb-ranched polyesteramides as new polymeric admixtures for cement" App. Poly. Science, Vol. 121, 2011, 309-320.

[15] Chung, D.D.L., "Use of polymers for cement-based structural materials", J. Mater. Sci., 39, 1, 2004, 2973-2978.

[16] Ramachandran, V.S.; Feldman, R.F.; "Concrete Admixtures Handbook, Properties, Science and Technology", 2nd Edn.; Noyes Publications: New Jersey, 1995

[17] Jumadurdiyev, A; Ozkul, M. H.; Saglam, A. R. and Parlak, N. "The utilization of beet molasses as a retarding and water-reducing admixture for concrete", Cem. Concr. Res., 35, 5, 2005, 874-882.

[18] Abo-El-Enein,. S. A.; El-Ashry, S. H. ; El-Sukkary, M. M. A.; Hussain, M. H. M. and Gad, E. A. M. "Effect of some admixtures based on naphthalene or benzene on the mechanical properties and physicochemical properties of Portland cement pastes", Sil. Ind. 62, Nr. 3-4, 1997, 75-81.

[19] ASTM –Standards "Standard Test Method for Normal water of Consistency of Hydraulic Cement", C187-86, 1993, 148-150.

[20] ASTM –Standards "Standard Test Method for Setting Time of Hydraulic Cement", C191-92, 1993, 866-868.

[21] ASTM-Standards "Standard Test Method for Compressive Strength of Dimension Stone", C170-90, 1993, 828-830.

[22] Darweesh, H.H.M.; El-Alfi, E.A., "Effect of some water-soluble polymer admixtures on the hydration characteristics of Portland cement pastes",Sil. Ind., 71, Nr. 1-2, 2005, 27-32.

[23] Rixom, R.; Mailvaganam, N., "Chemical Admixtures for Concrete", 3rd Edn. E & FN Spon, 1999.

[24] Ayoub, M.M.; Nasr, H.E.; Darweesh, H.H.M., "Characterization and utilization of polystyrene and polyacrylamide-graft-methoxypolyethylene as cement Admixtures", Polym., Plast. Technol. & Eng., 45, 2, 2005, 1307-1315.

[25] Amin, A.; Abdel-Megied, A. E.; Darweesh, H. H.; Ayoub, M. M. and. Selim, M. S "New Polymeric Admixture for Cement based on Hyperbranched Poly amide-ester with Pentaerithritol Core", ISRN Materials Science, Hindaway Publishing Corporation, vol. 2013, 1-7, 2013.

[26] Knapen, E and Van Gemert, D. "Cement hydration and microstructure formation in the presence of water-soluble polymers", Cem. Concr. Res., 39, 1, 2009, 6-13.

[27] Kersener, Z.; Darweesh, H.H.M. and Routil, L. "Alkali-activated slag - rich cement pastes with sodium silicate and water glass" Cem. Hormigon, Spain, Vol. 4, No. 945, 2011, 18-26.

# 6

# Economic non Metallic Mineral Resources in Quaternary Sediments of Tehran and its Environmental Effects

Kaveh Khaksar[*]

Institute of Scientific Applied Higher Education of Jihad-e-Agriculture, Education and Extension Organization, Ministry of Agriculture, Department of Soil Science, Karaj, Iran
*Corresponding author: kavehkhaksar@gmail.com

**Abstract** Air pollution in Tehran is widely recognized as a serious environmental challenge, posing significant threats to the health of the resident population. As one of the most evident natural hazards and significant environmental issues, the dust phenomenon has raised significant concern within the research community. In light of the negative effect of dust mass in urban areas, robust and effective early warning systems are necessary; continuously enhanced monitoring of dust aerosols are a critical step in developing such systems. Tehran plain in general is of alluvial fan accumulation. The Quaternary of Tehran plain has up to 1100 m of sediments belongs to four lithostratigraphic units. They contain both non-metallic resources. Non-metallic resources of Quaternary strata include construction aggregates for the residential, industrial and transportation segments of the population, ceramic clays and laterites. Tehran region can be divided into 4 geological units as follows: Hezardareh F., Kahrizak F., Tehran alluvial F. and Holocene alluvium. The quaternary sand and gravel mines of Tehran have been exploited widely and this exploration is one of the causes of the environmental pollution. The over-exploitation of construction material mines in southern Tehran has become a source of dust particles, adding to air pollution problems in the Iranian capital.

*Keywords:* *quaternary, sand and gravel mines, Tehran plain, environmental pollution*

## 1. Introduction

Air pollution in Iran has significant natural and anthropogenic sources. Cold season and temperature inversion, dust, particles released from construction work in the city, emissions from old cars, underdeveloped public transportation, too many cars in the streets, drought, lack of enough rain and wind due to special location of Tehran are among the other elements lie behind air pollution in Tehran.

The primary anthropogenic sources are industrial activities and transportation, both of which are centered in urban areas, while the main natural source is wind-blown dust. Cars account for 70 to 80 percent of the normal air pollution in Tehran due to the large number of automobiles, heavy traffic congestion, and petrol with a sulfur content 2-3 times greater than legally permissible levels [1]. Tehran's air pollution is responsible for thousands of deaths and costs millions of dollars each year [2]. Mining activity is one of the new causes of pollution in Tehran. Yet the country is one of the most important mineral producers in the world, ranked among 15 major mineral-rich countries [3], holding some 68 types of

minerals, 37 billion tons of proven reserves and more than 57 billion tons of potential reservoirs. Mineral production contributes only 0.6 per cent to the country's GDP [4]. Add other mining-related industries and this figure increases to just four per cent (2005). Many factors have contributed to this, namely lack of suitable infrastructure, legal barriers, exploration difficulties, and government control over all resources.

The most important mines in Iran include coal, metallic minerals, sand and gravel, chemical minerals and salt. Khorasan has the most operating mines in Iran. Other large deposits which mostly remain underdeveloped are zinc (world's largest), copper [5], iron (world's ninth largest), [6] uranium (world's tenth largest) and lead (world's eleventh largest) [7-13].

Iran with roughly 1% of the world's population holds more than 7% of the world's total mineral reserves [14,15].

Tehran has developed on recent sediment and quaternary. Geological maps confirm that quaternary and Pliocene alluviums and moraine deposit have developed in Tehran desert.

The Quaternary deposits comprising semiconsolidated to unconsolidated gravel, sand, silt and clay occupy the greatest part of Tehran plain. These deposits use in construction aggregates for the residual, industrial and transportation segments of the population, ceramic clays, and laterites. The Construction sand and gravel can be classified into two types of deposits:

1-Alluvial fan sands and gravel

2-River sand and gravel

Much of the ground water essential to agriculture and human existence emanates from aquifers in quaternary sedimentary environments.

Air pollution in Tehran is widely recognized as a serious environmental challenge, posing significant threats to the health of the resident population. Improving air quality will be difficult for many reasons, including climate and topography, heavy dependence on motor vehicles for mobility, and limited resources to reduce polluting emissions.

# 2. Materials and Methods

## 2.1. Study Area

Tehran Province is one of the 31 provinces of Iran. It covers an area of 18,909 square kilometers and is located in the north central plateau of Iran.

Coordinates of Tehran in decimal degrees

The study area consists of the 22 municipal districts of Greater Tehran (Figure 1). This area is about 1100 km$^2$ wide and is located between 51° 15′ and 51° 33′ Eastern longitude and 35° 32′ and 35° 49′ Northern latitude. Tehran plain starts near the city center with an elevation of 1250 meters above sea level, and stretches over to the southern parts of Rey in a mild slope. The elevation declines in a mild west-east slope.

Tehran is located on relatively recent alluvial deposits extending toward the south from the foothills of Alborz Mountains range. These deposits are the result of river activity and seasonal inundations.

Environmentally, the climate of Tehran province in the southern areas is warm and dry, but in the mountain vicinity is cold and semi-humid, and in the higher regions is cold with long winters. The hottest months of the year are from mid-July to mid-September when temperatures range from 28°C (82°F) to 30°C (86°F) and the coldest months experience 1°C (34°F) around December–January, but at certain times in winter it can reach −15°C (5°F). Tehran city has moderate winters and hot summers. Average annual rainfall is approximately 200 millimeters (7.9 in), the maximum being during the winter season. On the whole, the province has a semiarid, steppe climate in the south and an alpine climate in the north.

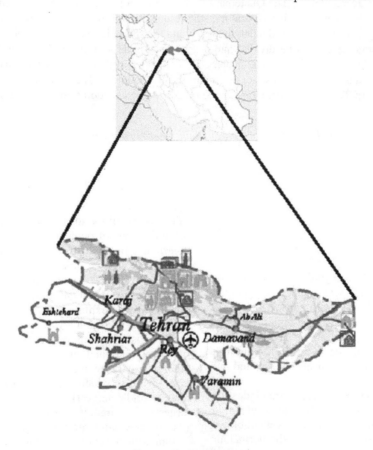

**Figure 1.** Study area in Iran

## 2.2. Methods

The main propose of this paper is the study of the new source of air pollution in Tehran, due to increase licensing of mining activities. New environmental pollution source in Tehran is dust particles, produced from mine exploration. Dust particle pollution have been produced from mine activities in the south and southeast of Tehran city.

Growing of dust pollution is due to the increasing number of mines in the area.

The Quaternary stratigraphy of Tehran is discussed in some detail. The following procedures were adopted in the data analysis:

-Preparation of geological maps.

-Field studies: The base of this study was the original geological information's of the upper plains including the properties of structure, petrology of geological outcrops in studied areas.

## 2.3. Bibliotheca Studies

The collection of data for this study was carried out using resources review.

# 3. Results and Discussion

**Stratigraphy**

Tehran has developed on recent sediment and quaternary. Geological maps confirm that quaternary and Pliocene alluviums and moraine deposit have developed in Tehran desert.

Bed rock: The bed rock of Tehran is the Tertiary formations, mostly Eocene lava, showing in the mountainous areas in the north of the city. The younger sediment has formed on this bed rock. The bed rock of the eastern heights in Sehpayeh and Bibi Shahrbanoo mountains have been formed from the dolomite limestone from Triassic and Cretaceous ages.

The Quaternary sediments of Tehran region were studied during the recent 15 years. On the basis of the stratigraphical and sedimentological development of the Quaternary deposits, the region can be divided into 4 Formations as follows [16,17,18]:

**1- Hezardareh Formation**

The name of the formation, literally "Bad land morphology" has been inspired by the geomorphologic properties of its surface and the existence of the multitude of its erosional valleys of great density. The oldest and most important of the four units. The Hezardarreh formation consists of materials that had source like the overlaying B and C units, mainly in the Eocene formation. The average grain size is between one centimeter to one decimeter. A characteristic of this ancient alluvium is the regularity of its bedding. Unregulary it displays beds of sandy gravel which is cemented by lime carbonate. An angular unconformity was observed by Rieben [16]. Further study by revealed three unconformity in A (Hezardarreh ) formation. This unit is folded and dissected by numerous faults.

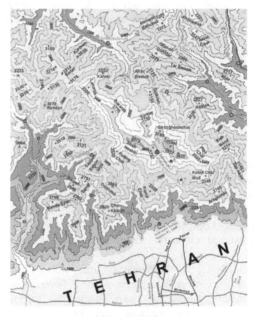

**Figure 2.** Topographical map of Tehran plain

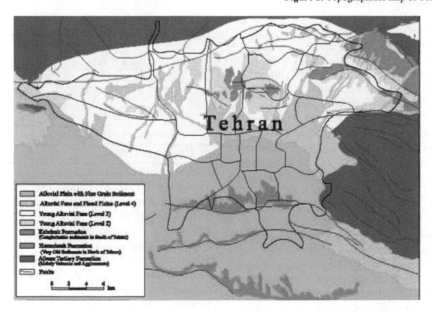

**Figure 3.** Geological map of Tehran [21]

The distinguished characters of Hezardareh Formation are:

-High thickness, Homogeneous, Regular stratification, Pebble with medial size, Advanced grade of weathering, Grey in color, High slope of layers.

-There are semi rounded grain results from Karaj Formation (90%) and other rocks (10%).

Which indicate during of Hezardareh Formation sedimentation uplift and erosion of Alborz Mountains. Hezardareh alluvial Formation overlaying Upper red Formation (Pliocene), gradually in the Ivanaky region. In the upper part of Upper red Formation, which composed of marl and red sandstone, with much conglomeratic layers, which indicated existence of torrential environments and came up Alborz Mountain and deposition of Hezardareh Formation begun. Hezardareh Formation is divided in two members. The first member of Hezardareh Formation has much little porosity, high erosion of pebble and strongly cementation. The second member of it has highly porosity respect of first member and less grade of erosion of pebble. Hezardareh Formation after folding, faulting and intense alternation, overlaying by Kahrizak Formation.

Age: Plio-Pleistocene

## 2- Kahrizak Formation

Kahrizak Formation in the type section overlaying Hezardareh Formation unconformitily that corresponding to tectonic phase of Passadenian and has been covered by Tehran alluvium. This unit is made of sandy or clayey loam. the colore is pale brown and overlay the A beds with a angular unconformity. In contrast to A beds it is not consolidated. The size of pebels in a matrix of sand and clay differ considerably in dimensions.

This heterogeneous formation overlay the surface of strongly eroded A-formation. The heterogenity of this formation was for Riviere [12] an indication for to take it as a moraine. The tilting of this unit exceed max 15 degree. The observation of an angular unconformity in B-beds leaded Rieben [5] to distinguish a lower and a upper B-formation.

Kahrizak Formation has different characters from layer to layer because of heterogeneous, mechanical resistance and changeable porosity.

Distinguished characters from Hezardareh Formation were:

-Unsolidificate, heterogeneous and poorly sorted conglomerate, Gravel size ranging from several cm to several meters which has been situated in sandstone cement, Low feeble cement and little mechanical resistance, Slope of layers (maximum of 15), Darkness color of Formation.

Age: Middle Quaternary.

## 3- Tehran alluvial Formation

Formation (Tehran plain alluvium) - it forms from gravels, pebbles and sands in the cement of sands and silts. The permeability and the strength of the formation are high. The volume of Tehran alluvium fans consists mainly of C formation. As to more thickness and permeability, it forms a suitable groundwater resource in Tehran.

Tehran alluvium Formation includes younger alluvial fans which from southern pediments of Alborz Mountains continue to south and spread part of Tehran city building on it.

This formation in general created by alluvial and streams sedimentation. The thickness of them is up to 50 m. The homogeneous sediments were composed of gravel, sand, silt and boulders. On the unconsolidated sediments species of stratification were seen. The presence of red conglomeratic layers laterizaed and weathering surface indicated the suspension of sedimentation. This formation is divided into two different stratigraphical alluvial plain units:

A) Khorramabad alluvial Formation. This formation has covered the formation of the older alluvial fans in the south of Tehran, forming a fairly smooth plain. In the northern and eastern parts, the sand is richer. The southwest of the plain is dominated by fine grain material such as clay and silt.

B) River alluviums unit. The most recent alluviums of alluvial levels in Tehran are formed by the deposits of the flood plain of rivers consisting Grave land pebble size clasts. These materials have an alluvial origin and are not very hard.

The layers of this Formation had horizontal aspect, which did not support tectonic movements. Tehran alluvium Formation overlaid Kahrizak Formation and covered by Holocene alluvium. These alluvium sediments were formed of erosion and resedimentation of Kahrizak Formation.

Age: Upper Quaternary

## 4- Holocene alluvium (Recent alluvium)

Formation (Recent alluvium) - the most recent riverbeds, alluvium terraces and the young fans consist of the D formation. It includes gravel and pebble, by the weak cements in the north, which gradually convert to silt and clay toward the south. Due to weathering, a layer with the thickness of 2-3 m is formed on the surface of the formation (Figure 4 and Table 1)

Age: Holocene

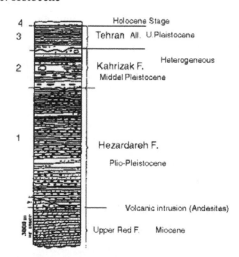

**Figure 4.** General stratigraphical section of Tehran alluvium, by [16]

**Table 1. Alluvial groups in Tehran [17]**

| Bed | Geological name | Thickness (m) |
|-----|-----------------|---------------|
| D | Recent Alluvium | Up to 20 |
| C | Tehran Formation | Up to 100 |
| B | Heterogenous Formation (Kahrizak F.) | Up to 300 |
| A | Hezardarreh Formation | Up to 500 |

## Mineral resource potential

Tehran province structurally belongs to two structural zone of Iran namely, Alborz and central Iran zones. Based on its geological and structural characters such as existence of magmatism (intrusive and extrusive), alteration, thrusting, and faults and hydrothermal, it contains diverse type of mineral resources like:

- Robatkarim manganese deposits 45 NW of Tehran.
- Sabou refractory clay deposit.
- Hovir feldspar and dolomite (Damavand).
- Keripton-Ah kaolin deposit.
- Silica sand and refractory sand in Masklu, Mobarakabad, Sarneza, Mosha and Damavand.
- Celestite in Mikabad.
- Talc and asbestos 48 north of Karaj (Sierra).
- Magnesite and huntite 75 km SE of Varamin.

Other deposit of Tehran province are sand and gravel (110 areas), limestone (14 areas), Gypsum (4 areas), marl

(1 area), industrial clay (4 areas), evaporitic salt (16 areas), copper (2 areas), iron, lead- zinc and coal (3 areas), bauxite (1 area), phosphate (1 area), barite (4 areas), dolomite (2 areas), ornamental stones (41 areas), bentonite (2 areas), dense limestone (2 areas), salt (2 areas) and travertine (1 area).

A large volume of silty clay and clay of economic importance was found lower most of the mapped area.

Aggregate sand and gravel, for construction activities, which are now widely developed in Tehran region. There are abundant sand and gravel deposits along the rivers throughout the western and eastern part of Tehran. Alluvial fan sand and gravel can be excavated from alluvial fan deposits. There are three large alluvial fan deposits in the western (Karaj), central (Kan) and eastern (Jajerud) part of Tehran plain, which are huge suppliers of sand and gravel for construction.

The total numbers of mines in Iran are up to 3200. The numbers of sand & gravel mines throughout Iran are 1174, which totally are open mines Table 2.

**Table 2. Number of employees (persons)**

| Sand and gravel, Ballast and Mineral cartridge (m³) | Minerals except sand and gravel, ballast and mineral cartridge (Tons) | Contractor | Without wages and salaries | With wages and salaries | Sum | Number of Mines | Province |
|---|---|---|---|---|---|---|---|
| 15747811 | 16531893 | 265 | 78 | 4029 | 4372 | 204 | Tehran |

**Figure 5.** Tehran particle dust pollution

There are 204 different mines in Tehran, with ~4372 employees. The numbers of sand and gravel mines are 110, which produce totally 36.500.000 tons annually evaluated more than 109.5 million US dollars. Annual production of sand and gravel in Iran is 63707157 m3. The whole value of annual production is more than 191 million US dollars. The Sand and Gravel business in Tehran shows a growth rate of about 17, 5% in 2013.

Than 110 mines of sand and gravel plays an important role in the economy of the Tehran province. The amount of certain resources of sand & gravel in Tehran evaluated more than one milliard m³, which in these mines up to 2500 persons are as employees. Sand and gravel extraction in the river areas is 80 percent.

**Environmental impact**

From an environmental standpoint sand and gravel mines near the Tehran, are one of the pollution sources, which increases the density of air (Figure 5), causing respiratory diseases. From the point of view environmental pollution from mining and its impact on people's quality of life is also very important, mines are indiscriminate exploitation leads to environmental degradation, and land settlement, pollution, etc. are actually some provincial official's dust the storm also caused the famous capital Tehran air pollution from mines knew every day.

Extracted gravel and sand are washed and processed by the factories near the river and transported to the consuming centers. Washed gravel and sand suspend fine sediments like clay and slit in the river. These fine sediments settle in the river and make impermeable layers in the bed and prevent seepage of surface water to groundwater and increase the dangers of flooding and damaging to the adjacent regions of the river and destroy the environment [18].

When we entered the adhesion of particles with diameters less than 2.5 microns in drought periods, and are a breeze to get up from the ground and in space. According to reports, the particles can stay suspended in the air for 15 years. Of course this in mind that external sources of pollution by about 30 percent and 70 percent dust the threat posed by these particles are related to internal resources. Deserts around Tehran as you mention the problem of drought, the main source of air pollution have been Tehran. The head of Iranian Parliament's Environment Commission says the city has been issuing too many licenses for mining construction materials in southern Tehran, and that is now "making Tehran out of breath [22].

# 4. Conclusion

The Quaternary sediments of Tehran area are drawn considering the geology and other physical processes occurring in the investigated area. Tehran region can be divided into 4 Formations as follows: Hezardareh F., Kahrizak F., Tehran alluvial F. and Holocene alluvium.

-The mapped area consisted of four quaternary geological units. They were:

-Floodplain deposits covered the major part of total mapped area.

-Four geological units (Formation) with different types of vertical section of sediments.

-All the deposits in the area were fluvial in origin.

-The economic non-metallic mineral resources of Quaternary strata include construction materials; consist of clay, sand and gravel.

The main reason for the air pollution in the capital is the vehicles. Tehran is wedged between two mountains that trap the fumes of its bumper-to-bumper traffic. The storm brought with him a message to Tehran: recently painted the sky with a layer of "toxic dust" from standard gasoline vehicles over the sprawling capital had already elusive yellow, composed of particles produced by sand and gravel mine activities, which also appear turned Inhaled air is several times more dangerous than the old infection. It seems that one way to reduce pollution these licences should be immediately rescinded.

# References

[1] Abrishamchi, A., 2013. Overview of Key Urban Air Pollution Problems in Iran and its Capital City, Tehran, PROCEEDINGS OF THE U.S.-IRAN SYMPOSIUM ON AIR POLLUTION IN MEGACITIES, Irvine, California, September 3-5, 2013.

[2] Bayat, R.; Torkian, R.; Najafi, M.; Askariyeh, M.H.; Arhami, [2012] M. Source apportionment of Tehran's air pollution by emission inventory, Paper presented at the 2012 International Emission Inventory Conference of EPA. 13-16 Aug 2012. Tampa, FL, U.S.A.

[3] "Mining in Iran – Country Mine". Info Mine. Retrieved 18 October 2011.

[4] "Iran's mineral exports up 39 percent". Press TV. 17 January 2011. Retrieved 18 October 2011.

[5] World's ninth largest - revised from second largest in 2010.

[6] US Geological Survey, Mineral commodity summary: iron ore, 2013.

[7] http://www.tehrantimes.com/index.php/economy-and-business/92737-irans-copper-output-will-increase-35-fold.

[8] http://www.irtp.com/howto/partner/partner/chap2/chap2iii.htm.

[9] "Iran's Economy". Iraniantrade.org. Retrieved 18 October 2011.

[10] http://www.earthstonegroup.com/blog/?p=709.

[11] "advantageaustria.org – official business portal Austria (B2B, import, export, Austrian products, investment)".

[12] Austriantrade.org. Retrieved 18 October 2011.

[13] "Iran's Islamic Revolution and Its Future – Harvard – Belfer Center for Science and International Affairs".

[14] "Iran's Islamic Revolution and Its Future – Harvard – Belfer Center for Science and International Affairs".

[15] Belfercenter.ksg.harvard.edu. Retrieved 18 October 2011.

[16] RIEBEN, H. 1966 - Geological observations on alluvial deposits in northern Iran. Geol. Survey of Iran, Rep. No. 9, 40 p., 10 figs., 1 pl. (map).

[17] Khaksar, K. and Khaksar, K., (2012). Correlation between quaternary stratigraphy units in different geological zones of Iran, International Research Journal of Geology and Mining (IRJGM) (2276-6618) Vol. 2(6) pp. 141-147, August.

[18] Khaksar K., Rahmati M. and Haghighi S., (2011). Quaternary stratigraphy of northern Alborz range – Iran, Proceedings of the VII All – Russian Quaternary Conference "The Quaternary in all of its variety. Basic issues, results, and major trends of further research", 12-17 September 2011.

[19] Ghayomian, J., Fatemi Aghda, S. M., Maleki, M. and Shoaei, Z., (2006). Engineering Geology of Quaternary Deposits of Greater Tehran, Iran, The Geological Society of London 2006, IAEG2006 Paper number 248.

[20] Hashemi Monfared, S. A. 2008, Environmental effects of irregular extracting of gravel from river beds, EE08 Proceedings of the 3rd IASME/WSEAS international conference on Energy & Environment P. 213-218.

[21] Tehran municipality, Public & International Relations Department, (1014), http://www.tehran.ir/.

[22] Shahrvand daily, Mining in southern Tehran adds to pollution in capital, July. 5. 2014.

# Geological Analysis of Zakiganj Upazila and Feasibility Study of Available Geo Resources

**Mohammad Masudul Alam[1],\*, Mir Raisul Islam[2], Md. Ashraful Islam Khan[2]**

[1]Department of Petroleum and Mineral Resources Engineering, Bangladesh University of Engineering Technology, Dhaka, Bangladesh
[2]Department of Petroleum and Mining Engineering, Shahjalal University of Science and Technology, Sylhet, Bangladesh
*Corresponding author: masud_pge@yahoo.com

**Abstract** This paper has been being written in pursuit of analyzing geology, evaluating existing mineral, oil & gas reserves and feasibility study of prevailing geo-resources in Zakiganj Upazila, one of the resourceful Upazilas in Sylhet region. Surma & Kushiyara River along with the Zakiganj deposit a large amount of sand every year which is usually used as raw materials of the nearest bricks factory. However, sand deposit is not extracted commercially but it would be prospective resources of our country to minimize the construction cost. This study usually entails sufficient information on extracting, processing, metallurgical, economic and other relevant factors.

*Keywords:* geology, feasibility study, Zakigonj, sand deposit

## 1. Introduction

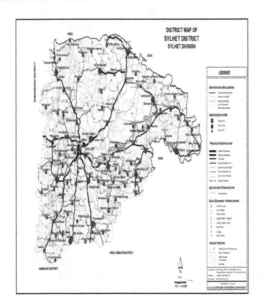

**Figure 1.** Sylhet district [1]

Sylhet district is located at 24.8917°N 91.8833°E 24.8917; 91.8833, in the north eastern region of Bangladesh. Geologically, this region is complex having diverse geomorphology; high topography of Plio-Miocene age such as Khasi and Jaintia hills and small hillocks along the border. At the centre there is a vast low laying flood plain, locally called Haors. Available limestone deposits in different parts of the region suggest that the whole area was under the ocean in the Oligo-Miocene.

Because of the geologic Structure most of natural resources such as gas, oil, hard rock, and limestone and so on are accumulated in Sylhet. According to these geologic positions it's highly recommended that Zakiganj is highly probable of geo resources contents.

### 1.1. Geography of Zakiganj

It is located between 24°51' and 25°00' North latitudes and between 92°13' and 92°30' East longitudes. The Upazila is bounded on the North by India and Kanaighat Upazila, on the East and South by India and on the west by Beanibazar and Kanaighat Upazilas.

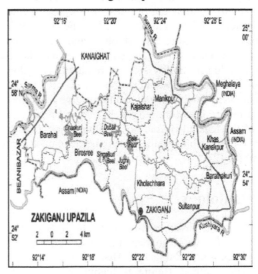

**Figure 2.** ZakiganjUpazila [3]

### 1.2. Some Features of Zakiganj Upazila

Zakiganj (Town) consists of 10 wards and 23 mahallas. The area of the town is 13.32 km². It has a population of 10465; male 50.31%, female 49.69%; density of population is 786 per km². Literacy rate among the town people is 34.1%.

**Administration**: Zakiganj Thana was established on 27 August 1947 and was turned into an Upazila in 1983. The Upazila consists of one municipality, 9 union Parishads, 119 Mouzas and 286 Villages.

**Population**: 174038; male 50.55%, female 49.45%; Muslim 86.48%, Hindu 13.47%, Buddhist, Christian and others 0.05%.

**Land use**: Arable land 20743.47 hectares, fallow land 604.21 hectares, grassland 1097.39 hectares; single crop 25%, double crop 60% and triple crop land 15%.

**Communication facilities**:

**Roads**: Pucca 124 km and mud road 1005 km; waterways 11 Nautical Mile.

**River**: Kushiara & Surma are along the border of the Zakiganj & divided India & Bangladesh which would be the probable reserve of sand deposit.

**Figure 3.** River Bed of Zakiganj Upazila [4]

## 1.3. Feasibility

A Project Feasibility Study is an exercise that involves documenting each of the potential solutions to a particular business problem or opportunity. Feasibility studies aim to objectively and rationally uncover the strengths and weaknesses of the existing business or proposed venture, opportunities and threats as presented by the environment, the resources required to carry through, and ultimately the prospects for success. In its simplest terms, the two criteria to judge feasibility are cost required and value to be attained. A feasibility study allows project managers to investigate the possible negative and positive outcomes of a project before investing too much time and money. The results of the feasibility study determine which, if any, of a number of feasible solutions will be developed in the design phase [2].

## 2. Methodology

The comprehensiveness of the study is to evaluate feasibility of the geo-resources amongst the available all geo-resources in Zakiganj Thana was done by questionnaire method. Before evaluating the feasibility it was important to know the geology of Zakiganj Upazila and this is done by reviewing different paper and books about the geology of Bangladesh. Study of feasibility of

prevailing resources in the area of interest begins with locating the geo-resources, identifying the geo-resources, performing qualitative study on the identified resources, introducing different possible methods of extraction, calculating & performing a qualitative study of the economic value of the studied geo-resources and finally assessing the feasibility of the prevailing resources. Geological concepts and concise basics about rocks and minerals helps identify the located geo-resources. Qualitative study on the identified geo-resources involves the characteristics study of the identified geo-resources. When a huge body of geo-resources is located and identified then the one thing one should come up with is the proper method to extract the identified resources. Evaluating the economic value of the identified resources basically focuses on the market demand, market value and other relevant factors. For a preliminary feasibility study on prevailing geo-resources, doing a qualitative cost/benefit analysis on the proposed project helps to obtain the preliminary feasibility. If the total project cost outweighs the total project benefit we call it Economically not feasible, otherwise this is feasible.

## 3. Geological Condition of Zakiganj

### 3.1. Sedimentation and Tectonics of the Sylhet Trough

The Sylhet trough, a sub-basin of the Bengal Basin in North-Eastern Bangladesh, contains a thick fill (12 to 16 km) of late Mesozoic and Cenozoic strata that record its tectonic evolution. The Sylhet trough occupied a slope/basinal setting on a passive continental margin from late Mesozoic through Eocene time. Subsidence may have increased slightly in Oligocene time when the trough was located in the distal part of a foreland basin paired to the Indo-Burman ranges. Oligocene fluvial-deltaic strata (Barail Formation) were derived from incipient uplifts in the eastern Himalayas. Subsidence increased markedly in the Miocene epoch in response to Western encroachment of the Indo-Burman ranges. Miocene to earliest Pliocene sediments of the Surma Group was deposited in a large, mud-rich delta system that may have drained a significant proportion of the eastern Himalayas.

Subsidence rates in the Sylhet trough increased dramatically (3-8 times) from Miocene to Pliocene-Pleistocene time when the fluvial Tipam Sandstone and DupiTila Formation were deposited. This dramatic subsidence change is attributed to south-directed overthrusting of the Shillong Plateau on the Dauki fault for the following reasons. (i) Pliocene and Pleistocene strata thin markedly away from the Shillong Plateau, consistent with a crustal load emplaced on the northern basin margin. (ii) The Shillong Plateau is draped by Mesozoic to Miocene rocks, but Pliocene and younger strata are not represented, suggesting that the massif was an uplifted block at this time. (iii) South-directed over thrusting of the Shillong Plateau is consistent with gravity data and with recent seismotectonic observations. Sandstone in the Tipam has a marked increase in sedimentary lithic fragments compared to older rocks, reflecting uplift and erosion of the sedimentary cover of the Shillong Plateau. If the Dauki fault has a dip similar to that of other Himalayan overthrusts, then a few tens of kilometers of horizontal tectonic transport would be required to carry the Shillong Plateau to its present elevation. Uplift of the Shillong Plateau probably generated a major (~300 km) westward shift in the course of the Brahmaputra River.

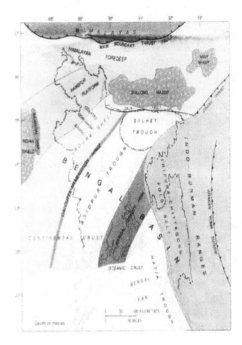

**Figure 4.** Geologic map of Bangladesh[5]

## 3.2. Structural Framework

The tectonic framework of Bangladesh is broadly divided into two main units : i) Stable platform in the North-West and ii) Deep (geosyncline) basin to the South-East. The deep geosynclinal basin is further subdivided into two parts-

- Fold belt in the East
- Foredeep in the West.

Our study area Zakiganj is located in the Eastern fold belt which is the most prolific natural gas province and has been the center of exploration activities in Bangladesh. The fold belt is characterized by series of meridional to submeridional folds and extends into the Indian territory of Assam, Tripura, and Mizoram to the east. Also known as frontal fold belt, this province represents the western and outermost part of the Indoburman origin. The fold belt shows sign of diminishing intensity of structures towards the West in which direction it gradually fade away and merge with the central foredeep province. Its boundary with the foredeep is therefore gradational, indistinct and arbitrary [6].

### 3.2.1. Characteristics of Eastern Fold Belt

**Source Rock:** The source of natural gas found in the province is believed to be the shale beds of Jenum Formation (Barail Group) of the Oligocene age. The shale have total organic carbon (TOC) of 0.6% to 2.4% and have attained thermal maturity with vitrinitereflectance of 0.65% at total depth indicating marginal maturity but would be fully mature in adjacent generative depression. It has been suggested that gas has been generated at depths

between 6000 to 8000 meters below the surface and migrated up through multi kilometer sand-shale sequence for a long vertical distance before being accumulated in the Mio-Pliocene sand reservoir [6].

Some geologists believe that lower Miocene shale in the lower part of Surma group may also have generated some gas. Generally Miocene shale have low (<0.5%) TOC content and are thermally immature to generate gas in the drilled structures. But some shale in the lower Miocene Bhuban Formation may have the required TOC and thermal maturity (when within generative depression) to generate some gas [6].

**Reservoir Rock:** The reservoirs of the gas in the fold belt province are all of Mio-Pliocene age sandstones generally occurring in the depth between 1000 to 3400 meters. These sand layers belong to Bokabil and Bhuban Formations. The sandstone reservoirs are generally excellent in quality with respect to porosity-permeability values.. The sands are generally medium to fine grained, sublitharenite in composition and texturally mature with little clay content. The sand reservoirs originated in the shallow marine to deltaic depositional conditions [6].

**Trap:** The traps of fold belts are formed by the anticlinal folds with shale seal. The anticlinal structures provide excellent traps for gas accumulation rendering the fold belt a rich natural gas province. The anticline in the fold belt range from simple gentle and concealed undulation in the subsurface in the western part of the fold belt to high amplitude strongly faulted ones with highly rugged surface topography towards east. Accordingly, the fold belt is divided into two zones-

- The western zone consists of the low to moderate amplitude simple anticline structures. The structures are not very strongly affected by major faults and show simple four way closures. The intensity of deformation and folding gradually decrease to the west in the zone. In fact the anticlines bordering the foredeep lack surface expression. Some of the large gas fields like Titas, Bakhrabad, Bibiyana etc. are located in this zone.
- The eastern zone is characterized by high relief, tighter folds with thrust faults as expressed in the surface by rugged topography. The lengths of the individual folds range up to 150 km. The intensity of faulting generally increases towards east. Recent LANDSAT imageries and field surveys suggest that the eastern zone is part of the fold-thrust belt related to the east dipping Bengal subduction. Many of these anticlines in the eastern zone have been intruded by clay diapirs associated with overpressured shale in the deeper subsurface. Because of thrusting and massive faulting associated with intense tectonic deformation many of these structures are breached and their hydrocarbon prospects have been downgraded [6].

## 3.3. Stratigraphy of Deep Basin

Stratigraphy of the deep basin including foredeep and foldbelt to the South-East is characterized by enormous thickness of Tertiary sedimentary succession. This is a record of rapid subsidence and sedimentation. It has been suggested that Bangladesh has the thickest accumulation of sedimentary deposit in the world [7].

The sequence of the rocks encountered in the deep basin area is Oligocene to Recent in age as described below.

**Tertiary (2 to 65 million years):** During Tertiary period of the Bengal Basin as it is started to take shape. It is during this time that the major part of the very thick sedimentary succession of Bangladesh has been deposited. In the deep basin area no rocks older than Oligocene age are encountered in the surface or by drilling although their presence is suggested beneath the drilled section. Rocks older than Oligocene age are deeply buried [6].

**Oligocene:** The Oligocene is represented by the Barail Group, named by Evans (1992) after the Barail range in nearby Assam, India where the unit has its type locality. The Barail Group is composed of alternating sandstone, shale, siltstone and occasional carbonaceous rich layers. In the neighboring Assam, about 3000 meter of Barail sediments are recorded and the unit is divided into three units from bottom upward, an arenaceousLaisong Formation, an argillaceous Jenum Formation and an arenaceousRenji Formation [6].

**Miocene:** The Surma Group of Miocene-Pliocene age overlies the Baril Group with an unconformity. The Surma Group, named after Surma valley by Evans (1932), has a thickness of about 3500 to 4500 meter and is composed of monotonous alteration of subequalportion of sandstone and shale with siltstone and some conglomerates.

The Surma Group is divided into two formations, a lower sandier Bhuban Formation and an upper more argillaceous (clayey) Bokabil Formation. Both the Bokabil and Bhuban Formations show extensive lateral facies change as well as vertical variations in sand to shale ratio from place to place. The Surma sediments are poor in diagnostic fossil and therefore their age designation is often difficult.

The Surma Group is the most important stratigraphic unit in Bangladesh in terms of thickness and economic importance. It is represented by great thickness in all the wells drilled in the deep basin and also forms the backbone of the eastern hilly areas of the country including Sylhet and Chittagong hill where it is extensively exposed. The unit is traditionally believed to be deposited in Deltaic to shallow marine environment (Holotrop & Keiser 1970, Jonson & Alam 1991). All the reservoirs discovered so far in the Bangladesh are housed in the Surma Group [6].

**Pliocene-Pleistocene:** Following the filling up of the basin by deltaic deposits, a broad front of river plain environment was established under which sand dominating (arenaceous) units was deposited. This is called **Tipam Group** which is divided into three formations from bottom upward, Tipam sandstone Formation, Girujan Clay Formation and Dupitila Formation ( Holotrop& Keiser 1970) [6].

**Tipam Sandstone Formation**, of Pliocene age, is typically consists mainly of gray brown medium to coarse grained, cross bedded to massive sandstone with minor intervals of clay beds. It unconformably overlies the Surma Group in marginal part of the basin, but in the basin center the contact is conformable. Like the Surma Group it is extensively exposed along the hill ranges of the fold belt. The Tipam Sandstone Formation is about 1200 to 2500 meter thick and indicates deposition under river plain environment [6].

The overlying **Girujan Clay Formation** is a clay unit with thickness of 100 to 1000 meter. The unit has local extent and represents deposition in Lake Environment. The unit is conformably overlain by sand dominating DupiTila Formation; however where the Girujan Clay unit is missing, the Tipam Sandstone Formation would not be distinguished from the DupiTila Formation and their boundary became vague [6].

The **DupiTila Formation** of Pliocene-Pleistocene age, has been named after Dupigaon hills in Sylhet (Evans 1932), the only place in Bangladesh given a type locality status in Tertiary stratigraphy. It is a sand dominating unit with minor interbedded clay stone. The sandstone is red to brown, medium to coarse grained, loosely compacted, cross bedded, occasionally pebbly and contain petrified wood in several places. The unit was deposited under fluvial/river plain environment. The thickness of the unit varies from 500m in Sitakund anticline to over 3000m in Sylhet trough (Reiman 1993). DupiTila Formation is the major groundwater aquifer in Bangladesh [6].

**Quarternary (Present to 2 million years ago):** The Quarternary rocks are represented by Modhupur Clay Formation of Pleistocene age. This unit is composed of reddish to brownish clay with subordinate silt and typically occurs in the uplifted terraces as well as in the subsurface (Morgan and McIntyre 1959, Monsurrt al 2003). The above is covered with about 100 meter of sandy, silty and clayey sediment of Bengal Alluvium of Recent (Holocene) age [6].

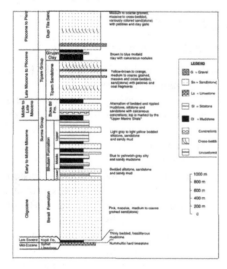

**Figure 5.** Stratigraphy of SylhetTough [6]

# 4. Location Observing Data

### Surma & Kushiyara Sand Deposit

The Surrna and Kusliiyara Rivers are the bifurcated channels of the Barak River from Amalshid. The Surma has an average slope of 50 mm/km from Amalshid to Kanaighat. The Surma River is approximately 150m wide at bankfull stage. It is 11m deep at Amalshid, and then decreases to 8.3m at the junction of the Lubha. From Lublha, the river bed falls for the next 16km downstream, lowering to 13.8m below the bank level. The Surma above the Lubha towards Amalshid is not significant and carries only a small portion of the Barak flow. The Kushiyara

River is approximately 150m wide when full and the average depth is about 12m.

**Soils**

Soils in the project area were developed in alluvial sediments laid down by the Surrma and Kushiyara Rivers. Because both rivers originate from the Barak River, parent materials of the soil are similar. Eight soil series have been identified so far.

Heavy clay soils occur in the deeply flooded basins and cover about 5916 ha (20 percent) of the cultivated area. Silty clay soils occur on low, smoothed out ridges and edges of basins, and cover about 7277 hectares (26 percent). Silty clay loams are found primarily on ridges on about 10.462 hectares (37 percent) while medium texture soils (loam to silt loam) occupy the highest topographical positions and cover about 4400 hectares (16 percent).

Fine texture soils (silty clays and clays) are poorly to very poorly drained, grey to dark grey in color and have low available moisture holding capacity. Moderately fine textured (silty clay loam) and medium textured (silt loam) soils are olive brown to grey in color, imperfect to poorly drained and have high to moderately high available moisture holding capacity. The natural fertility of these soils is moderate and they are capable of producing fairly good crops with very little fertilizer. Agricultural production, however, can be increased by applying mixed fertilizers.

# 5. Result

We got that the percentage of sand both in Surma & Kushiyara is high enough with respect to other minerals. This sand is mostly used in the bricks factory.

Besides this the thing is to say that most of the sand extracted from the sand deposit of Surma&Kushiyara river bed is used in the bricks factory near to Zakiganj as raw materials.

# 6. Conclusion

The main thing we have to locate & consider to analysis the feasibility of any area that the proven or already found geo-resources. But the main fact is that there has not arranged yet any survey or extraction data through we could determine or calculate the feasibility of my respective area. On the other hand another possibility is highly appreciable that if we analyze the structural framework of the Sylhet region, we find Zakiganj is also be a probable area for geo-resources as we have already found different types of geo-resources in different location of all over the Sylhet which is lying in the same geological area. As of now the current situation clearly state that Zakiganj does not possess any kind of mineral resource which is extractable or has any prospect of contributing to the economy of Bangladesh. All the survey data readily available also point us to the same direction. The unavailability of resources also indicates to the point that inadequate survey in the Zakiganj area, which may be the main reason of not finding any kind of geo-resources in the corresponding area. Beside this the region is on the Surma Basin of Sylhet which geologically and stratigraphically possesses a significant amount of features suitable for probable reserve of available geo-resources. Though the area has good prospect; from my investigation, collected data & all other present estimate of the respective upazilla there has not yet been discovered any reserve of geo-resources. Accuracy & inaccuracy determination of the presently collected data is highly important for further analysis and development of the feasibility test.

# Acknowledgement

The authors would like to thank almighty for empowering and guiding throughout the completion of this research paper. They would also like to thank and remember their parents for their continued support. The authors are also grateful to the local community of the research area for their support.

# References

[1]    Sylhet District digital map: Auther- Local Govt. Eng. Dept.; viewed on 14/06/2012 http://www.lged.gov.bd/ViewMap.aspx

[2]    Georgakellos, D. A. &Marcis, A. M. (2009). Application of the semantic learning approach in the feasibility studies preparation training process.*Information Systems Management* 26 (3) 231-240.

[3]    ZakiganjUpazila Map: Auther- Banglapedia; viewed on 12/06/2012.
       http://www.banglapedia.org/httpdocs/Maps/MZ_0008.GIF.

[4]    Capture on 05/06/12 at Zakiganj.

[5]    Map from Alam *et al.* 1990, permitted from GeolSurvy Bangladesh

[6]    Imam, B. (2005) *"Energy resources of Bangladesh"*, University Grant commission of Bangladesh Dhaka, pp.19-37 & 84-104.

[7]    Curray, J.R. 1991, possible green schist metamorphism at the base of a 22 kilo meter sedimentary section, Bay of Bengal. *Geology*, v.19. p. 1097-1100.

# 8

# Precambrian Stratigraphy of Central Iran and Its Metallogenic

Kaveh Khaksar[1,*], Keyvan Khaksar[2], Saeid Haghighi[3]

[1]Institute of Scientific Applied Higher Education of Jihad-e-Agriculture, Department of Soil Science, Education and Extension Organization, Ministry of Agriculture, Karaj, Iran
[2]Faculty of Basic Sciences, Qom Branch, Islamic Azad University, Qom, Iran
[3]Department of Soil Sciences, Rudehen Branch, Islamic Azad University, Rudehen, Iran
*Corresponding author: kavehkhaksar@gmail.com

**Abstract** Iran has been divided into several structural units, each characterized by a relatively unique record of stratigraphy, magmatic activities, metamorphism, orogenic events, tectonics, and overall geological style. It is imagined that Precambrian Rocks has extent in Iran and some of them are characterized more than 1.5 billion years. However it is characterized that many rocks are pertaining to Precambrian but it has an age young than 900 million years. Some of these complexes and some of formations are pertaining to Precambrian and Infracambrian in old reports and they are: kahar Formation, Tashk, Morad series, Rhizo Series and Dezu in central Iran. Stratigraphy and fossil studies in last years, results from changes in stratigraphy position in some of mentioned rocks. These deposits constitute the most largest and important economical deposits of Iran. The Chador Malou, Choghart, Golgohar, Sechahoun and Gelmandeh Iron Ore deposits, Kushk lead sphalerite mines, Saghand and Narygan Uranium deposits and Esfordy phosphate, deposits are the some of these examples. Based on new investigation it is suggested that the separation of ore rich melt and the ensuing hydrothermal processes dominated by alkali metasomatism were both involved to different degrees in the formation of ore deposits in Central Iran. Because of high concentration of various and largest deposits in this limited area of Precambrian age we can call the Precambrian of Central Iran as metallogenic province and metallogenic epoch.

*Keywords: Precambrian rocks, central Iran, stratigraphy and metallogenic*

## 1. Introduction

Precambrian complexes are exposed in northern and central Iran. Central Iran Located as a triangle in the middle of Iran (Figure 1), Central Iran is one of the most important and complicated structural zones in Iran. In this zone, rocks of all ages, from Precambrian to Quaternary, and several episodes of orogeny, metamorphism, and magmatism can be recognized. Micro continent of central Iran is a part of middle Iran that is bounded with ophiolithic suture zones in Sistan, Naiin, Bafgh, Doruneh fault and Kashmar–Sabzevar ophiolites and is classifiable into Lut block, Shotori upland, Tabas subduction, Kalmard upland, Posht Badam block, Biaze-Bardsir basin and Yazd block by means of long faults which are dextral strike–slip faults and have westwards inclination.

In past, micro continent of central Iran was known as a part of Central Iran zone but according to Stöcklin, this motioned area, after hardening of Precambrian bed rock, had characteristics of platform in Paleozoic and has been turned in an active zone in Mesozoic and Cenozoic. Nevertheless it should be said that major structural pattern

in this micro continent is type of separated blocks with main faults which each of them have different characteristics and dynamic of micro continent is not same in everywhere [1].

Between the bounding ranges, the central high plateau is a major feature of Iran. It is a territory of depressions, low rises, playas, dune fields, broad alluvial fans, and isolated mountain chains.

## 2. Study Area

Central Iran consists of the Alborz Mountains in the north, the Zagros Mountains in the south and west, and the scattered mountains ofKhorasan in the east.

It includes the provinces of Esfahan, Yazd, Chahar Mahal and Bakhtiari, Markazi, Qazvin, Alborz, Tehran, Qum and Semnan of Iran.

Most of the region has a warm and dry weather with a milder climate in the mountain areas.The central unit is interpreted as an assemblage of marginal Gondwana fragments that were united with the mother-continent and separated from the N (Eurasian) continent in the Paleozoic, but detached from Gondwana and attached to Eurasia in the Mesozoic, and finally rejoined by Gondwanic Afro-

Arabia in the Late Cretaceous. It comprises central Iran and the Alborz.

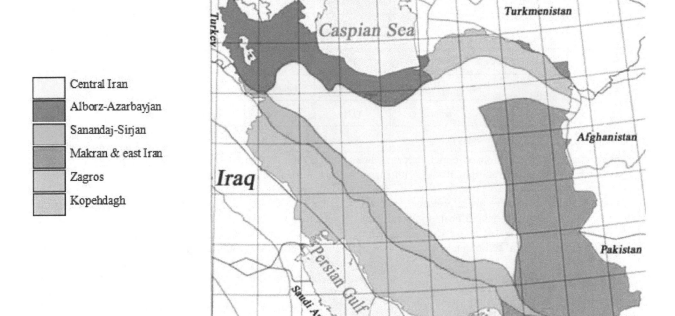

**Figure 1.** Central Iran geographic situation

## 3. Method

The data used in this paper are from all compiled studies that includes of all Precambrian stratigraphical sections in the Central Iran. The most important metallic deposits of Iran occurred in the Precambrian rocks of Central Iran.In this limited area of Precambrian age we have high concentration of various and largest deposits. In this paper we present some characteristic of the Central Iran and metallurgic deposits.

## 4. Results and Discussion

### 4.1. Metamorphic Rocks Complexes

Some of these complexes are attributed to Precambrian: Chapdoni Complex, Sarkuh Complex, Boneh Shureh Complex, Ney Baz Complex in Central Iran.

### 4.2. Kahar Formation

The Kahar Formation, synonymous with the "non-metamorphic green shales" of [2], is extensively exposed in the Qara Dagh and Molla Dagh in the southeastern part of the Soltanieh Mountains and reappears in their northwestern part in smaller outcrops at Chavarzad and Shahbolagh. The Formation consists of a rather uniform sequence of argillaceous to slightly siliceous slaty shale, most commonly of green-grey to violettish-green color but including also local red and green-red-banded varieties. Many of the shales are of the "flagstone" type and show a characteristic silk-like luster due to orientated sericite. Yellow-brown ankeritic dolomite, grey quartzite, and dark-green to violet tuffaceous shale form subordinate intercalations. These lithologically very characteristic rocks are apparently identical in lithology and stratigraphic position with the Kahar Formation of the central Alborz [3], and therefore have been given the same rock-unit name. As has been stated above, no sharp limit can be drawn between the Kahar Formation and the metamorphic complex. In its most typical development the Kahar Formation is practically nonmetamorphic, but all transitional stages to the phyllitic schists that constitute the bulk of the metamorphic complexes of the Soltanieh Mountains can be seen. The distinction between the Kahar and metamorphics on the map is, therefore, very arbitrary. It is well possible, that the phyllites of the Talesh Mountains largely represent the metamorphosed Kahar Formation.

[2] Compare the Kahar Formation with the Morad "Series" and the Kalmorz Formation of eastern central Iran. Precambrian rocks of very similar lithology have been observed in the quadrangles of Takab (Azerbaijan), Golpaygan, Yazd, and Ardekan (Central Iran). Another lithologically very similar Formation and possible stratigraphic equivalent of the Kahar Formation is the Taknar Formation of northeastern Iran, (Table 1) [4].

### 4.3. Sedimentary Formations

Some of formations are pertaining to Precambrian and Infracambrian in old reports and they are: Kahar, Baiandor, Soltanieh, Barut, ZaguaFormations, in Alborz and Azerbaijan, Tashk and Morad series, Kalmard Formation and Taknar Series in central Iran, stratigraphy and fossil studies in last year, results from changes in stratigraphy position in some of mentioned rocks.

**Table 1. Precambrian geological formations of Central Iran and its character**

| Sedimentary-Structural Zone | Formation | Location Type section | Presented by: | Inferior boundary | Superior boundary | Coverd by: |
|---|---|---|---|---|---|---|
| Central Iran | Kahar Formation | southeastern part of the Soltanieh Mountains | [2] | Non exposed | Contact tectonically disturbed | Soltanieh Formation |
| | Morad series | 20 Km northwest of Kerman | [10] | Is not seen | Unconformity | Rizu series (Late Precambrian-Cambrian) |
| | Kalmard Formation | Kalmard Anticlinal | [11] | Is not seen | Unconformity | Shigesht F. (Ordovician) |
| | Upper Tashk | Saghand | [12] | Is not seen | Unconformity | Rizu series (Late Precambrian-Cambrian) |
| | Taknar Formation | South of Saqbzevar | [4] | Is not seen | Unconformity | Soltanieh F. (Late Precambrian-Cambrian) |

In central Iran, characterization of upper Precambrian Rocks is accompanied with Cambrian Rocks with many difficulties.

Based on geological evidences in Bafgh Region and we may attribute volcanic rocks to Rhizo Formation and they are with two times and Dezu Rocks are located on them and Tashk Formation / detrital sedimentary rocks.

## 4.4. Taknar Series Rocks

[4] Has attributed to Precambrian arrow from uniform schists with an alternation of rhyolite and metarhyolite which are around Taknar Mine, around Kashmar located to NE Iran / with of Precambrian age and it is believed that these series is covered with unconformity by Paleozoic sediments.

[5] And [6] has compared Taknar Rocks under title of Taknar Formation to Kahar Formation and so Precambrian age for these rocks.

Hamdi quotes from Houshuand Zadeh that Taknar Rocks and a set from volcanic rocks – early to late. These rocks are located on Mila Formation and in Taknar Mine with spore and Pollen fossils.

## 4.5. Upper Tashk Formation

A monotonous sequence of dark green-grey phyllites, quartzites and slates with rare rhyolite layers, locally passing into micaschists, exposed in the mountains east of the Saghand-Posht e Badam road. The unit is at least 2000 m thick and overlies conformably the metamorphic Boneh Shurou Formation, from which it is persistently separated by a dolomite-marble key bed. The Tashk Formation is eastwards overlain by a unit of cherty dolomites, gypsum and volcanic rocks attributed to the Infracambrian Rhizo Series and this farther southeast by the Lalun Sandstone and fossiliferous Cambrian beds.

## 4.6. Kalmard Formation

No metamorphic and fossiliferous Precambrian deposits of thick shale, sandstone and sandy limestone of the central Iran that have been folded due to Katanganeventand have been covered by Ordovician deposits (Shirgesht Formation) by an angular unconformity that is evidence of first long sedimentary gap.

## 4.7. Morad Series

Morad Series: the old sediments of Central Iran are called Morad Series. Type section of this series is in core of an anticline at 20 km of northwestern Kerman. This formation includes uniform evaporate sediments of slate shales-sand shale with 500 m thickness. A Yugoslavian group found another section of this formation in 80 km of northwestern Zarand.

[7] Is as a row from detrital sediments/including silt shales, sandy shale to fine grained sandstone and finally arkosic mica-bearing sandstones and fine–grained sandstone and finally mica-bearing arkosic sandstone and quartzite sandstone with thickness more than 500 m in Ab Morad Regions / North west Kerman and North Zarand (Rud shur Area and Godgal Mine) as Morad Series with Precambrian Age.

In late surveys / in was obvious that relation between Morad series to Rhizo series is as fault in Morad Area.

Series is located on Morad Series [8,9] has compacted Morad series to Kahar Formation and it has assumed Rhizo series with Dezu Gypsum–beating Formation. As a complex and then we know that equivalent to Soltanieh Formation to North Iran.

## 4.8. Metallogenic of the Central Iran

The most important metallic deposits of Iran occurred in the Precambrian rocks of Central Iran. The main part of these deposits occurred in the metamorphosed rocks of Saghand Chadormalou Regions. These deposits constitute the most important and economical deposits of Iran.The Chador Mmalou, Choghart, Golgohar, Sechahoun and Gelmandeh Iron Ore deposits. Koshk lead sphalerite mine, Saghand and Narygan Uranium deposits, Esfordi phosphate and Salt Diapers deposits are the some of these examples, the banded mine of Narygan Manganese also can be added to these deposits. Therefore the mineralization of this epoch and province can be categorized in to (1) Iron ore, for example, Choghart was a prominent iron oxide deposit in the Bafgh mining district of Iran. 800 meters length and 300 meters width, standing 150 m above the surrounding plain and 1257 m above sea level (2).

Lead Sphalerite deposit (3) Uranium and (4) Salt Diapers deposits without considering the Narygan Manganese in which the last two. Each of these deposits is one of the largest and important deposits in Iran (5) The Esfordi apatite-magnetite deposit is situated in the Bafgh district of central Iran is the most P rich deposits in that Iran and is hosted by a sequence of early Cambrian rhyolitic volcanic rocks and intercalated shallow-water sediments. Bafgh Mishidovan refractory is a potential of the refractory group minerals [14].

## 5. Conclusion

Between periods of Upper Precambrian-Lower Cambrian Iran was under extensional period of pan-African orogenic phase materialized by alkaline magmatism and associated metasomatism.

Mineral resources related to this phase are iron, manganese, apatite, magnetite-apatite, REE minerals, Uranium, thorium, lead-zinc and Central Iranian (Figure 2) and Hormoz type evaporites. Granitoids and migmatite of Central Iran typified by Kalmard, Posht-e-Badam and Saghand bodies, and granite bodies like Zarigan, Narigan in Bafgh Central Iran having calc-alkaline and alkaline overall chemistry [15] as well as Doran granite of Zanjan, Taknar, Muteh and Golpaigan (with gold mineralization) are all developed in Late Proterozoic.

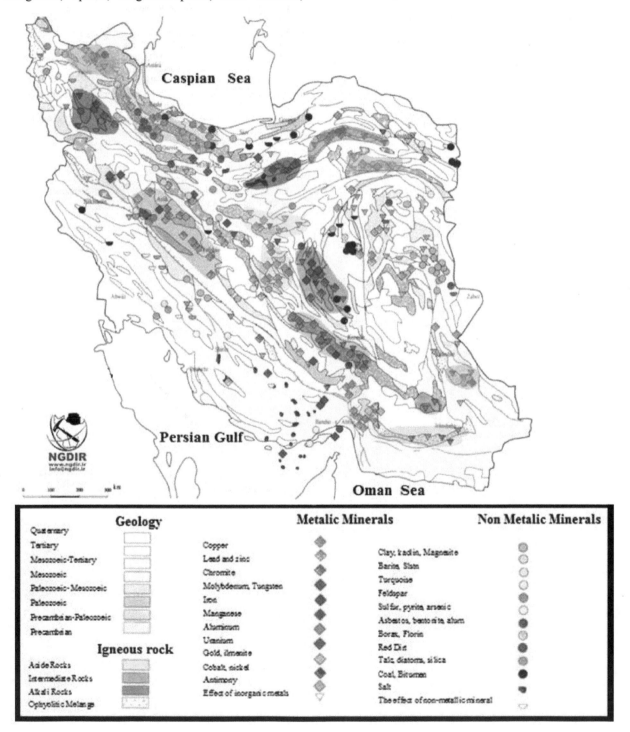

**Figure 2.** Mineral Distribution Map of Iran and rocks extension[13]

The host rocks of these deposits various from sedimentary to high grad metamorphic rocks (Meta greywacke slate, shale, quartzite, amphibolites and gneiss) and also extrusive and intrusive rocks. Most of them have been cut by late magmatic activities especially this is very common phenomena in the Bafgh iron ore deposit. The various mineralization in the Precambrian rocks of Central Iran indicate the potential resources of different geological activities and mineralization processes which have been occur in these areas. From the events of the above four categorized ore deposits, one can easy conclude that, the Precambrian age in Central Iran is one of the most metallogenic epochs and Province in Iran.

# References

[1] Manoochehr Farboodi, Kaveh Khaksar 2013, brief Precambrian–Infracambrian geological description of North-Central Iran, 3rd International Conference on Precambrian Continental Growth and Tectonism (PCGT-2013), Bundelkhand University, Jhansi – 284 128, India 23-26 November.

[2] Stöcklin, J., A. Ruttner and M. Nabavi 1964. New data on the lower Paleozoic and pre-Cambrian of north Iran. Geological Survey of Iran, Report no. 1, p. 1-29.

[3] Dedual, E., 1967. Zur Geologic des mittleren und unteren Karaj-Tales, Zentral-Elburz (Iran): Mitt. Geol. Inst ETH u. Univ. Zurich, n.s., 76, 123 pp.

[4] Razzaghmanesh B. (1968) Die Kupfer-Blei-Zink-Erzlagerstatten von Taknar und ihr geologischer Rahmen (NE-Iran).Diss. Aachen, 131 p, Aachen.

[5] Stöcklin, J. (1972): Iran central, septentrional et oriental (en collaboration avec les géologues du Service Geologique de l'Iran). – Lexique Stratigr. Intern. 3 (fasc. 9b), Iran: 1–283, Paris.

[6] Muller R., Walter R. "Geology of the Precambrian-Paleozoic Taknar inliers northwest of Kashmar, Khorasan province, NE Iran", GSI. Rep. No. 51, (1983)165-183.

[7] Huckriede, R., M. Kursten, and H. Venzlaff, 1962, Zur Geolog, Des Gebietes Zuischen Kerman and Saghakd (Iran): Beih. Geo. Jarb, Report.51: 197p.

[8] Stöcklin, J. 1971. Stratigraphic lexicon of Iran. Part I: Central, North and East Iran. Geological Survey of Iran, Report No. 18:338.

[9] Stöcklin, J. 1986. The Vendian-lower Cambrian salt basins of Iran, Oman and Pakistan: stratigraphy, correlations, paleogeography. Science de la Terre Memoire 47, p. 329-345.

[10] Ganser, A., 1955, New aspects of the geology in Central Iran. Proc.4th World Petrol, eongr. Rome sect. I/A/5. [a] er 2, pp. 280-300, 5 figs.

[11] Stocklin, J. 1968. Structural history and tectonics of Iran, A review. AAPG Bulletin 52, 7:1229-1258.

[12] Haghipour, A., Iranmanesh, M.H. Takin, M., 1972. The Ghir earthquake in Southern Persia (a field report and geological discussion). Geol. Sur. Iran, Int. Rep. 52.

[13] Aghanabati, A., 2004. Geology of Iran. Ministry of Industry and Mines, Geological Survey of Iran, 582 p.

[14] Mollaei, H., 2010. Geology of Precambrian Rocks in Central Iran, as Evidence of Metallogenic province and Metallogenic Epoch for Metallic Deposits. The 1st International Applied Geological Congress, Department of Geology, Islamic Azad University - Mashad Branch, Iran, 26-28 April.

[15] Berberian, M. and King, G. C. P., 1981. Towards a Paleogeography and tectonic evolution of Iran. Canadian Journal of Earth Science, 18 210-265.

# 9

# Ventilation Air Methane of Coal Mines as the Sustainable Energy Source

Junjie Chen[*], Deguang Xu

School of Mechanical and Power Engineering, Henan Polytechnic University, Jiaozuo, China
*Corresponding author: comcjj@163.com

**Abstract** Underground coal mines emitting large quantities of methane to atmosphere is one of the sources of methane. Approximately 70% of the methane emitted from coal mines is released as the ventilation air methane (VAM). Unfortunately, due to the low methane concentration in ventilation air, its effective utilization is considerably low. However, the global warming potential of methane can be reduced up to 95% by oxidizing the methane. Energy recovery may be possible as the products of oxidization. In this work, the existing and developing methods, based on the oxidation of methane, are introduced with a discussion of the features of the methods of the mitigation and utilization of VAM. The main operational parameters of the methods such as combustion method, technical feasibility and engineering applicability were also discussed.

***Keywords:*** *coal mines, ventilation air, greenhouse gases, reversal reactor, gas turbine*

## 1. Introduction

Greenhouse gases are released mainly from the activities such as burning of fossil fuels, industrial processes, transportation, agricultural facilities and waste management processes. Accumulation of greenhouse gases in atmosphere has led to the increase of earth's temperature [1]. As a result of the increases in global temperatures, it is expected that important changes affecting socioeconomic sectors, ecological systems and humans' life would come into existence [2].

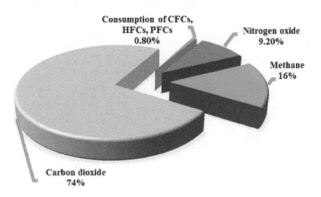

**Figure 1.** Contribution of gases to anthropogenic greenhouse gas emissions

Carbon dioxide, methane, nitrogen oxide and chlorofluorocarbons have vital importance on the global warming and the related environmental problems. Carbon dioxide has solely a rate of 74% in total anthropogenic greenhouse gas emissions [3] and [4]. Methane, nitrogen

oxide and high global warming potential gases follow carbon dioxide (Figure 1).

Methane can trap the heat about 20 times of $CO_2$. In spite of the fact that methane is the second biggest contributor to anthropogenic greenhouse gas emissions, it affects climate changes at least as carbon dioxide does [5] and [6]. The variation of non-$CO_2$ greenhouse gases has been given in Figure 2 between 1995 and 2010. Methane emissions have been increased gradually in 1995-2010 period and these increases are expected to keep their trend in the future. Estimated increase is about 12-16% for major contributing sectors as coal mining and agriculture by 2020 [7].

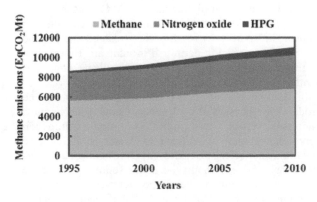

**Figure 2.** Changes of non-$CO_2$ greenhouse gases between 1995 and 2010 (EqCO2Mt)

Methane is released mainly from agriculture, energy, industry and waste processing sectors. Energy sector is the second biggest contributor (30%) to anthropogenic methane emissions [8]. Activities causing methane emissions in energy sectors include oil and natural gas systems; coal mining, stationary and mobile combustion and biomass combustion [9]. Emissions from coal mining account for 22% of emissions from energy sector (Figure 3).

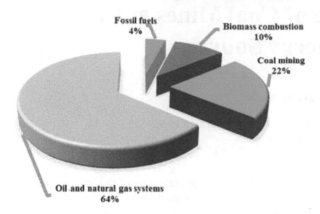

**Figure 3.** Contribution of activities to methane emissions from energy sector

The quantity of gas emitted from mining operations is a function of two primary factors as coal rank and depth of seam. Coal rank is a measure of the carbon content of the coal. Higher coal ranks mean to higher carbon content and generally the higher methane content. Coals such as anthracite and semianthracite have the highest coal ranks, while peat and lignite have the lowest. The importance of the depth of coal seam is related with the pressure over the coal. Pressure increases with depth, preventing methane from migrating to the surface. Thus, underground mining operations typically emit more methane than surface mining [5] and [10].

Methane content of coal seams increases with depth. It is 0.02 $m^3$ per tonne of coal for coal beds under 100 m of surface. The methane concentrations increase up to 7.069 $m^3$ per tonne of coal at seams at 2000 m depth below the surface [11]. Singh et al. [12] developed a correlation for Indian coal seams. They found that the methane content of the coal seam increases 1.3 $m^3$ by 100 m increase of depth.

Methane is liberated during coal extraction and diluted through ventilation fans. The diluted methane is discharged to atmosphere via mine exhausts [13]. This gas is called "ventilation air methane (VAM)" and has very low methane concentration. The low concentration of methane in VAM is the result of the threshold limits of the methane concentration permitted in mine air. However, it is responsible for 60-70% of total methane emissions related to coal mining [14] and [15].

In addition to its effect on the continuity of the production processes of coal mines, methane has its most important effect in global warming. Mitigation opportunities may be utilized to overcome its adverse environmental effects. Besides, methane can be utilized as an energy source. Among the major sources of methane emissions, coal mining has important share in methane emissions. In U.S. coal industry 42 bcf of methane content of coal mines was liberated through drainage systems [16] and [17]. About 100 bcf of methane liberated as ventilation air. This is the most identical case for the most of the coal mining practices in all countries. Therefore, the

ventilation air can be considered as a primary methane emitting source.

This work discusses the available and/or theoretically possible methane mitigation and utilization methods suitable for ventilation air methane projects. The methods are also compared with respect to their main operational parameters such as combustion method, technical feasibility and engineering applicability. Additionally, various VAM projects from all over the world are presented.

## 2. Mitigation and Utilization of VAM

Methane in coal seams has to be recovered in order to both maintain the safety in working environment during the production process and use the captured gas in diverse areas of industries. The recovered gas may contain methane up to 95% [6]. A general classification of coal mine methane (CMM) mitigation and utilization methods is illustrated in Figure 4.

The gas captured by the drainage methods has methane concentrations over 30% and this gas may be used in industry. However, the utilization of ventilation air containing very low methane is difficult owing to the fact that the air volume is large and variable in concentration [18]. In order to use in industry, the methane concentration of ventilation air has to be increased. Effective technology to increase methane concentration is yet not available but is being developed and majority of the efforts has been concentrated on the oxidation of methane in ventilation air. Methane is transformed to carbon dioxide by oxidation and energy production can be possible with the heat produced. As a result of the oxidation, the effect of methane on climate change can be reduced almost 20 times [19] and [20].

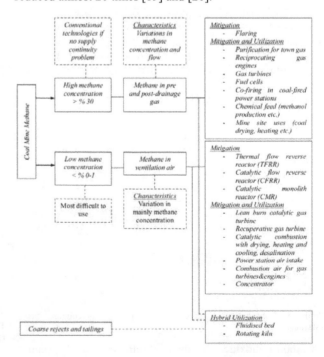

**Figure 4.** Classification of coal mine methane mitigation and utilization methods

Oxidation methods for methane may be classified as thermal and catalytic oxidation from the standpoint of the kinetic combustion mechanisms. VAM is used as ancillary and principal fuel in these oxidation technologies [21] and [22].

## 3. Methane Oxidation Mechanisms

The overall combustion mechanism of methane may be represented by the following equation:

$$CH_4 + 2O_2 = CO_2 + 2H_2O$$
$$\Delta_r H_m(298K) = -802.7 \text{ kJ/mol}$$

However, this is a gross simplification as the actual reaction mechanism involves many radical chain reactions [16] and [23]. The combustion of methane may produce CO or $CO_2$ depending on the methane ratio by the reactions below:

$$CH_4 + 2O_2 = CO_2 + 2H_2O$$
$$CH_4 + 3/2O_2 = CO + 2H_2O$$

Other reactions may also be present as following:

$$CH_4 + H2O = CO + 3H_2$$
$$2H_2 + O_2 = 2H_2O$$
$$CO + H_2O = CO_2 + H_2$$

Catalytic combustion mechanism of methane is more complicated especially heterogeneous reactions taken into consideration. Possible mechanism for methane catalytic oxidation is shown in Figure 5.

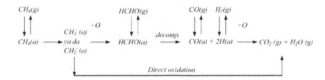

**Figure 5.** A possible mechanism for methane catalytic oxidation

## 4. Ancillary Uses of Ventilation Air Methane

The captured ventilation air can be used as an ancillary fuel to increase the combustion performance in combustion processes. Basic applications utilizing the ventilation air methane as ambient air are pulverized coal-fired power stations, hybrid waste/coal methane combustion unit, gas turbines and internal combustion engines.

An assessment of ancillary uses technologies of ventilation air is presented in Table 1 with respect to the main operational parameters such as combustion method, technical feasibility and engineering applicability.

**Table 1. Mitigation and utilization technologies of ventilation air methane as an ancillary fuel**

| Technology | Oxidation mechanism | Principal | Application status |
|---|---|---|---|
| Combustion air for conventional power station | Thermal | Combustion in power station boiler furnace | Mitigation/utilization. In a pilot scale unit but, large scale unit studies under consideration |
| Combustion air for gas turbines | Thermal | Combustion in conventional gas turbines combustor | Mitigation Utilization - studied |
| Combustion air for gas engine | Thermal | Combustion in a gas engine combustor | Mitigation Utilization - demonstrated |
| Hybrid waste coal/methane combustion in a kiln | Thermal | Combustion inside a rotating combustion chamber | Mitigation Utilization - being tried in a pilot scale unit |
| Hybrid waste coal/methane combustion in a fluidised bed | Thermal | Combustion inside a fluidised bed | Mitigation Utilization - being proposed as a concept |

Energy recovery using these methods may be certain. The expected success is dependent mainly on the safe connection of these units to mine shafts. But, this is a site specific issue and has not yet been fully examined [16].

Table 2 compares the methods of the ancillary use of ventilation air methane from the standpoint of main operational parameters.

**Table 2. A comparison of methods of the ancillary use of ventilation air methane**

| Technology | Feature | Combustion temperature (°C) | Technical feasibility and engineering applicability | Potential issues |
|---|---|---|---|---|
| Combustion air for conventional power station | Pulverized coal-fired furnace | 1400-1650 | Technically: yes Engineering: not demonstrated at a mine site | Limited sites Potential operational problems to existing boilers |
| Hybrid waste coal/methane combustion in a fluidised bed | Fluidised bed | 850-950 | Technically: yes Engineering: not demonstrated at a mine site | Minimum requirement for coal/tailings quality Proving tests needed for methane oxidation |
| Hybrid waste coal/methane combustion in a kiln | Rotating kiln | 1200-1550 | Technically: yes Engineering: not demonstrated at a mine site | Self-sustaining combustion Minimum requirement for coal/tailings quality |
| Combustion air for gas turbines | Gas turbine | 1400-1650 | Technically: yes Engineering: not demonstrated at a mine site | Small percentage of turbine fuel A lot of methane is emitted in by-passing air for a single compressor machine. If two compressor are used, there is increasing system complexity and decreasing capacity using ventilation air |
| Combustion air for gas engine | Motor | 1800-2000 | Technically: yes Engineering: demonstrated at a mine site | Small percentage of engine fuel Using a small percentage of ventilation air |

## 4.1. Pulverized Coal-Fired Power Station

Captured ventilation air can be utilized as ambient air at large power stations for the available combustion processes. A pilot scale study for this application has been conducted at a power station in Australia and the results have shown that the application of the method is technically possible especially if a power plant is already built near the mine exhaust shafts [14].

Power stations in general are not convenient to all gassy mines. This does not allow keeping the suitability of the method. Variation of methane concentration in ventilation air may affect the equipments' operational performances negatively depending on the methane concentration and flow rate. It also increases the complexity of power station operation. For instance, methane concentration in ventilation air could increase during combustion and it can result in damages to the equipment (boiler furnace, etc.) and slagging and residuals [16].

## 4.2. Hybrid Waste/Coal/Tailings/Methane Combustion Units

When considered the methane oxidation mechanism, it can be recognized that the use of ventilation air in hybrid waste/coal/tailings/methane combustion units in either rotating kiln or fluidised bed has similarities with the use of ventilation air in pulverized coal boilers. However, there is a need for additional regulations to organize the combustion process and provide the stability.

Some rotating kilns have been developed by several companies to hybridize waste/coal which is low quality. Studies carried out with these kinds of kilns have shown that high quality gas or fuel is required to maintain the stability of combustion process [23]. In a study conducted by Cobb [24], low performances were obtained in case of using hard coal wastes having high quality in these kinds of kilns.

Even though a wide range of pilot scale plant using VAM as an ancillary fuel in fluidised bed combustion units is available, there has been no experimental study proving the methane will be fully oxidized in a fluidised combustion unit [25].

## 4.3. Internal Combustion Engines

Internal combustion engines commonly use medium quality gas to generate electric and; therefore, they are suitable for using VAM as ambient air in combustion processes. It is an option requiring low capital cost to reduce the methane in ventilation air if it has advantage in transportation. Due to the higher temperatures reached during combustion, it produces more $N_2O$ gases than other methods do [14]. Despite the fact that the method has low capital cost; only, a small percentage of methane in ventilation air can be used in this application.

## 4.4. Conventional Gas Turbines

Conventional gas turbines have similarities with internal gas engines and a small percentage of methane in ventilation air meets the gas turbine's fuel needed. On the other hand, using of ventilation air to dilute the combustion process and cool the turbine results in the methane passing through the turbine without combustion. To avoid this, not only more complex turbine systems requiring compressed air from other sources are required but also compressed ventilation air is needed [16] and [26].

# 5. Principal Uses of VAM

Ventilation air can be used as principal fuel in combustion processes for mitigation and utilization of methane in vented air. However, the principal uses of ventilation air may not be possible for some methods in terms of methane concentration for the operational requirement. Mitigation and utilization methods of VAM as a principal fuel are presented in Table 3. Ventilation air could be used in thermal and catalytic flow reverse reactors, catalytic-monolith reactors, lean burn gas turbines, concentrators. Detailed descriptions of these technologies are below.

**Table 3. Mitigation and utilization technologies of ventilation air methane as principal fuel**

| Technology | Oxidation mechanism | Principal | Application status |
|---|---|---|---|
| Thermal flow reverse reactor (TFRR) | Thermal | Flow reverse reactor with regenerative bed | Mitigation: demonstrated<br>Utilization: not demonstrated yet |
| Catalytic flow reverse reactor (CFRR) | Catalytic | Flow reverse reactor with regenerative bed | Mitigation: demonstrated<br>Utilization: not demonstrated yet |
| Catalytic-monolith combustor | Catalytic | Monolith reactor with a recuperator | Mitigation: demonstrated<br>Utilization: not demonstrated yet |
| Catalytic lean burn gas turbine | Catalytic | Gas turbine with a catalytic combustor and recuperator | Mitigation: combustion demonstrated<br>Utilization: being developed in a lab scale unit |
| Recuperative gas turbine | Thermal | Gas turbine with a recuperative combustor and recuperator | Mitigation: combustion demonstrated<br>Utilization: demonstrated in a pilot scale unit but needed further modification |
| Concentrator | N/A, adsorption | Multi-stage fluidised/moving bed using adsorbent, and a desorber | Mitigation and utilization: under development |

## 5.1. Thermal Flow Reversal Reactor Technology (TFRR)

Thermal flow reversal reactors are the equipments used for thermal oxidation processes of organic components, and their operating principles have been described [26] and [27]. Basically, a TFRR consists of a bed of a silica gravel or ceramic heat-exchange medium with a set of electric heating elements in the center. Mine ventilation air containing methane enters to the reactor through valves or channel available on the equipment [28]. A schematic illustration of TFRR is shown in Figure 6.

Ambient temperature is required for autoignition of methane in ventilation air. Electrical heating elements in medium pre-heat the center of reactor to start the process with the aim of autoignition of methane. Ventilation air containing methane entering to reactor at ambient temperature, flows through the reactor in one direction and its temperature increases till oxidation of methane occurs near the center of the bed. The hot products of oxidation continue through the bed losing heat to the far side of the bed in the process. When the far side of bed is sufficiently heated and the near side has cooled because of the inflow of ambient-temperature ventilation air, the reactor automatically reverses the direction of ventilation airflow [21] and [22]. Ventilation air (VA) enters now the far side (heated) of the bed and then oxidation occurred. The hot gases transfer heat to the near side of the bed and exit the reactor at a temperature just modestly above ambient. Then, the process again reverses. Oxidized methane produces $CO_2$ and heat [29] and [30]. If meets demands, energy production can be possible from the oxidation products. Thermal flow reversal reactors oxidize $\geq 95\%$ of methane in ventilation air mine (VAM) [31].

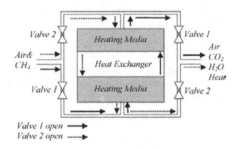

**Figure 6.** A schematic illustration of thermal flow reversal reactor

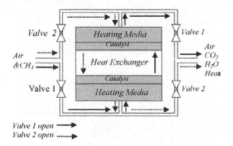

**Figure 7.** Cyclic air flow alternatives in thermal flow reversal reactors

Thermal flow reversal reactors have alternative designs according to their cyclic of airflow and valves. The schematic illustrations of alternatives of TFRR are shown in Figure 7 [31].

## 5.2. Catalytic flow reversal reactor (CFRR)

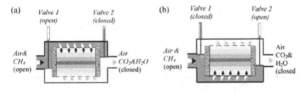

**Figure 8.** A schematic illustration of catalytic flow reversal reactor

Catalytic flow reversal reactors differ from the TFRR in terms only of a catalyst used in CFRR method (Figure 8).

CFRR decreases the autoignition temperature of methane in ventilation air and keeps the durability of system reaction during combustion [32]. Over heating or over cooling may be prevented by adding air or air-water mixer to the heat-exchanger [5] and [33]. These kinds of reactors have some advantages such as working at low temperatures, releasing low amount of $N_2O$ (can be omitted), low production and engineering costs, requiring small equipments [34] and [35].

A study reported by Sapoundjiev and Aube [35], has shown that CFRR would be used for generating electric at coal mines being far away to thermal energy users. Discussing on the benefits from CFRR in spite of using in coal plants for mitigation of methane in VA, they stated that this method could be applied for reducing the methane emissions released from different fields. Figure 9 shows the advantages of CFRR method using methane (0.5%) in VAM. One can understand from the figure that the greenhouse effect of VAM would be decreased and the products of oxidation can be used in diverse areas.

CFRR oxidizes approximately 90% of methane in VA. Thus, it allows obtaining an important energy source, reducing the greenhouse effect of VAM [36].

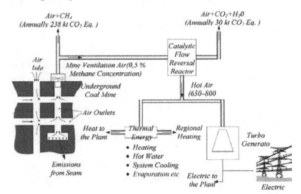

**Figure 9.** Advantages of catalytic flow

## 5.3. Catalytic-monolith reactors (CMR)

Somewhat similar principles which exist for the other methods, apply in operational and design characteristics of catalytic-monolith reactors. Catalytic-monolith reactor uses a monolithic reactor like a honeycomb. It is widely utilized because of its outstanding characteristics as very low pressure drop at high mass flows, high geometrical area and high mechanical strength [37]. Monoliths used consists of a structure of a parallel channels with walls coated by a porous support containing catalytically active particles. Thus, compared with the other reactor designs (TFRR or CFRR), it can be seen that the CMR unit is more compact in terms of processing the same amount of ventilation air, but will require a recuperator to pre-heat the ventilation air. Table 4 compares the different reactor designs (TFRR, CFRR and CMR) in terms of main performance and design characteristics.

VAM, captured at mine exhaust, not only has high volume and low concentration of methane but also its methane concentration is variable. These drawbacks can result in inefficient working of mitigation and utilization methods. MEGTEC, a firm which produced TFRR, has reported that TFRR unit can continue its function at concentration of 0.08% methane. However, a simulation results carried out, have indicated that temperatures would drop below the minimum required if the methane concentration drops below 0.35% [14]. To sustain CFRR operation, the minimum methane in the ventilation air should be above 0.08%. It is unclear how long CFRR unit can be operated on 0.08% methane in air [33]. As a result

of the experimental catalytic combustion work aimed to define the minimum methane concentration needed for operation of CMR, it was found that it can be operated when the methane concentration is greater than 0.4% in VAM [16].

**Table 4. Comparison of the reactor technologies (TFRR, CFRR and CMR) from the standpoint of their outstanding features**

| Features | Thermal flow reversal reactor (TFRR) | Catalytic flow reversal reactor (CFRR) | Catalytic-monolith reactor (CMR) |
|---|---|---|---|
| Principles of operation | Flow reversal | Flow reversal | Monolithic reactor |
| Catalyst | No | Yes | Yes |
| Auto-ignition temperature | 1000 °C | 350-800°C | 500°C |
| Cycle period length | Shorter | Longer | Continuously |
| Minimum $CH_4$ concentration | 0.2% | 0.08% | 0.4% |
| Applicability | $CH_4$ mitigation | $CH_4$ mitigation | $CH_4$ mitigation |
| Possibility of recovering heat to generate power | May need additional fuel to increase $CH_4$ concentration and maintain it constant | May need additional fuel to increase $CH_4$ concentration and maintain it constant | May need additional fuel to increase $CH_4$ concentration and maintain it constant |
| Variability of $CH_4$ concentration | Variable | Variable | Variable |
| Plant size | Huge | Larger | Compact |
| Operation | More complicated | More complicated | Simple |
| Life time | N/A | N/A | >8000 h for catalyst |
| $NO_x$ emission | N/A | Low | Low (<1 ppm) |
| $CO_x$ emission | Low | Low | Low (~0 ppm) |

## 5.4. Lean-burn Gas Turbines

In recent years, many lean-burn gas turbines are being developed such as EDL's recuperative gas turbine, CSIRO lean-burn catalytic gas turbine, Ingresoll-Rand micro turbine with a catalytic combustor [38]. VAM is mostly used in recuperative gas turbine using heat from the combustion process to pre-heat the air having methane. Comparison of some important features of lean-burn gas

turbines is presented in Table 5. The methane concentration in ventilation air should be above 2.0%. Thus, it may require the addition of substantial quantities of methane to the ventilation air to reach the adequate methane concentration in order to be used as an ancillary fuel. Not only low concentration methane in ventilation air can be used in these kinds of turbines but also the methane captured from pre and post mining may be used [16].

**Table 5. Comparison of some features of Lean-Burn gas turbine technologies**

| Features | EDL recuperative turbine | CSIRO catalytic turbine | IR catalytic micro turbine |
|---|---|---|---|
| Principles of operation | Air heater inside combustion chamber | Monolith reactor | Monolith reactor |
| Catalyst | No | Yes | Yes |
| Auto-ignition temperature | 700-1000 °C | 500°C | N/A |
| Experience | Pilot-scale | Bench-scale study on combustion | Conventional micro turbine development |
| Cycle period length | Continuously | Continuously | Continuously |
| Minimum $CH_4$ concentration | 2.0% | 0.08% | 0.08% |
| Applicability | $CH_4$ mitigation and power generation and need additional fuel to increase $CH_4$ concentration | $CH_4$ mitigation and power generation and need additional fuel to increase $CH_4$ concentration | $CH_4$ mitigation and power generation and need additional fuel to increase $CH_4$ concentration |
| Possibility of recovering heat to generate power | Feasible | Feasible | Feasible |
| Variability of $CH_4$ concentration | Constant | Constant | Constant |
| Operation | Simple and stable | Simple and stable | Simple and stable |
| Life time | May be shorter due to high temperature combustion heat exchanger | >8000 h for catalyst, 20 years for turbine | N/A |
| $NO_x$ emission | Higher | Low (<3 ppm) | Low |
| $CO_x$ emission | Low | Low (~0 ppm) | Low |

A technical and economic assessment has been carried out on the implementation of 0.8% and 2.0% methane in gas turbines on the basis of real methane emission data from two Australian gassy coal mines [39]. As a result of this study, it has been concluded that 50-60% of the fuel for firing 1% methane catalytic turbine is the methane from ventilation air. On the other hand, it can be seen that 30-60% of the fuel for firing 2.0% methane catalytic turbine is the methane from ventilation air. Additionally, it

has been determined that while almost 100% ventilation air was used for the turbines using 0.8% methane concentration, approximately 30-50% ventilation air was used for turbines using 2.0% methane concentration.

## 5.5. Concentrators

Concentrators are used to capture volatile organic compounds by a number of industries. These types of concentrators can enrich the methane in ventilation air and they provide the gas (methane) concentration required for the lean-burn gas turbines. Ventilation air containing 0.08-0.8% methane enters to these concentrators and leaves at a concentration of greater than 20% methane. If methane concentration is/or greater than 30% as a result of enrichment, ventilation air can be used to generate power using conventional gas turbines as well [39].

# 6. Conclusions

Coal mining is responsible for 7% of antrophogenic methane emissions. 70% of these emissions come from ventilation air at underground coal mines. Methane discharged into atmosphere may be considered as a contribution for the adverse effects of GHGs. Additionally, an energy source can be wasted. As a GHG, ventilation air methane is a contribution for the ineffective use of the world's energy produced. Mitigation and utilization methods may help to reduce methane content in atmosphere. In addition, a wasted source of energy could be utilized.

Effective method for mitigation and utilization of ventilation air methane is not yet available but many efforts have been devised in recent years. Majority of works has been concentrated on the oxidation of methane in ventilation air. The following conclusions can be drawn from the analysis of the current technological possibilities and the theoretical bases:

- Methane gas captured by drainage methods can be utilized depending on the methane concentration. However, it is so difficult to use mine ventilation air since it has high volume and contains low and variable concentration of methane.
- VAM can be used as ancillary and principal fuel in combustion processes for the purposes of mitigation and utilization. Ancillary uses of ventilation air methane serve mainly in reducing the greenhouse effect of methane. In principal use, both the energy production and the greenhouse effect reducing goals may be achieved.
- In case of inadequate methane concentration in ventilation air to meet the demands for the mitigation and utilization processes, ventilation air should be enriched to increase methane concentration. Concentrators are suitable devices for enrichment of low methane in ventilation air. After enrichment if methane concentration is greater than 30%, it can be possible to use it for power generation in conventional gas turbines.
- The applicability of mitigation and utilization methods for VAM at any mine site depends mainly on site specific conditions. It is very important to investigate any possible safety issue when any type of technologies is connected to the mine site (mine exhaust).
- Globally the effect of methane from underground coal mines via ventilation air on climate changes could be reduced about 95% by using oxidation methods. On the other hand emissions from coal mining could be reduced to 67%.

# References

[1] D. Mira Martinez, D.L. Cluff, X. Jiang, Numerical investigation of the burning characteristics of ventilation air methane in a combustion based mitigation system, *Fuel*, 133 (2014), pp. 182-193.

[2] Y.X. Zhang, E. Doroodchi, B. Moghtaderi, Utilization of ventilation air methane as an oxidizing agent in chemical looping combustion, *Energy Conversion and Management*, 85 (2014), pp. 839-847.

[3] G. Hu, S. Zhang, Q.F. Li, X.B. Pan, S.Y. Liao, H.Q. Wang, C. Yang, S. Wei, Experimental investigation on the effects of hydrogen addition on thermal characteristics of methane/air premixed flames, *Fuel*, 115 (2014), pp. 232-240.

[4] K. Baris, Assessing ventilation air methane (VAM) mitigation and utilization opportunities: A case study at Kozlu Mine, Turkey, *Energy for Sustainable Development*, 17 (1) (2013), pp. 13-23.

[5] K. Gosiewski, A. Pawlaczyk, Catalytic or thermal reversed flow combustion of coal mine ventilation air methane: What is better choice and when?, *Chemical Engineering Journal*, 238 (2014), pp. 78-85.

[6] I. Karakurt, G. Aydin, K. Aydiner, Mine ventilation air methane as a sustainable energy source, *Renewable and Sustainable Energy Reviews*, 15 (2011), pp. 1042-1049.

[7] S. Su, A. Jenny, Catalytic combustion of coal mine ventilation air methane, *Fuel*, 85 (2006), pp. 1201-1210.

[8] C.O. Karacan, Modelling and prediction of ventilation methane emissions of U.S. longwall mines using supervised artificial neural networks, *International Journal of Coal Geology*, 73 (2008), pp. 371-387.

[9] P.-h. Hu, X.-j. Li, Analysis of radon reduction and ventilation systems in uranium mines in China, *Journal of Radiological Protection*, 32 (2012), pp.289-300.

[10] Y.V. Kruglov, L.Y. Levin, A.V. Zaitsev, Calculation method for the unsteady air supply in mine ventilation networks, *Journal of Mining Science*, 47 (2011), pp.651-659.

[11] L. Chen, T. Feng, P. Wang, Y. Xiang, B. Ou, Catalytic properties of Pd supported on hexaaluminate coated alumina in low temperature combustion of coal mine ventilation air methane, *Kinetics and Catalysis*, 54 (2013), pp.767-772.

[12] A. Setiawan, J. Friggieri, E.M. Kennedy, B.Z. Dlugogorski, M. Stockenhuber, Catalytic combustion of ventilation air methane (VAM) - long term catalyst stability in the presence of water vapour and mine dust, *Catalysis Science & Technology*, 4 (2014), pp.1793-1802.

[13] A. Widiatmojo, K. Sasaki, N.P. Widodo, Y. Sugai, J. Sinaga, H. Yusuf, Numerical simulation to evaluate gas diffusion of turbulent flow in mine ventilation system, *International Journal of Mining Science and Technology*, 23 (2013), pp. 349-355.

[14] K. Gosiewski, A. Pawlaczyk, M. Jaschik, Combustion of coal-mine ventilation air methane in a thermal flow reversal reactor, *Przemysl Chemiczny*, 90 (2011), pp. 1917-1923.

[15] M.C. Suvar, C. Lupu, V. Arad, D. Cioclea, V.M. Pasculescu, N. Mija, Computerized simulation of mine ventilation networks for sustainable decision making process, *Environmental Engineering and Management Journal*, 13 (2014), pp. 1445-1451.

[16] S. Torno, J. Torano, M. Ulecia, C. Allende, Conventional and numerical models of blasting gas behaviour in auxiliary ventilation of mining headings, *Tunnelling and Underground Space Technology*, 34 (2013), pp. 73-81.

[17] J. Cheng, S. Yang, Data mining applications in evaluating mine ventilation system, *Safety Science*, 50 (2012), pp. 918-922.

[18] N. Sahay, A. Sinha, B. Haribabu, P.K. Roychoudhary, Dealing with open fire in an underground coal mine by ventilation control techniques, *Journal of the Southern African Institute of Mining and Metallurgy*, 114 (2014), pp. 445-453.

[19] H. Dello Sbarba, K. Fytas, J. Paraszczak, Economics of exhaust air heat recovery systems for mine ventilation, *International Journal of Mining Reclamation and Environment*, 26 (2012), pp.185-198.

[20] P. Zapletal, V. Hudecek, V. Trofimov, Effect of natural pressure drop in mine main ventilation, *Archives of Mining Sciences*, 59 (2014), pp. 501-508.

[21] B. Kucharczyk, W. Tylus, Effect of promotor type and reducer addition on the activity of palladium catalysts in oxidation of methane in mine ventilation air, *Environment Protection Engineering*, 37 (2011), pp. 63-70.

[22] B.P. Kazakov, A.V. Shalimov, A.S. Kiryakov, Energy-saving mine ventilation, *Journal of Mining Science*, 49 (2013), pp. 475-481.

[23] D. Cioclea, C. Lupu, I. Toth, I. Gherghe, C. Boanta, F. Radoi, Fast network connections for ensuring decision operativity in mining ventilation, *Environmental Engineering and Management Journal*, 11 (2012), pp. 1225-1228.

[24] N.I. Alymenko, A.V. Nikolaev, Influence of mutual alignment of mine shafts on thermal drop of ventilation pressure between the shafts, *Journal of Mining Science*, 47 (2011), pp. 636-642.

[25] E. Rusinski, P. Moczko, P. Odyjas, D. Pietrusiak, Investigation of vibrations of a main centrifugal fan used in mine ventilation, *Archives of Civil and Mechanical Engineering*, 14 (2014), pp. 569-579.

[26] S. Bluhm, R. Moreby, F. von Glehn, C. Pascoe, Life-of-mine ventilation and refrigeration planning for Resolution Copper Mine, *Journal of the Southern African Institute of Mining and Metallurgy*, 114 (2014), pp. 497-503.

[27] D. Xie, H. Wang, K.J. Kearfott, Modeling and experimental validation of the dispersion of Rn-222 released from a uranium mine ventilation shaft, *Atmospheric Environment*, 60 (2012), pp. 453-459.

[28] A.V. Shalimov, Numerical modeling of air flows in mines under emergency state ventilation, *Journal of Mining Science*, 47 (2011), pp. 807-813.

[29] S.A. Kozyrev, A.V. Osintseva, Optimizing arrangement of air distribution controllers in mine ventilation system, *Journal of Mining Science*, 48 (2012), pp.896-903.

[30] J. Du, H. Li, L. Wang, Phase equilibria and methane enrichment of clathrate hydrates of mine ventilation air plus tetrabutylphosphonium bromide, *Industrial & Engineering Chemistry Research*, 53 (2014), pp. 8182-8187.

[31] M.M. Konorev, G.F. Nesterenko, Present-day and promising ventilation and dust-and-gas suppression systems at open pit mines, *Journal of Mining Science*, 48 (2012), pp. 322-328.

[32] T. Krings, K. Gerilowski, M. Buchwitz, J. Hartmann, T. Sachs, J. Erzinger, J.P. Burrows, H. Bovensmann, Quantification of methane emission rates from coal mine ventilation shafts using airborne remote sensing data, *Atmospheric Measurement Techniques*, 6 (2013), pp. 151-166.

[33] J.C. Kurnia, A.P. Sasmito, A.S. Mujumdar, Simulation of a novel intermittent ventilation system for underground mines, *Tunnelling and Underground Space Technology*, 42 (2014), pp. 206-215.

[34] R. Patterson, K. Luxbacher, Tracer gas applications in mining and implications for improved ventilation characterisation, *International Journal of Mining Reclamation and Environment*, 26 (2012), pp. 337-350.

[35] S.S. Erdogan, C.O. Karacan, E. Okandan, Use of reservoir simulation and in-mine ventilation measurements to estimate coal seam properties, *International Journal of Rock Mechanics and Mining Sciences*, 63 (2013), pp. 148-158.

[36] J. Torano, S. Torno, M. Menendez, M. Gent, Auxiliary ventilation in mining roadways driven with roadheaders: Validated CFD modelling of dust behaviour, *Tunnelling and Underground Space Technology*, 26 (2011), pp. 201-210.

[37] B. Kucharczyk, Activity of monolithic Pd/Al$_2$O$_3$ catalysts in the combustion of mine ventilation air methane, *Polish Journal of Chemical Technology*, 13 (2011), pp. 57-62.

[38] B. Kucharczyk, W. Tylus, Removal of methane from coal mine ventilation air by oxidation over monolithic palladium catalysts, *Przemysl Chemiczny*, 89 (2010), pp. 448-452.

[39] H. Bystron, The influence of dry mine air chemical composition on the safety state of the ventilation system of a deep mine using underground main fan, *Archives of Mining Sciences*, 55 (2010), pp. 377-387.

# 10

# Operational Dependence of Galvanized Steel Corrosion Rate on Its Structural Weight Loss and Immersion-Point pH in Sea Water Environment

C. I. Nwoye[1,*], E. C. Chinwuko[2], I. E. Nwosu[3], W. C. Onyia[4], N. I. Amalu[5], P. C. Nwosu[6]

[1]Department of Metallurgical and Materials Engineering, Nnamdi Azikiwe University, Awka, Nigeria
[2]Department of Industrial and Production Engineering, Nnamdi Azikiwe University, Awka, Nigeria
[3]Department of Environmental Technology, Federal University of Technology, Owerri, Nigeria
[4]Department of Metallurgical and Materials Engineering, Enugu State University of Science & Technology Enugu, Nigeria
[5]Project Development Institute Enugu, Nigeria
[6]Department of Mechanical Engineering, Federal Polytechnic, Nekede, Nigeria
*Corresponding author: nwoyennike@gmail.com

**Abstract**  The operational dependence of galvanized steel corrosion rate on its structural weight loss and immersion-point pH (pH of stagnant sea water trapped in holes and grooves of galvanized steel made structures or equipment) in sea water environment was studied. SEM analysis of the surface structure of the corroded steel revealed that the adherent and compact nature of the white rust layers absorbed on the zinc surface affected the level of corrosion attacks on the zinc and invariably on the steel structure. The corrosion rate of the galvanized steel decreased with increase in the steel weight loss and immersion-point pH. Formation and presence of $(ZnOH)_2$ in corrosion medium retarded the corrosion process because of its alkaline nature. A two-factorial model was derived, validated and used for the predictive evaluation of the galvanized steel corrosion rate. The validity of the model was rooted on the core model expression $\zeta + 5 \times 10^{-5} \ln\gamma + 6.166 \times 10^{-5} = -1.5 \times 10^{-5} \vartheta^2 + 0.0001\vartheta$ where both sides of the expression are correspondingly approximately equal. The standard errors incurred in predicting the corrosion rate for each value of the weight loss & immersion-point pH considered as obtained from experiment, derived model and regression model-predicted results were $1.516 \times 10^{-7}$, $5.415 \times 10^{-7}$ and $2.423 \times 10^{-9}$ & $1.39 \times 10^{-7}$, $4.529 \times 10^{-7}$ and $2.548 \times 10^{-8}$ % respectively. Deviational analysis indicates that the derived model operates most viably and reliably within a deviation range of 0-15.38% from experimental results. This translated into about 84% operational confidence and response level for the derived model as well as 0.84 reliability response coefficient of the corrosion rate to the collective operational contributions of weight loss and immersion-point pH in the sea environment.

*Keywords:* galvanized steel, corrosion rate, immersion-point ph, weight loss, sea water environment

## 1. Introduction

The growing usefulness of galvanized steel as an effective material for building and water distribution system has raised the need for research and development aimed at improving the coating of zinc on the steel substrate. Galvanized steel has dual properties; mechanical, due to steel and chemical, resulting from its enhanced corrosion resistance due to presence of zinc. [1]

Observation has shown [1] that the nature of the corrosion product deposited on galvanized steel surface following its exposure in water is waxy and white. This is referred to as white rust. The research indicates that if the white-rust corrosion product is kept wet, it often feels waxy; if it dries it usually feels hard and brittle.

It has been revealed [1] that white rust is a rapid, localized corrosion attack on zinc that usually appears as a voluminous white deposit. Results of this work also show that there exists beneath the white deposit a localized area where the zinc has been attacked. This area appears as a shallow pit at its early formation stages.

Studies [2,3,4] have shown that the good corrosion resistance of zinc coating could be significantly improved by alloying zinc with other metal (e.g. Co, Ni, Mn, Al). Some scientists have therefore suggested that the higher protective ability of zinc results from the presence of zinc hydroxide salts $(Zn(OH)_2)$, which is formed due to interaction with the corrosion medium [5]. In addition, corrosion current and corrosion potential are significantly sensitive to the zinc surface conditions as well as the environmental factors (pH of the solution, dissolved oxygen concentration, Cl⁻ ion concentration, temperature, etc). These factors are related to the presence of oxidized species (oxide, hydroxide and carbonate) due to the contact with aqueous solution, and to the contribution of the cathodic reduction of dissolved oxygen.

Investigations [6,7,8] on atmospheric corrosion of galvanized steel have indicated that the composition of the rust layer on galvanized steel depends on the exposure conditions, type and level of the pollutants, as well as the number of the wet- dry cycles.

It has been proposed [9] that zinc ion dissolved from the rust layer on galvanized steel prevents further corrosion of the steel substrate. The researcher [9] demonstrated the contribution of the zinc-containing rust layer to the corrosion retardation for the Fe substrate, as well as the high sacrificial anodic effect of the metallic zinc.

Galvanized steel has found application [10,11] in areas such as building, automotive body parts and water distribution systems because of its good resistance to environmental corrosion. The protection proffered by zinc coating is due to barrier and galvanic double protective effect [12,13]. However, many cases of heavy damage of galvanized pipes and tanks have been reported as being due to corrosion processes in water hanging system, as clearly evidenced by the production of rust layer in those systems after an unexpectedly short service life [14].

The aim of this research is to evaluate galvanized steel corrosion rates based on the steel's structural weight loss and immersion-point pH while serving in natural sea water. A model would be derived, validated and used for the evaluation. Structures and equipment made of galvanized steel and used in sea water evaporating system are known to have series of holes and grooves which entrap water. It is strongly believed that the corrosion rates of these areas could be predicted by substituting into the derived model, values of the galvanic steel structural weight loss (resulting from corrosion) and immersion pH (the pH of the trapped sea water).

## 2. Materials and Methods

Materials used for the experiments are galvanized steel pipes obtained from oil fields in Port Harcourt, Nigeria. The other materials used were acetone (analytical grade), distilled water, graduated pyrex beakers and Erlenmeyer flasks. The equipment used were Micro drilling machine (Model H), analytical digital weighing machine (Mettle 4900) and pH meter (SeaFET™ Ocean pH Sensor).

### 2.1.    Specimen    Preparation    and Experimentation

The galvanized steel pipes were cleaned using 0.5M picric acid to remove any existing trace of rust. These pipes were then washed in running water, distilled water and acetone before air-drying at room temperature. The dried steel pipes were cut into test samples of cross-sectional area: 12 cm² and weight: 14 g. Each sample piece was drilled to 0.5 mm diameter to provide hole for

the suspension of the strings and submersion of the sample in the sea water.

The method adopted for this phase of the research is the weight loss technique. The test pieces were weighed and exposed to 200 cm³ of sea water contained in a beaker for 250 hrs after which they were withdrawn. The pH of the sea water was measured as each test piece was withdrawn. The withdrawn test pieces were washed with distilled water, acetone and then dried in open air before weighing to determine the final weight. The experiment was repeated with 270, 280, 290 as well as 300 hrs exposure time and the corresponding sea water pH measured.

**Figure 1.** Galvanized steel pipe

**Figure 2.** Corroded pieces of galvanized steel cut and exposed to sea water environment

## 3. Results and Discussion

### 3.1. Surface Structural Analysis of Corroded and un-corroded Galvanized Steel

The corrosion processes of the galvanized steel changed differently under the same sea water Cl⁻concentration. Figure 3 shows the SEM images at different sampling time intervals. Figure 3 (a) presented the SEM images of the as-received sample of galvanized steel before immersion in sea water. This image (un-corroded steel) was for comparison with those of corroded samples. Evidently, the zinc coatings were compact, smooth and completely covering the substrate surface. No corrosion was found before the immersion of the test piece. Loose white rust and corrosion products were absorbed on the zinc surface after 250 hrs (Figure 3 (b)), and as time elapsed, through 270 hrs (Figure 3 (c)), designating localized corrosion attack on the zinc covering the steel

[1]. These results further support the assumption of the oxygen diffusion control step. With time, the rust layer absorbed on the zinc coating was gradually damaged under the erosion of Cl⁻. However, lots of needle-like white rust layers (Figure 3 (d)) were adherent and compactly absorbed on the zinc surface, and so reduced corrosion was observed on the zinc coating after 280 hrs. This indicates drop in corrosion attack. Figure 3 (e) also shows a reduction in the white waxy rust at 290 hrs (compared with other exposure times) due to decrease in the diffusion of oxygen. This resulted from the compact nature of the protective film formed at this particular exposure time. Corrosion attack was correspondingly reduced as a result of the highlighted hindrance to oxygen diffusion. At an exposure time of 300 hrs (Figure 3 (f)), the white rust deposit further deceased due to much significant decrease in the diffusion of oxygen, resulting from an increased compact nature of the formed protective film. This resulted to much decrease in the corrosion attack on the galvanized steel.

**Figure 3.** The SEM images of galvanized steel in sea water environment with different weight loss: (a) before immersion; (b) 250 hrs; (c) 270 hrs; (d) 280 hrs, (e) 290 hrs; (f) 300 hrs (50μm)

**Table 1. Variation of corrosion rate $\zeta$ of galvanized steel with its exposure time $\Phi$ (hr) weight loss $\vartheta$, and immersion-point pH $\curlyvee$**

| $(\zeta)$ (mm/yr) | $(\Phi)$ (hr) | $(\curlyvee)$ | $(\vartheta)$ (g) |
|---|---|---|---|
| $1.3010 \times 10^{-5}$ | 250 | 6.00 | 3.50 |
| $1.2740 \times 10^{-5}$ | 270 | 6.08 | 3.70 |
| $1.1261 \times 10^{-5}$ | 280 | 6.12 | 3.83 |
| $1.1248 \times 10^{-5}$ | 290 | 6.16 | 3.90 |
| $1.2360 \times 10^{-6}$ | 300 | 6.20 | 4.00 |

## 3.2. Variation of Corrosion Rates with Immersion-point pH and Weight Loss

Table 1 shows that the corrosion rate of the galvanized steel decreases with increase in the weight loss and immersion-point pH. It was believed that the protective film on the zinc grew, and its hardness, adherence and coherency enhanced with increased exposure time thereby reducing corrosion through oxygen diffusion.

$$Zn(s) + 1/2O_2 + H_2O \rightarrow Zn(OH)_2 \quad (1)\,[15]$$
$$\rightleftarrows ZnO + H_2O$$

$$1/2\,O_2 + H_2O + 2e \rightarrow 2OH^- \quad (2)\,[15]$$

Decrease in the corrosion rate of galvanized steel with increase in immersion-point pH stems on the formation

and presence of $Zn(OH)_2$ in the corrosion medium (which reversibly gives $ZnO + H_2O$) from the reaction between $OH^-$ and $Zn^{2+}$. This reaction resulted to the initial rust. Based on the foregoing, presence of $Zn(OH)_2$ or $ZnO$ in aqueous solution (around the immersed galvanized steel) during the corrosion process increases the immersion-point pH and invariably the corrosion resistance in line with past findings [15]. $OH^-$ was formed as result of oxygen reduction at the cathodic zone as shown in equation (2).

**Table 2. Variation of $\zeta + 0.00005 \ln\curlyvee + 6.166 \times 10^{-5} = -1.5 \times 10^{-5}\vartheta^2 + 0.0001\vartheta$**

| $\zeta + 0.00005 \ln\curlyvee + 6.166 \times 10^{-5}$ | $-1.5 \times 10^{-5}\vartheta^2 + 0.0001\vartheta$ |
|---|---|
| $1.6425 \times 10^{-4}$ | $1.6625 \times 10^{-4}$ |
| $1.6465 \times 10^{-4}$ | $1.6465 \times 10^{-4}$ |
| $1.6485 \times 10^{-4}$ | $1.6297 \times 10^{-4}$ |
| $1.6504 \times 10^{-4}$ | $1.6185 \times 10^{-4}$ |
| $1.6525 \times 10^{-4}$ | $1.6000 \times 10^{-4}$ |

Computational analysis of experimental results shown in Table 2, gave rise to Table 3 which indicate that;

$$\zeta + K \ln\curlyvee + S = -N\vartheta^2 + N_e\,\vartheta \quad (3)$$

Introducing the values of K, S, N and $N_e$ into equation (11) reduces it to;

$$\zeta + 0.00005 \ln\curlyvee + 6.166 \times 10^{-5} = -1.5 \times 10^{-5}\,\vartheta^2 + 0.0001\vartheta \quad (4)$$

$$\zeta = -0.00005 \ln\curlyvee - 1.5 \times 10^{-5}\,\vartheta^2 + 0.0001\vartheta - 6.166 \times 10^{-5} \quad (5)$$

Where
K = 0.00005, S = 6.166 × 10⁻⁵, N = 1.5 × 10⁻⁵ and
$N_e$ = 0.0001 are empirical constants (determined using C-NIKBRAN [16]
$(\zeta)$ = Corrosion rate (mm/yr)
$(\vartheta)$ = Weight loss (g)
$(\curlyvee)$ = Immersion-point pH (pH of stagnant sea water trapped in holes and grooves of galvanized steel made structures or equipment)

The derived model is equation (5). Computational analysis of Table 1 gave rise to Table 2. The derived model is two-factorial in nature, being composed of two input process factors: weight loss and immersion-point pH. This implies that the predicted corrosion rate of galvanized steel in the sea water environment is dependent on just two factors: weight loss and galvanized steel immersion-point pH.

## 3.3. Boundary and Initial Conditions

Consider short cylindrically shaped galvanized steel exposed to sea water environment, interacting with some corrosion-induced agents. The sea water is assumed to be affected by undesirable dissolved gases. The range of the exposure time was considered: 250-300 hrs, range of galvanized steel immersion point-pH considered: 6.0-6.2.

The boundary conditions are: aerobic environment for zinc coating (covering galvanized steel) oxidation (since the atmosphere contains oxygen. At the bottom of the exposed steel, a zero gradient for the gas scalar are assumed. The exposed steel is stationary. The sides of the solid are taken to be symmetries.

## 3.4. Model Validity

The validity of the model is strongly rooted on equation (4) (core model equation) where both sides of the equation are correspondingly approximately equal. Table 2 also agrees with equation (4) following the values of $\zeta$ + 0.00005 ln$\gamma$ + 6.166 x $10^{-5}$ and - 1.5 x $10^{-5}$ $\vartheta^2$ + 0.0001$\vartheta$ evaluated from the experimental results in Table 1. Furthermore, the derived model was validated by comparing the corrosion rate predicted by the model and that obtained from the experiment. This was done using some statistical tools, graphical comparison, comparison with regression model, computational and deviation analysis.

### 3.4.1. Statistical Analysis

#### 3.4.1.1. Standard Error (STEYX)

The standard errors incurred in predicting the galvanized steel corrosion rate for each value of weight loss & immersion-point pH considered as obtained from experiment and derived model were 1.516 x $10^{-7}$ and 5.415 x $10^{-7}$ & 1.39 x $10^{-7}$ and 4.529 x $10^{-7}$ % respectively. The standard error was evaluated using Microsoft Excel version 2003.

#### 3.4.1.2. Correlation

The correlation coefficient between galvanized steel corrosion rate and weight loss & immersion-point pH were evaluated (using Microsoft Excel Version 2003) from results of the experiment and derived model. These evaluations were based on the coefficients of determination $R^2$ shown in Figure 4- Figure 7.

$$R = \sqrt{R^2} \qquad (6)$$

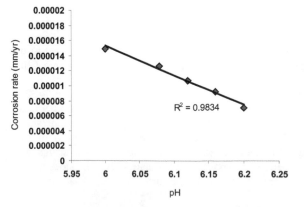

**Figure 4.** Coefficient of determination between galvanized steel corrosion rate and immersion- point pH as obtained from the experiment

**Figure 5.** Coefficient of determination between galvanized steel corrosion rate and immersion-point pH as predicted by derived model

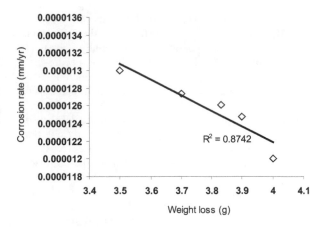

**Figure 6.** Coefficient of determination between galvanized steel corrosion rate and weight loss as obtained from the experiment

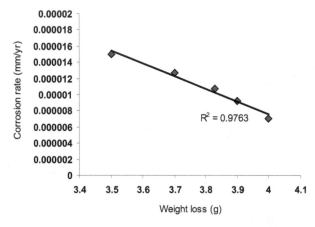

**Figure 7.** Coefficient of determination between galvanized steel corrosion rate and weight loss as predicted by derived model

**Table 3. Comparison of the correlations between corrosion rate and weight loss as evaluated from experimental (ExD) and derived model (MoD) predicted results**

| Analysis | Based on weight loss | |
|---|---|---|
| | ExD | D-Model |
| CORREL | 0.9350 | 0.9881 |

The evaluated correlations are shown in Table 3 and Table 4. These evaluated results indicate that the derived model predictions are significantly reliable and hence valid considering its proximate agreement with results from actual experiment.

**Table 4. Comparison of the correlations between corrosion rate and immersion-point pH and as evaluated from experimental and derived model predicted results**

| Analysis | Based on immersion-point pH | |
|---|---|---|
| | ExD | D-Model |
| CORREL | 0.9457 | 0.9917 |

### 3.4.2. Graphical Analysis

Figure 8 and Figure 9 show curves from derived model and experiment. Comparative analysis of these figures shows a high degree of curves alignment which indicates proximate agreement between ExD and MoD predicted results.

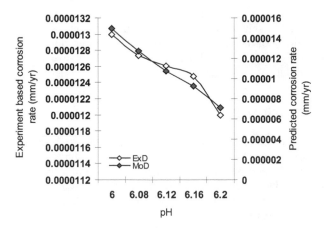

**Figure 8.** Comparison of the galvanized steel corrosion rates (relative to immersion-point pH) as obtained from experiment and derived model

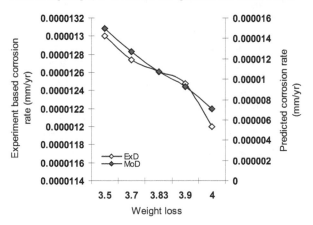

**Figure 9.** Comparison of the galvanized steel corrosion rates (relative to weight loss) as obtained from experiment and derived model

### 3.4.3 Comparison of derived model with standard model

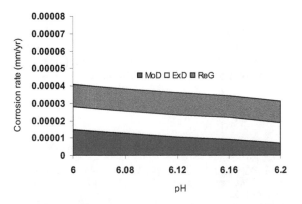

**Figure 10.** Comparison of the galvanized steel corrosion rates (relative to immersion-point pH) as obtained from experiment, derived model and regression model

The validity of the derived model was also verified through application of the regression model (ReG) (Least Square Method using Excel version 2003) in predicting the trend of the experimental results. Comparative analysis of Figure 10 and Figure 11 shows close dimensions of shaped areas of corrosion rates, which precisely translated into significantly similar trend of data point's distribution for experimental (ExD), derived model (MoD) and regression model-predicted (ReG) results of corrosion rates. Furthermore, the calculated correlations (from

Figure 10 and Figure 11) between galvanized steel corrosion rates and weight loss & immersion-point pH for results obtained from regression model were 1.0000 & 0.9980 respectively. These values are in proximate agreement with both experimental and derived model-predicted results. The standard errors incurred in predicting steel corrosion rates for each value of weight loss & immersion-point pH considered as obtained from regression model were 2.423 x $10^{-9}$ and 2.548 x $10^{-8}$ % respectively.

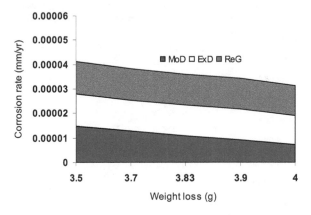

**Figure 11.** Comparison of the galvanized steel corrosion rates (relative to weight loss) as obtained from experiment, derived model and regression model

### 3.4.4 Deviational Analysis

Comparative analysis of the corrosion rates precisely obtained from the experimental data and derived model shows deviation on the part of model-predicted results. This was attributed to the fact that the effects of the surface properties of the galvanized steel which played vital roles during the corrosion process were not considered during the model formulation. This necessitated the introduction of correction factor, to bring the model-predicted corrosion rate to those of the corresponding experimental values.

The deviation Dv, of model-predicted corrosion rate from the corresponding experimental result was given by

$$Dv = \left( \frac{\zeta_P - \zeta_E}{\zeta_E} \right) \times 100 \qquad (7)$$

Where $\zeta_E$ and $\zeta_P$ are corrosion rates evaluated from experiment and derived model respectively.

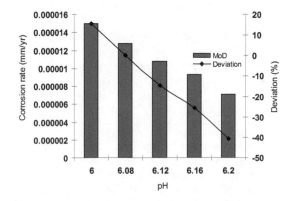

**Figure 12.** Variation of model-predicted corrosion rate with associated deviation from experimental results (relative to immersion-point pH)

Critical analysis of Figure 12 and Figure 13 show that the derived model operates most viably and reliably within a deviation range of 0-15.38%. This translates into over 84% operational confidence and response level for the derived model as well as over 0.84 reliability response coefficient of corrosion rate to the collective operational contributions of the weight loss and galvanized steel immersion-point pH (under service) in the sea water environment. This deviation range corresponds to corrosion rates in the range: $1.0762 \times 10^{-5}$ - $1.5 \times 10^{-5}$ mm/yr, exposure time: 250-280 hrs as well as galvanized steel immersion pH: 6 - 6.12 and range of weight loss: 0 - 3.5 g respectively.

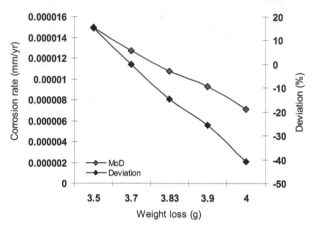

**Figure 13.** Variation of model-predicted corrosion rate with associated deviation from experimental results (relative to weight loss)

Correction factor, Cf to the model-predicted results was given by

$$Cf = \left( \frac{\varsigma_P - \varsigma_E}{\varsigma_E} \right) \times 100 \qquad (8)$$

Analysis of Table 5 as well as Figure 12 and Figure 13 shows that the evaluated correction factors are negative of the deviation as shown in equations (7) and (8).

The correction factor took care of the negligence of operational contributions of the effects of surface properties of the galvanized steel which actually affected the corrosion process. The model predicted results show deviation from those of the experiment because these contributions were not considered during the model formulation. Introduction of the corresponding values of Cf from equation (8) into the model gives exactly the corresponding experimental corrosion rate

**Table 5. Variation of correction factor with weight loss and immersion-point pH**

| $(\vartheta)$ (g) | $(\gamma)$ | Cf (%) |
|---|---|---|
| 3.50 | 6.00 | 15.38 |
| 3.70 | 6.08 | 0 |
| 3.83 | 6.12 | - 14.66 |
| 3.90 | 6.12 | - 25.56 |
| 4.00 | 6.20 | - 40.75 |

Table 5 indicates that the most reliable range of correction factors to the model-predicted corrosion rates is 0-15.38%. This range correction factor also corresponds to corrosion rates in the range: $1.0762 \times 10^{-5}$ - $1.5 \times 10^{-5}$ mm/yr, exposure time range: 250-280 hrs as well as galvanized steel immersion pH range: 6 - 6.12 and range of weight loss: 0 - 3.5 g respectively.

It is important to state that the deviation of model predicted results from that of the experiment is just the magnitude of the value. The associated sign preceding the value signifies that the deviation is a deficit (negative sign) or surplus (positive sign).

## 4. Conclusion

The operational dependence of the galvanized steel corrosion rate on its structural weight loss and immersion-point pH in sea water environment has been evaluated. It was concluded that the adherent and compact nature of the white rust layers absorbed on the zinc surface affected the level of corrosion attacks on the zinc and invariably on the steel structure. The corrosion rate of the galvanized steel decreased with increase in the steel weight loss and immersion-point pH. This was so because the protective film on the zinc grew, and its hardness, adherence and coherence enhanced with increased exposure time, and so resisted corrosion process by hindering of oxygen diffusion. Furthermore, formation and presence of $(ZnOH)_2$ in corrosion medium retarded the corrosion process because of its alkaline nature. A two-factorial model was derived, validated and used for the predictive evaluation of the galvanized steel corrosion rate. The validity of the model was rooted on the core model expression $\varsigma + 5 \times 10^{-5} \ln\gamma + 6.166 \times 10^{-5} = -1.5 \times 10^{-5} \vartheta^2 + 0.0001\vartheta$ where both sides of the expression are correspondingly approximately equal. Standard errors incurred in predicting the corrosion rate for each value of the weight loss & immersion-point pH considered as obtained from experiment, derived model and regression model-predicted results were $1.516 \times 10^{-7}$, $5.415 \times 10^{-7}$ and $2.423 \times 10^{-9}$ & $1.39 \times 10^{-7}$, $4.529 \times 10^{-7}$ and $2.548 \times 10^{-8}$ % respectively. Deviational analysis indicates that the derived model operates most viably and reliably within a deviation range of 0-15.38% from experimental results. This translated into about 84% operational confidence and response level for the derived model as well as 0.84 reliability response coefficient of the corrosion rate to the collective operational contributions of weight loss and immersion-point pH in the sea environment.

## References

[1] Engineering Bulletin (Evapco). (2009). White Rust on Galvanized Sheet, No. #34A,

[2] Zhang, X., Russo, S. L., Miotello, A., Guzman, L., Cattaruzza, E., Bonora, P.L. and Benedetti, L. (2001). Surf. Coat. Technol., 141:187-193.

[3] Volovitch, P., Allely, C., Ogle, K. (2009). Corros. Sci., 5: 1251-1562.

[4] Boshkov, N., Petrov, K., Vitkova, S., Nemska, S., and Raichevsky, G. (2002). Surf. Coat. Technol., 157: 171-178.

[5] Oritz, Z. I., Diaz-Arista, P., Meas, Y., Ortega-Borges, R. and Trejo, G. (2009). Corros. Sci., 51: 2703-2715.

[6] Zapponi, M., Perez, T., Ramos, C., and Saragovi, C. (2005). Corros. Sci., 47: 923-936.

[7] Azmat, N. S., Ralston, K. D., Muddle, B. C., and Cole, I. S. (2011). Corros. Sci., 53: 1604-1615.

[8] Yadav, A. P., Nishikata A., and Tsuru, T. (2004). Corros. Sci., 46: 169-181.

[9] Tsuru, T., (2010). Corros. Eng. 59: 321-329.

[10] El-Sayed M. Sherif, A. A. Almajid, B. Bairamov, A. K., and Eissa, A. (2012). Int. J. Electrochem. Sci., 7: 2796-2810

[11] Lin, B. L., Lu, J. T., and Kong, G. (2008). Corros. Sci., 4: 962-967.

[12] Bajat, J.B., Stankovic, S., Jokic, B. M., Stevanovic, S. I., (2010). Surf. Coat. Technol., 204: 2745-2753.

[13] Kartsonakis, I. A., Balaskas, A. C., Koumoulos, E. P., Charitidis, C. A., and Kordas, G. C., (2012). Corros. Sci., 57: 30-41.

[14] Zhang, X. G. (2005). Corrosion of zinc and zinc alloys, Corrosion: Materials, ASM Handbook, p. 402-406.

[15] Hamlaoui,Y., Tifouti, L., Pedraza, F.(2010). Corros. Sci., 52: 1883-1888.

[16] Nwoye, C. I. (2008). C-NIKBRAN Data Analytical Memory (Software).

# Study of Corrosion and Corrosion Protection of Stainless Steel in Phosphate Fertilizer Industry

**Rajesh Kumar Singh**[*], **Rajeev Kumar**

Department of Chemistry, Jagdam College, J P University, Chapra, India
*Corresponding author: rks_jpujc@yahoo.co.in

**Abstract** Phosphate industries use bulk amount of concentration $H_2SO_4$ during production of phosphate fertilizers. Stainless steel is major supporting metal for completion of several processing operational works. This acid produces corrosive effect for stainless steel. It develops corrosion cell on the surface of base metal and it changes its internal morphology as well as physical, chemical, mechanical properties. $H_2SO_4$ behaves like diabetes for this industrial metal and industries face economical. The eradication of corrosion problems used organic inhibitors like 1-(2-chlorophenyl)methanamine and 1-(2-bromophenyl)methanamine and its inhibition effect and surface coverage area studied at different temperatures $333^0K$, $343^0K$ and $353^0K$ in presence of 15% $H_2SO_4$ and 15mM concentration of inhibitors. The corrosion rate of metal was determined by weight loss experiment and potentiostat techniques. The surface adsorption and surface thin film formation were analyzed by application of activation energy, heat of adsorption, free energy, enthalpy and entropy. The inhibition efficiencies and surface coverage areas were shown that the used inhibitors produced anticorrosive effect in acidic medium.

*Keywords:* stainless steel, inhibitors, weight loss, potentiostat, surface coverage area

## 1. Introduction

The Corrosion of metal is not fully control but its effect can be minimized by application suitable methods e.g. give proper design and shape of operational metals [1], take care of surrounding operating temperatures and atmosphere [2], used different types of coatings [3], addition of inorganic and organic inhibitors as cathodic and anodic protection [4] and applied nanocoating [5]. When operational equipments came in contact of acidic environment [6], they exhibited several types of corrosion problems like galvanic, pitting crevice, stressed, intergranular, blistering, embritlement. Different types of coatings methods used for corrosion alleviation of metal such coatings were metallic coating [7], inorganic coating [8], organic coating [9], painting coating, polymeric coating, and nanocoatings [10]. These coating did not provide good support for metal in acidic environment because porosities were developed on the surface of base metal during coatings. The $H^+$ ions entered into porosities of coating materials by the process of diffusion and it developed corrosion cell on the surface base metal.

Various types of inhibitors like inorganic, organic and mixed types used to control the corrosion of metal. Organic inhibitors which possessed nitrogen, oxygen, sulphur, silicon, phosphorous, methyl, phenyl, primary, secondary and tertiary alkyl groups whereas these organic compounds have high electron rich functional groups and they have capacity to produce thin film on surface of metal. Inhibitors were bonded with metals by physical-chemical adsorption. Aromatic and heterocyclic organic compounds containing above mentioned functional groups produced anticorrosive effect in acidic medium.

Nanocoatng of $Zn_3(PO_4)_2$ [11], $Mg_3(PO_4)_2$ [12] and $AlPO_4$ [13] in presence of DLC (diamond like carbon) controlled high temperatures corrosion and minimize hydrogen ions attack. Plasma and composite coating gave corrosion protection of metal in acidic environment. Inhibitor 1-(2-chlorophenyl)methanamine and 1-(2-bromophenyl)methanamine used for this work. These inhibitors contained electron rich functional which minimize the attack of $H^+$ ions and forming thin film and it also increased surface coverage area and efficiency. The thermodynamical results noticed that these inhibitors have good adsorption capability.

## 2. Experimental Procedure

Stainless steel coupons were cut into size of (5 x 3) $cm^2$. Its surface was rubbed with emery paper and samples were washed with double distilled water. Finally it was rinsed with acetone and dried with air dryer and kept into desiccator. Test sample dipped into 250ml biker with support glass hook and corrosion rate metal determined absence and presence of inhibitors 1-(2-chlorophenyl)methanamine and 1-(2-bromophenyl)methanamine at different temperatures $333^0K$, $343^0K$ and $353^0K$ and 15mM concentration and thermostat used to mention temperature. The corrosion rate was measured by gravimetric method.

The corrosion current density and corrosion rate were calculated by potentiostatic polarization technique with help of an EG & G Princeton Applied Research Model 173 Potentiostat. A platinum electrode was used as an auxiliary electrode and a calomel electrode was used as reference electrode with stainless steel coupons. The used inhibitors structure mentioned as:

1-(2-chlorophenyl)methanamine
IH(I)

1-(2-bromophenyl)methanamine
IH(II)

## 3. Results and Discussion

The corrosion rates of stainless steel without and with inhibitors 1-(2-chlorophenyl)methanamine and 1-(2-bromophenyl) methanamine were determined by equation1 and its results were mentioned in Table 1.

$$K(mmpy) = 87.6W / DAt \qquad (1)$$

where W = weight loss of test coupon expressed in gm, A = Area of test coupon in square centimeter, D = Density of the material in g/cm$^3$.

The surface coverage areas ($\theta$) and the inhibition efficiencies (IE) occupied by inhibitors were calculated equation 2 and 3 and their results were also written in Table 1.

$$\theta = (1 - K / K_o) \qquad (2)$$

where $\theta$ = Surface coverage area, $K_o$ = corrosion rate without inhibitor, K = corrosion rate with inhibitor

$$IE = (1 - K / K_o) X 100 \qquad (3)$$

where $K_o$ is the corrosion rate without inhibitor, K= corrosion rate with inhibitor

**Table 1. Corrosion of stainless steel at different temperatures without and with inhibitors in 15% H$_2$SO$_4$**

| Inhibitors | Temperatures | 333$^0$K | 343$^0$K | 353$^0$K | C (m M) |
|---|---|---|---|---|---|
| IH(0) | $K_o$ | 391 | 565 | 836 | 00 |
|  | $\log K_o$ | 2.592 | 2.752 | 2.922 |  |
| IH(I) | K | 118 | 192 | 230 | 15 |
|  | $\log K$ | 2.071 | 2.283 | 2.518 |  |
|  | $\log(\theta/1-\theta)$ | 0.363 | 0.290 | 0.185 |  |
|  | $\theta$ | 0.698 | 0.661 | 0.605 |  |
|  | IE (%) | 69.80 | 66.10 | 60.50 |  |
| IH(II) | K | 61 | 155 | 201 | 15 |
|  | $\log K$ | 1.785 | 2.191 | 2.303 |  |
|  | $\log(\theta/1-\theta)$ | 0.729 | 0.421 | 0.498 |  |
|  | $\theta$ | 0.843 | 0.725 | 0.759 |  |
|  | IE (%) | 84.30 | 72.50 | 75.90 |  |

The results of Table 1 observed that corrosion rate increased in acidic medium without addition of inhibitors but its values decreased after addition of addition of inhibitors. The results of surface coverage area and inhibition efficiency with 1-(2-chlorophenyl)methanamine and 1-(2-bromophenyl)methanamine enhanced at different temperatures and it looked in Figure 1 plot between $\theta$ (surface coverage area) versus T$^0$K and Figure 2 IE (inhibition efficiency) versus T$^0$K. These results indicated that used inhibitors produced anticorrosive effect against acid.

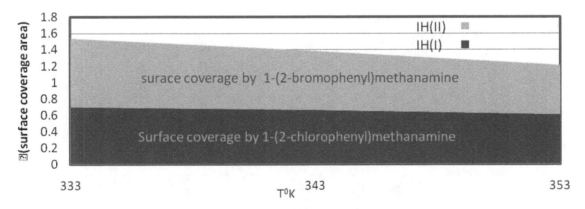

**Figure 1.** $\theta$ (surface coverage area) versus T$^0$K for stainless steel at different tenpertaures

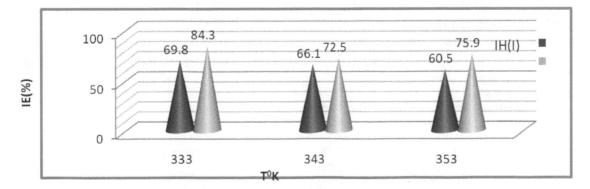

**Figure 2.** IE(%) versus T$^0$K for stainless steel at different temperatures

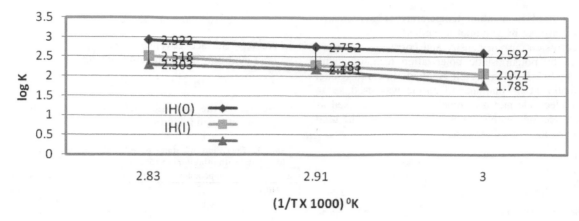

**Figure 3.** log K versus 1/T for stainless steel at different temperatures

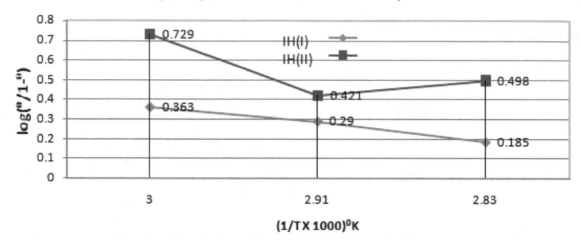

**Figure 4.** log(θ/1-θ) versus 1/T for stainless steel at different temperatures

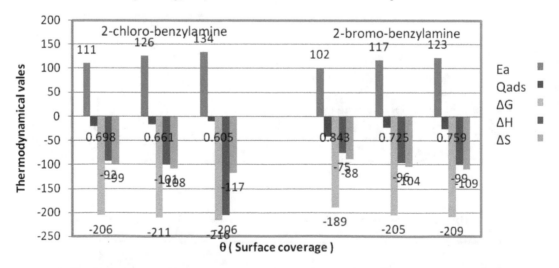

**Figure 5.** Thermodynamical values of different inhibitors versus θ(surface coverage area) at different temperatures

The activation energy, heat of adsorption, free energy, enthalpy and entropy of inhibitors 1-(2-chlorophenyl)methaneamine and 1-(2-bromophenyl)methaneamine were calculated by equation4, equation 5, equation 6 and equation 7 and their values were recorded in Table 2. The activation energy increased without inhibitors and its values decreased with inhibitors and its values were determined by plot of logK versus 1/T in Figure 3. It indicated that inhibitors bonded with base metal. Heat of adsorption found to be negative which indicating that inhibitors adhered with metal by physio-

chemio adsorption. Its values were calculated by plot of log(θ/1-θ) versus 1/T in Figure 4. The results of free energy, enthalpy and entropy values were shown negative sign which depicted that adsorption occurred on the surface of metal and the graph of all thermodynamical values (Ea, $Q_{ads}$, ΔG, ΔH and ΔS) versus θ (surface coverage area) were presented in Figure 5 .

$$d/dt\left(logK\right) = E_a /RT^2 \qquad (4)$$

where T is temperature in Kelvin and $E_a$ is the activation energy

$$\log(\theta/1-\theta) = \log(A.C) - (Q_{ads.}/RT) \quad (5)$$

where T is temperature in Kelvin and $Q_{ads.}$ heat of adsorption

$$\Delta G = -2.303RT \left[ \log C - \log(\theta/1-\theta) + 1.72 \right] \quad (6)$$

where T is temperature in Kelvin and $\Delta G$ free energy

$$K = RT/N\, h \log(\Delta S^{\#}/R) \, X \log(-\Delta H^{\#}/RT) \quad (7)$$

where N is Avogadro's constant, h is Planck's constant, $\Delta S^{\#}$ is the change of entropy activation and $\Delta H^{\#}$ is the change of enthalpy activation.

**Table 2. Thermodynamical values of inhibitors in 15% $H_2SO_4$ for stainless steel**

| Inhibitors | Temperatures | $333^0$K | $343^0$K | $353^0$K |
|---|---|---|---|---|
| IH(0) | $E_a(o)(kJmol^{-1})$ | 148 | 153 | 157 |
| IH(I) | $E_a(kJmol^{-1})$ | 118 | 126 | 134 |
| | $Q_{ads.}(kJmol^{-1})$ | -21 | -16 | -10 |
| | $\Delta G(kJmol^{-1})$ | -206 | -211 | -216 |
| | $\Delta H(kJmol^{-1})$ | -92 | -101 | -206 |
| | $\Delta S(JK^{-1})$ | -99 | -108 | -117 |
| IH(II) | $E_a(kJmol^{-1})$ | 102 | 117 | 123 |
| | $Q_{ads.}(kJmol^{-1})$ | -42 | -23 | -26 |
| | $\Delta G(kJmol^{-1})$ | -189 | -205 | -204 |
| | $\Delta H(kJmol^{-1})$ | -75 | -96 | -99 |
| | $\Delta S(JK^{-1})$ | -88 | -104 | -109 |

The corrosion current density determined in the absence and presence of inhibitor with the help of equation 8 and their values were recorded in Table 3.

$$\Delta E/\Delta I = \beta_a \beta_c / 2.303\, I_{corr}(\beta_a + \beta_c) \quad (8)$$

where $\Delta E/\Delta I$ is the slope which linear polarization resistance ($R_p$), $\beta_{a \text{ and }} \beta_c$ are anodic and cathodic Tafel slope respectively and $I_{corr}$ is the corrosion current density in mA/cm$^2$.

The metal penetration rate (mmpy) was determined by equation9 in absence and presence of inhibitors.

$$C.R(mmpy) = 0.327\, I_{corr}(mA/cm^2) \times Eq.Wt(g)/\rho(g/cm^3) \quad (9)$$

where $I_{corr}$ is the corrosion current density $\rho$ is specimen density and Eq.Wt is specimen equivalent weight.

The results of Table 3 indicated that corrosion current increase without inhibitors and its values reduced after addition of inhibitors because these inhibitors enhanced cathodic current so corrosion current and corrosion rate minimized. Tafel graph was plotted in Figure 6 between electrode potential and corrosion current density in the absence and presence of inhibitors.

**Table 3. Potentiostatic polarization of inhibitors in 15mM concentration and 15% of $H_2SO_4$**

| $\Delta E(mV)$ | $\Delta I$ | $\beta_a$ | $\beta_c$ | $I_{corr.}(mA/cm^2)$ | K(mmpy) | C(mM) |
|---|---|---|---|---|---|---|
| -650 | 550 | 350 | 250 | 55.79 | 4.379 | 0.00 |
| -600 | 475 | 225 | 315 | 45.17 | 0.354 | 15 |
| -575 | 350 | 175 | 325 | 30.10 | 0.236 | |

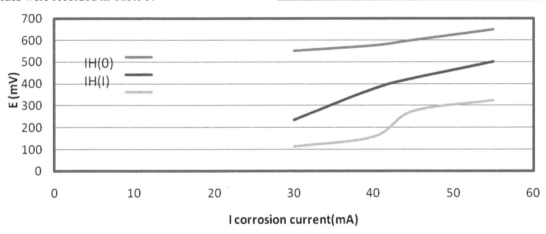

**Figure 6.** Plot of E (mV) Vs. I (mA) for stainless steel with 15mM concentration of inhibitors

## 4. Conclusion

These inhibitors possessed electron releasing functional which had capability to enhance electron charge density towards corred metal and protected base metal by formation of thin film. The results of surface coverage area and inhibition efficiency for both inhibitors indicated that both inhibitors adhered with the surface of metal. The results of activation energy, heat of adsorption, free energy, enthalpy and entropy were shown both inhibitors bonded with base metal physical-chemical adsorption.

## Acknowledgement

I am thankful to UGC, New Delhi for providing me financial support for this work. The author is thankful to Professor Sanjoy Misra, Department of chemistry, Ranchi University, Ranchi who provide me suggestion and guidance. I am also thankful to the department of chemistry, Ranchi University, Ranchi and the department of applied Chemistry Indian school of Mines, Dhanbad for providing laboratory facilities.

## References

[1] Khramov AN, Voevodin NN (2004), Hybrid organo-ceramic corrosion protection coating with encapsulated organic corrosion inhibitors. Thin solid films, 447, 549-557.

[2] Seth A, van Ooij (2004), Novel water based high-performance primers that can replace metal pretreatments and chromate-containing primers. J. Mater. Eng. Perform. 13, 468-474.

[3] Ianmuzzi M, Frankel GS, (2007) Mechanisms of corrosion inhibition of AA 2024-T3 by vanadates. Corros. Sci., 49, 2371-2391.

[4]   Moutarlier V, Neveu B, Gigandet MP (2008), Evaluation of corrosion protection for sol-gel coatings doped with inorganic inhibitors. Surf Coat Technol, 202, 2052.

[5]   Code A, Arenas M A, de Frutos (2008), Effective corrosion protection of 8090 alloy by cerium conversion coatings. Electrochim Acta, 53, 7760-68.

[6]   Shem M, Schmidt T, Gerwann J et al (2009), CeO2-filled sol-gel coatings for corrosion protection of AA2024-T3 aluminium alloy, Corrosion Sci. 51, 2304.

[7]   Pulvirenti A L, Bishop E J, Adel-Hadai M A, Barkatt A, (2009), Solubilisation of nickel from powders at near-neutral pH and the role of oxide layers. Corros. Sci. 51, 2043-2054.

[8]   Glezakou V A, Dang LX, Mc Grail BP,(2009) Spontaneous activation of CO2 and possible corrosion pathways on the Low-index iron surface Fe (100), J. Phys. Chem. 113, 3691-3696.

[9]   Farias MCM, Santos CAL, (2009) Friction behavior of lubricated zinc phosphate coating, Wear, 266, 873-877.

[10]  Cuevas-Arteaga C, (2008) Corrosion study of HK-40m alloy exposed to molten sulphate|vandate mixtures using the electrochemical noise technique, Corros. Sci. 50, 650-663.

[11]  Singh R K, Misra Sanjoy,(2013) Corrosion protection of stainless steel in CO2 environment by nanocoating of zinc phosphate with DLC filler, Journal of Metallurgy & Materials Science, 55, 149-156.

[12]  Singh R K, Misra Sanjoy, (2013) Corrosion protection of rebar steel in marine atmosphere by nanocoating, Journal of Metallurgy & Materials Science, 55, 313-321.

[13]  Singh R K, Misra Sanjoy,(2013) Corrosion protection of mild steel in SO2 environment by nanocoating with DLC filler, Journal of Metallurgy & Materials Science, 55, 227-234.

# Groundwater Quality and Hydrogeochemistry of Toungo Area, Adamawa State, North Eastern Nigeria

**J.M. Ishaku[1], B.A. Ankidawa[2,*], A.M. Abbo[1]**

[1]Department of Geology, School of Pure and Applied Sciences, Modibbo Adama University of Technology, PMB 2076, Yola, Nigeria
[2]Department of Agricultural and Environmental Engineering, School of Engineering and Engineering Technology, Modibbo Adama University of Technology, PMB 2076, Yola, Nigeria
*Corresponding author: ankidawa03@yahoo.com

**Abstract** Analytical results revealed that the water from various sources in the study area is unfit for human consumption due to bacteriological pollution. The water quality for agricultural practice indicated that water is good for agricultural practice. The Sodium Absorption Ratio (SAR) values range from 0.04 meq/l to 0.60 meq/l with an average of 0.13 meq/l. The Residual Sodium Carbonate (RSC) values range from 0.35 meq/l to 2.64 meq/l with an average of 1.43 meq/l as to of the water of generally safe. Total dissolved solid (TDS) values ranges from 95 to 285 mg/l hence the water is good for irrigational purposes. The study also revealed that the water may be unsuitable for some industries due to high iron concentration in some places, the values range from 0.21 to 3.97 mg/l with an average of 0.5 mg/l which indicate values above the recommended limits of 0.2 mg/l for industrial purpose. The water has total hardness ranging from 19.64 mg/l to 92.01 mg/l with a mean value of 50.97 mg/l, hence suitable for some industrial activities. PCA, HCA and rock-water interaction diagrams identified diffused form of contamination, leaching of bed rock geochemistry, salinity, natural mineralization, anthropogenic contamination, silicate weathering and oxidation as the major processes controlling the groundwater geochemistry. The groundwater is of calcium magnesium bicarbonate facies which belong to the normal alkaline group.

**Keywords:** *bacteriological pollution, agricultural practice, water quality, groundwater, industries, geochemistry*

## 1. Introduction

The study area forms part of the Yola arm of the Upper Benue Trough North eastern Nigeria. The people living in this area are mostly farmers and cattle rearers. The area is inhibited by the Fulani who are traditionally livestock rearers of Northern Nigeria and the Chamba people who cultivate crops like millet, beans groundnut, maize, cassava and guinea corn.

In any hydrogeological setting surface water and groundwater are the main sources of water supply. Surface water in the area includes; water from River Kom, streams, and lakes, whereas groundwater is obtained from boreholes, hand-dug wells and springs. Both sources of water are prone to contamination and pollution. Hence, the need for water quality assessment for enhanced socio-economic growth and development [14].

Evaluation of water quality for human consumption, agricultural and industrial activities have not been given attention especially in developing countries like Nigeria. The chemical composition of water is an important factor to be considered before it is used for domestic or irrigation purpose [36]. Water is the primary need of every living thing on this planet earth which is essential for sustaining life. Hence, the need for the analysis of water hydro-

geochemically and bacteriologically in other to know the acceptable quality for various purposes which must be pursued. Groundwater is never chemically pure, dissolution of substance takes place in the course of its percolation through the rocks and groundwater which makes itacquire some of these dissolved chemical constituents.

The chemical composition of groundwater is important if the water is to be used for domestic, industrial and agricultural purpose. The dissolved geochemical constituents in water affect its usage for various purposes. The concentrations of the dissolved constituents are correlated with the standards of the World Health Organization (WHO) or other organizations charged with the responsibility of providing safe drinking water to the ever increasing population.

No work has been done precisely on the study area. Most of the previous works in the area are mainly regional in nature. The geological survey of Nigeria initiated the search for water as far back as 1926 by carrying out the hydrogeological investigations in parts of Northern Nigeria. This led to the exploitation of groundwater through hand-dug wells and boreholes. By 1938, a water well drilling section was set up in the geological survey of Nigeria.

Carter et al. [7], Du Preeze and Barber [11] and Consulting Nigerian Limited [8], respectively published some details about the geology, hydrogeology and water quality of some parts of the then northern Nigeria.

Another work was also done by Kiser [17] which involved detailed chemical analysis of groundwater in Northern Nigeria. Furthermore, the government of Northeastern States under the Ministry of Works and Housing carried out physical and chemical analysis of water in 1975.The work by Ntekim [25] reveals that the chemical characters of groundwater in Adamawa State are meteoric and has been influence by bedrock chemistry as well as atmospheric and environmental activities.

The work by Ishaku et al. [16], assessed water quality for enhanced rural water supply in Adamawa State Northeastern Nigeria. Similar work was done at Michael Opara University of Agriculture Umudike and Environs Southeastern Nigeria by Magnus [20]. Subsequently, Magnus [20] carried out hydro chemical evaluation of groundwater in the Blue Nile Basin, Eastern Sudan, using conventional and multivariate techniques.

This research work is aimed at evaluating the quality groundwater from hand-dug wells and boreholes in Toungo area thereby ascertaining the suitability of the water for domestic, agricultural and industrial activities.

The study is alsoaimed at assessing the geochemical processes affecting the groundwater chemistry.

## 1.1. The Study Area

The study area is Toungo town; it constitutes part of sheet 238 Toungo S.W on a scale of 1:50,000. It lies within latitudes 08° 05′ N to 08° 09′ N and longitudes 12° 03′ E and 12° 06′ E, and covers an area of about 57.7 km$^2$ (Figure 1). The area form part of Yola Arm of the Upper Benue trough.

The area is characterized by moderate to high relief; it is mark by isolated hills and valley, and towards the central part there are few hills which stand out with the general elevation, with boulders and minor hills surrounding it. The elevation rises from 1,400 to 1,750 m above mean sea level; the area is well drain by ephemeral rivers and streams which gather most of the runoff from the hills. The drainage system exhibits a dentritic drainage pattern and it is dominated by the River Kom which flows from northeast to southwest (Figure 1).

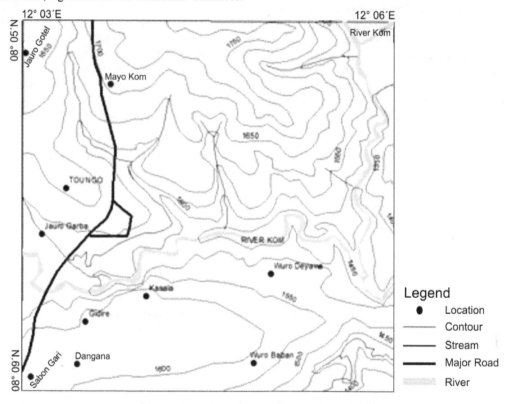

**Figure 1.** Topographic map of the study area

The study area is mark by two distinct seasons: the rainy season which start from April to October and the dry season which starts from November to March [38]. The annual rainfall and its variation were based on data recorded from existing rainfall gauging station at Upper Benue River Basin Authority Yola. The mean annual rainfall is about 700 to 1,200 mm. The rainfall begins in March and ends in November, with July, August and September recording the heaviest rainfall. The mean annual temperature range from 24 to 27°C with an average of 26°C recorded during the dry season. In the area humidity is generally low during harmattan period (December to January) due to the onset of the northeast trade wind from the continental interior. The increase in

humidity is noted during rainy season from April to October.

## 1.2. Geology and Hydrogeology of the Study Area

The study area is underlain by the Pan African granites which occupy about two third of the study area, and enclosed the migmatite – gneiss – quartzite complex. They are mostly located to the east and north central portion of the area. The granites have been identified as biotite granite, pegmatitic granites, granite gneisses and amphibolite (Figure 2). According to Rahaman [31], the older granites were emplaced during the Pan African thermotectonic event (450 ma).

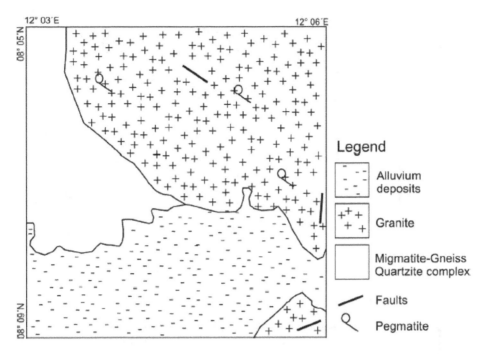

**Figure 2.** Geological map of the study area

The youngest deposits in the area include eluvium and the alluvial deposit. The eluvium is the weathered portion of the Precambrian basement and is mostly lateritic located in south-east and western portion of the study area, while the alluvial is characterized by sands, silts, clays and gravels and arefound along floodplains in the area.

Figure 3 indicates that groundwater flow takes place from mayo Kom in the north and flows towards the northeast and north west, respectively. Another flow zone takes place from Wuro Baban in the south and flows towards Wuro Deyawa in the south east and also flows towards Kasala. Groundwater flow also takes place from Sabon Gari in the south and flows towards the southwest. Recharge areas occur around Wuro Baban in the south, mayo Kom in the north and Sabon Gari area. Discharge areas occur around Wuro Deyawa, Jauro Garba and Jauro Gotel areas, respectively.

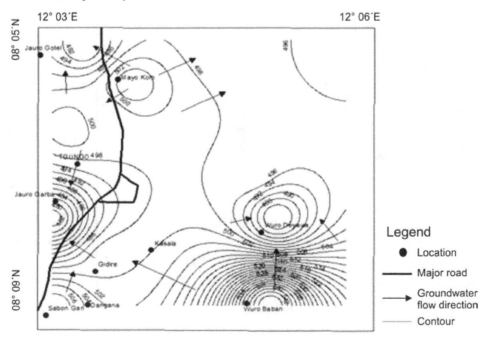

**Figure 3.** Hydraulic head distribution in unconfined aquifer in the study area

The boreholes in the study area generally attain average depth of 30 to 45 m tapping water largely from the unconfined aquifers. The weathered overburden aquifer generally has a thickness of about 9 m while the fracture basement aquifer is about 20 m thick (Figure 4). The water level is generally 4 m from ground surface and shallow wells yield is about 200 to 1,000 litres per minute in the study area. The hand-dug wells generally have depth of about 3 to 8 m at most locations [38].

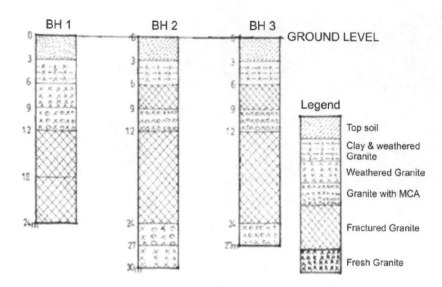

**Figure 4.** Lithologic log of some boreholes in the study area

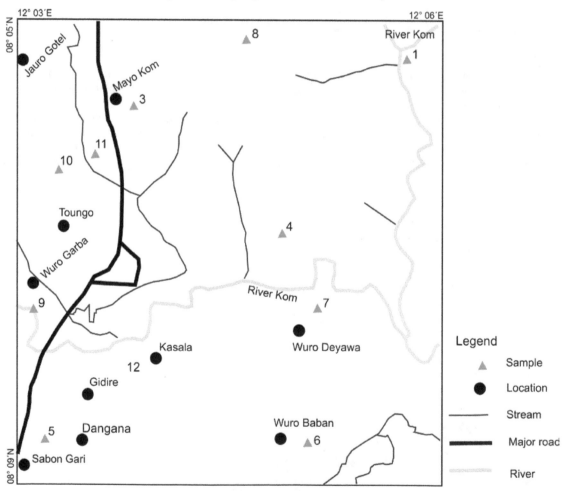

**Figure 5.** Sampling points in the study area

## 2. Methodology

The first of this work stage involved the collection of topographic map from Adamawa State Land and Survey (Figure 1). The second stage involved the collection of ten water samples from boreholes and hand-dug wells (Figure 5). The field work took place on 28th August 2012. The position of boreholes and hand-dug wells were determined using the Global Positioning System (GPS) and later transferred to the base map of the study area. Before collection, the water samples were collected in one-litre containers which were rinsed thoroughly with the water to be sampled according to Bercelona et al. [5]. Physical parameters such as temperature, pH, conductivity and total dissolved solid (TDS) were measured in the field using

TDS/conductivity meter (HACK KIT), while pH was measured using a portable pH meter. The chemical analysis was carried out using HACH digital spectrophotometer (model DR 2400), and samples were also analysed bacteriologically using the membrane filtration method according to WHO. The entire sample for chemical analysis was carried out within 24 hours of collection. While samples for bacteriological analysis was carried out within same period of time. The analysis was carried out at the Adamawa State Water Board Yola. The results were then subjected to statistical analysis using SPSS software version 15.0 and presented in for of minimum, maximum, mean and standard deviation. The geochemical processes were studied using the rock-water interaction diagrams.

The advanced multivariate statistical analysis techniques were adopted in this study using the Statistical software SPSS statistics, version 15.0. Multivariate analyses were employed because of its usefulness as a tool to reduce and organize large hydrochemical data sets into groups with similar characteristics and then relating them to specific changes in hydrological processes [28]. The multivariate statistical analyses such as PCA and Hierarchical Cluster Analysis (HCA) have been used to provide a quantitative measure of relatedness of water quality parameters and to suggest the underlying natural and anthropogenic processes in groundwater aquifers [22].

The Principal Component Analysis (PCA) is a multivariate statistical analysis for the purpose of data reduction with a view to determining the sources of elements and their controlling factors [15]. In PCA many factors can be obtained as there are many parameters/variables in the data set [24]. The total number of factors generated from a typical factor analysis indicates the total number of possible sources of variation in the data [4]. The multivariate statistical technique therefore relates variables into principal associations (factors) based on their mutual correlation coefficients and these associations may be interpreted in terms of mineralization, lithology and environmental processes [26]. Suleiman and Sameer

[35], classify factor loadings into 'strong', 'moderate' and 'weak' corresponding to greater than 0.75, 0.75-0.50 and 0.50-0.30, respectively. Significant attention is given to variables with strong positive or negative contribution to the factorial axis which is easier to understand the source of variability explained by the factor [24].

The Hierarchical Cluster Analysis (HCA) groups objects into classes or clusters on the basis of similarities within a class and dissimilarities [33]. A low distance shows the two objects are similar or close together whereas a large distance indicates dissimilarity [3]. The HCA according to Ward [40] with squared Euclidean distances was applied to detect multivariate similarities in groundwater quality. The results are presented as dendrogram of the groups and their proximity [27].

# 3. Results and Discussions

The standard for drinking water is based on two main criteria: the presence of objectionable taste, odours and colours and the presence of substances with adverse health effects [9]. The dissolved geochemical constituents in groundwater (Table 1) are correlated with the World Health Organization (WHO) standards (Table 2). Hydro geochemicalresults which are considered in the evaluation of water quality for domestic purposes reveal that the physical properties pH, EC and temperature are generally below the recommended limit of WHO [41,43]. The pH values range from 4.2 to 6.2 with an average of 5.3, which indicates the acidic condition of the water. TheElectrical Conductivity (EC) of the water samples range from 141 μS/cm to 424 μS/cm with a mean value of 244.7μS/cm which indicate low and fall within the recommended WHO [43] standards of 500 μS/cm. The temperature values range from 16.7°C to 21.7°C which fall below the recommended values of 30°C to 35°C of WHO [41]. The TDS values range from 95 mg/l to 285 mg/l, the values are within the recommended limit of WHO [43].

**Table 1. Chemical and bacteriological results of analysis from the different water sources in the study area**

| Sample location | Temp. (°C) | pH | Cond. (μs/cm) | TDS | HCO₃ (mg/l) | CO₂ (mg/l) | Total hardness (mg/l) | Fe²⁺ (mg/l) | F (mg/l) | Cl⁻ (mg/l) | NO₃ (mg/l) | Mg²⁺ (mg/l) | Na⁺ (mg/l) | Ca⁺ (mg/l) | SO₄ (mg/l) | K⁺ (mg/l) | Cu²⁺ (mg/l) | colifm count cfu/ml | RCS | SAR |
|---|---|---|---|---|---|---|---|---|---|---|---|---|---|---|---|---|---|---|---|---|
| BH1 | 19.80 | 4.6 | 162 | 110 | 171 | 0.00 | 52.93 | 0 | 0.25 | 10.31 | 12.11 | 28.47 | 1.02 | 30.18 | 21.39 | 2.06 | ND | 7 | 0.87 | 0.04 |
| BH2 | 21.70 | 5.4 | 198 | 130 | 174 | 0.00 | 23.03 | 0.11 | 0.04 | 49.77 | 43.38 | 14.52 | 0.14 | 10.91 | 23.26 | 1.1 | 0.01 | 76 | 1.92 | 0.00 |
| BH3 | 19.67 | 6.2 | 319 | 210 | 217 | 0.00 | 58.88 | 0.02 | 0.17 | 11.11 | 19.23 | 27.43 | 0.09 | 31.06 | 14.93 | 3 | 0 | 10 | 1.62 | 0.00 |
| BH4 | 21.00 | 6 | 221 | 150 | 236 | 0.00 | 92.01 | 3.97 | 0.012 | 87.68 | 68.1 | 45 | 6.41 | 48.91 | 27.63 | 7.61 | ND | 4 | 0.77 | 0.22 |
| BH5 | 20.37 | 5.3 | 159 | 105 | 211 | 0.00 | 21.93 | 0.17 | 0.19 | 15.73 | 37.86 | 11.96 | 0.18 | 13.03 | 27.33 | 3.96 | 0 | 16 | 2.64 | 0.00 |
| BH6 | 18.62 | 4.3 | 247 | 163 | 153 | 0.00 | 63.61 | 0 | 0.11 | 9.26 | 10.16 | 30.22 | 0.94 | 36.17 | 16.56 | 2.1 | ND | 13 | 0.35 | 0.04 |
| HDW1 | 18.00 | 4.2 | 424 | 285 | 131 | 2.00 | 79.74 | 0.43 | 0.06 | 23.83 | 54 | 14.61 | 2.96 | 19.27 | 29.39 | 4.4 | ND | 19 | 1.12 | 0.02 |
| HDW2 | 19.73 | 5.1 | 269 | 177 | 193 | 3.10 | 50.42 | 0.016 | 0.41 | 10.88 | 8.19 | 27.21 | 1.76 | 21.96 | 22.18 | 2.4 | 0.06 | 26 | 1.59 | 0.08 |
| HDW3 | 16.73 | 5.9 | 141 | 95 | 198 | 0.00 | 43.28 | 0.03 | 0.31 | 7.77 | 26.39 | 24.62 | 2.08 | 21.33 | 18.62 | 1.7 | ND | 28 | 1.61 | 0.10 |
| HDW4 | 21.00 | 5.7 | 307 | 201 | 168 | 1.70 | 30.12 | 0.09 | 0.09 | 16.91 | 14.67 | 15.88 | 3.61 | 17.21 | 16.97 | 3.31 | 0 | 40 | 1.74 | 0.20 |

BH=Borehole, HW=Hand-dugwel.

The evaluation of water quality revealed that parameters such as magnesium, sulphate, chloride, and total hardness are generally low. Magnesium concentration varies from 12 to 45 mg/l and falls within the recommended limit of WHO [43]. Sulphate varies from 14.9 to 29.4 mg/l with mean value of 21.8 mg/l, which is within the recommended limit of 400 mg/l of WHO [43]. Nitrate concentration values range from 8.2 mg/l to 68.1 mg/l

with mean value of 29.4 mg/l. About 30% of the water samples indicate values above WHO [42] recommended limit of 50 mg/l. High nitrate concentration is responsible for the blue baby syndrome (methamoglobiemia). The northeastern and northwestern parts of the study area show high nitrate values above WHO standards (Figure 6), these areas are therefore characterized by high risk of nitrate contamination.

Table 2. Summary of groundwater quality in the study area

| Parameters | Minimum | Maximum | Mean | Standard Deviation | WHO (2011) |
|---|---|---|---|---|---|
| Temperature (°C) | 16.70 | 21.70 | 19.66 | | 30 – 35* |
| pH | 4.20 | 6.20 | 5.27 | 0.712 | 6.5 - 8.5 |
| EC (µS/cm) | 141.00 | 424.00 | 244.7 | 88.08 | 500 |
| TDS (mg/l) | 95.00 | 285.00 | 162.600 | 58.487 | 500 |
| Sodium (mg/l) | 0.10 | 3.60 | 1.322 | 1.256 | 200 |
| Potassium (mg/l) | 1.70 | 7.60 | 3.25 | 1.778 | 12 |
| Magnesium (mg/l) | 12.00 | 45.00 | 23.99 | 10.0558 | 50 |
| Calcium (mg/l) | 10.90 | 48.90 | 25.01 | 11.642 | 75 |
| Bicarbonate (mg/l) | 131.00 | 236.00 | 185.20 | 31.762 | 500 |
| Chloride (mg/l) | 7.80 | 87.70 | 25.23 | 25.623 | 250 |
| Sulphate (mg/l) | 14.90 | 29.40 | 21.83 | 5.079 | 250 |
| Nitrate (mg/l) | 8.20 | 68.10 | 29.42 | 20.596 | 50** |
| Fluoride (mg/l) | 0.01 | 0.40 | 0.175 | 0.12713 | 1.5 |
| TH (mg/l) | 21.90 | 92.00 | 51.58 | 23.222 | 500** |
| Iron (mg/l) | 0.00 | 4.00 | 0.486 | 1.241 | 0.3 |
| Copper (mg/l) | 0.00 | 0.06 | 0.0167 | 0.02422 | 1.5 |
| Coliform | 4.00 | 122.00 | 37.0000 | 37.08957 | 0 - 3 |

* World Health Organization [41], **World Health Organization [42].

Iron concentration from the water samples range from 0 to 0.4 mg/l, these values are within the recommended limit of 0.3 mg/l of WHO [43]. The high iron concentration in BH4 of 4.0 mg/l, exceeded the recommended limit of WHO. Fluoride concentration ranges from 0.01 to 0.4 mg/l with an average of 0.22 mg/l which is below the recommended limit of WHO [43] The bacteriological analysis reveals high coliform number count ranging from 4 to 122 which indicate bacteriologically contaminated water. Bacteriological contaminated groundwater is associated with water borne diseases such as viral hepatitis, schistosomiasis and cholera. The quality of groundwater in some section of the study area is bacteriologically contaminated and therefore unfit for human consumption. Pockets of areas containing lessthan 10 coliform counts occur in the western and northern part of the study area (Figure 7).

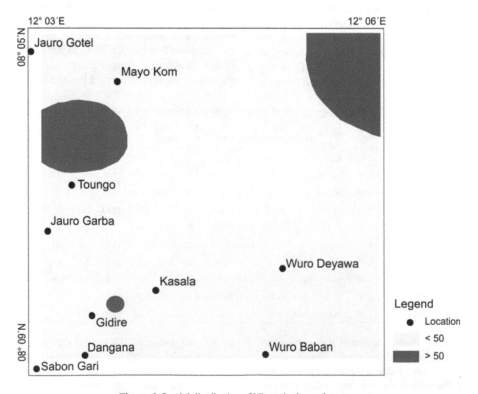

Figure 6. Spatial distribution of Nitrate in the study area

The electrical conductivity (EC) is a good measure of salinity hazard to crops, excess salinity reduces the osmotic activity of plant and thus interferes with the absorption of water and nutrients from the soil [32]. The EC values ranged from 141 to 424 µS/cm with mean value of 244.7 µS/cm. The EC values are within the range of excellent to good quality water for irrigation practice [39]. The values of TDS from the water samples range from 95 to 285 mg/l with mean value of 162.6 mg/l, all the values are less than 1000 mg/l, hence are within the non-saline class. The computed SAR values for all the water samples vary from 0.04 meq/l to 0.60 meq/l with an average of

0.40 meq/l. The results of analysis are correlated with the standard SAR values. The water samples from the study area can be utilized on all agricultural soil [21]. All the values are within the excellent class and can be applicable on all soils.

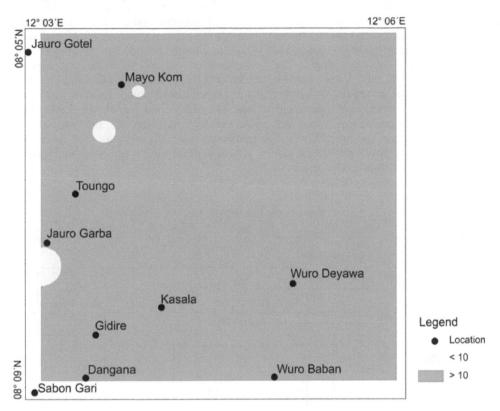

**Figure 7.** Coliform distribution in the study area

According to the California Fertilizer Committee [6], the Residual Sodium Carbonate (RSC) of less than zero is the most desirable for irrigational purposes, an RSC ranging from 0 to 1.24 mg/l is probably safe, RSC ranging from 1.25 to 2.50 mg/l is high and therefore suggest that water with such RSC should only be used on appropriate soil. The RSC values of the samples vary from 0.30 meq/l to 2.64 meq/l with an average of 1.43 meq/l, from the result in Table 2, 95% of the water is generally safe for irrigational practices, while the remaining 5% of the sample have potentials to cause salinity problems, hence if used on soil, it may likely develop salinity problem.

According to Okafor [29] water for industrial use should be odourless, colourless, free from suspended matter and micro-organism and should be of low iron and manganese contents. Water used in boilers should be soft and non-corrosive, while laundry water should be soft and colourless. Iron and manganese concentration above 0.2 mg/l precipitate upon oxidation and causes stain on plumbing features and foster growth in reservoir filters and distribution system [37]. Most industrial users object to water containing more than 0.2 mg/l of iron and manganese. From the results presented in Table 1, the iron concentration range from 0.00 mg/l to 4 mg/l, with an average of 0.5 mg/l. Based on the above standard 98% of the water samples reveal low concentration of iron except in BH4 which reveal value of 3.97 mg/l. The values of total hardness range from 21.9 to 92 mg/l with an average

of 51.6 mg/l. From the hardness classification of water after Vasanthavigar [39], only water samples from HDW1and BH4 arecharacterized by hard water. All the other water samples have total hardness <75 hence are classified as soft water, so the water is suitable for industrial usage.

**Table 3. Rotation Principal Component Analysis (PCA) loading matrix**

| Elements | Component | | |
|---|---|---|---|
| | 1 | 2 | 3 |
| Temperature | 0.514 | -0.225 | -0.396 |
| pH | 0.120 | 0.127 | -0.627 |
| EC | 0.198 | 0121 | **0.846** |
| TDS | 0.212 | 0.139 | **0.853** |
| Bicarbonate | 0.158 | 0.344 | -0.800 |
| TH | 0.252 | **0.873** | 0.369 |
| Iron | 0.699 | 0.639 | -0.250 |
| Fluoride | -0.744 | -0.047 | -0.119 |
| Chloride | **0.884** | 0.319 | -0.180 |
| Nitrate | **0.912** | 0.129 | -0.026 |
| Magnesium | -0.002 | **0.932** | -0.289 |
| Sodium | -0.142 | -0.124 | **0.684** |
| Calcium | 0.071 | **0.958** | -0.124 |
| Sulphate | **0.750** | -0.084 | 0.107 |
| Potassium | **0.793** | 0.489 | -0.020 |
| % Variance explained | 28.83 | 23.72 | 23.06 |
| % Cumulative | 28.83 | 52.55 | 75.61 |

**Multivariate Statistical Analyses**

Principal Component Analysis (PCA) on chemical data indicates three factors which explain about 75.61% of the total variance (Table 3). Factor 1 accounts for about 28.83% of total variance and is characterized by strong positive loading with respect to $NO_3^-$, $Cl^-$, $K^+$ and $SO_4^{2-}$. Factor 1 is interpreted as diffused form of contamination due to application of chemical fertilizer such as NPK, potash and manure [15]. Factor 2 accounts for about 23.72% of the total variance, and exhibits strong positive loadings with respect to TH, $Ca^{2+}$ and $Mg^{2+}$. The association of these elements to this factor may be attributed to leaching of bed rock materials, weathering and rock-water interaction. Hardness of water is caused by calcium and magnesium ions and can be tied to bed rock geochemistry [30]. Factor 3 accounts for about23.06% of the total variance, and has strong positive loadings with respect to EC, TDS and moderate loading with respect to

$Na^+$. This factor accounts for the temporary salinity of the water. The presence of $Na^+$ may be due to cation exchange by which $Ca^{2+}$ and $Mg^{2+}$ are replaced by $Na^+$ [28].

The results of cluster analysis are presented in Figure 8 and indicate two clusters. Cluster 1 is subdivided into two sub clusters, and sub cluster 1 comprises of EC, TDS and $Na^+$. This sub cluster is also related to factor 3. The sub cluster 1 is interpreted as salinity controlled by $Na^+$. The second sub cluster comprises of pH, and temperature with fluoride loosely bounded to the cluster. This cluster is ascribed to natural mineralization involving fluorite. Cluster 2 is also subdivided into two sub clusters; the first sub cluster shows close similarities between $Ca^{2+}$, $Mg^{2+}$ and TH, and second sub cluster shows close similarities between $Fe^{2+}$, $K^+$, $NO_3^-$, $Cl^-$ and $SO_4^2$and is interpreted as anthropogenic contamination.The first sub cluster is related to factor 1and controls the hardness of the water.

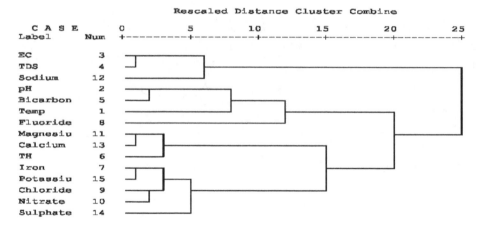

**Figure 8.** Dendrogram of hydrochemical data

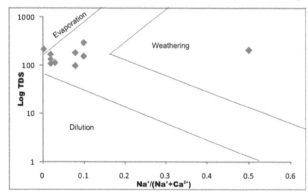

**Figure 9.** Plotting in Gibbs Diagram for Cations

**Rock-water Interaction**

The concentration of dissolved ions in groundwater depends on the hydrogeochemical processes that take place in the aquifer [23]. The authors added that the study of relative concentrations of the various major ions in groundwater is used in the identification of geochemical processes. Generally, different chemical processes occur during rock-water interaction, which include dissolution/precipitation, ion exchange processes, oxidation and reduction. The mechanisms controlling groundwater chemistry can be interpreted by Gibb's scatter diagrams and Piper diagrams and [23]. The plots of $Na^+/(Na^+ + HCO_3^-$ $Ca^{2+})$and $Cl^-/(Cl^-$ $+HCO_3^-)$ as a function of TDS are widely employed to

determine the sources of dissolved geochemical constituents. Gibbs [13] indicated a close relationship between water composition and the hydrochemical processes involving precipitation, water interaction and evaporation. Figure 9 and Figure 10 indicate that most points of the samples plotted in the region of rock dominance and weathering.

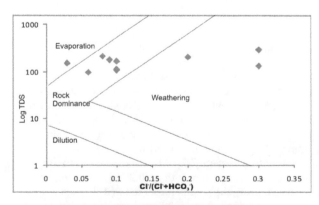

**Figure 10.** Plotting in Gibbs Diagram for Anions

*Calcium and Magnesium*

In the scatter diagram (Figure 11), about 70% of the samples plotted above the 1:1 equiline, thus representing carbonate weathering. Carbonate weathering may be caused by atmospheric water charged with $CO_2$ which further results in the formation of carbonic acid. This accelerates the dissolution of carbonate rocks such as dolomite, limestone and gypsum along groundwater flow path.

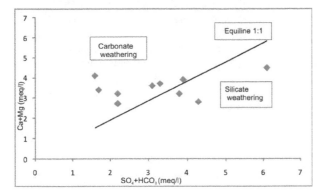

**Figure 11.** Scatter diagram of (Ca + Mg) vs (SO₄ + HCO₃)

The resultant is the release of $Ca^{2+}$ and $Mg^{2+}$ from the carbonate weathering into groundwater system through recharge. The weathering of calcite minerals which is responsible for the release of the ions is expressed through the following equations:

$$CO_2 + H_2O \rightarrow H_2CO_3 \quad (1)$$

$$CaCO_3 + H_2CO_3 \rightarrow Ca^{2+} + 2HCO_3^- \quad (2)$$
$$(\text{Calcite dissolution})$$

$$CaMg(CO_3) + 2H_2CO_3 \rightarrow Ca^{2+} + Mg^{2+} + 4HCO_3^- \quad (3)$$
$$(\text{Magnesium calcite dissolution})$$

*Sodium and Potassium*

When halite dissolution is prominent, Na vs Cl relationship gives a 1:1 ratio [23], however where there is increased concentration of Na than Cl is typically interpreted as Na released from silicate weathering [10]. Figure 12 indicates that most points plotted above the equiline of 1:1 suggesting no halide dissolution. Sodium is relatively less than Cl which indicates absence of much silicate weathering [12]. In the scatter plots of Na vs Ca (Figure 13) and Na vs Mg (Figure 14) scatter diagrams, most points plotted above the equiline of 1:1, thus indicating reduction in Na concentration in groundwater, due to ionic exchange. Further indication of reduction in sodium concentration is shown in Figure 15 where about 80% of the samples plotted below the equiline of 1:1, indicating the reduction of sodium from groundwater.

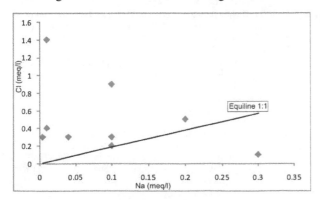

**Figure 12.** Scatter diagram of Na vs Cl

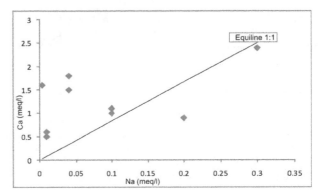

**Figure 13.** Scatter diagram of Na vs Ca

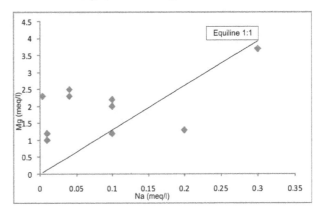

**Figure 14.** Scatter diagram of Na vs Mg

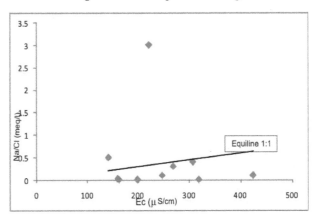

**Figure 15.** Scatter diagram of EC vs Na/Cl

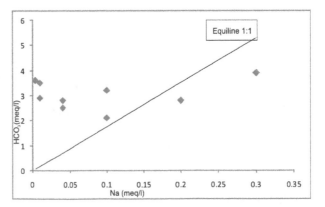

**Figure 16.** Scatter diagram of Na vs HCO₃

*Bicarbonate and Sulphate*

The Na vs HCO₃⁻ scatter diagram (Figure 16) indicates most plots occur above the equiline of 1:1, thus indicating increased $HCO_3^-$ compared to Na which resulted from silicate weathering. The weathering is given by the following equation:

$$2NaAlSi_3O_8 + 2H_2CO_3 + 9H_2O$$
$$\rightarrow Al_2Si_2O_5(OH)_4 + 2Na^+ + 4H_4SiO_4 + 2HCO_3^- \quad (4)$$

Figure 17 shows plot of $SO_4$ vs Cl, and indicate that most samples plotted above the equiline of 1:1, thus indicating low concentration of chloride. The sulphur concentration in groundwater may be due to oxidation of reduce sulphur gases and sulphate source [2].

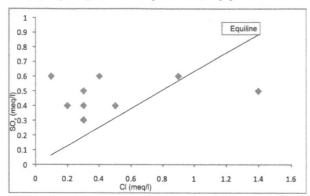

**Figure 17.** Scatter diagram of Cl vs SO₄

*Hydrogeochemical facies*

An iron can be considered as type facies when its concentration is within 50 to 100% domain. The characterization of the water in the study area using piper trilinear diagram (Figure 18) indicates that all samples plotted in the region of Ca-Mg-HCO₃.The Ca-Mg-HCO₃ facies is the dominant water type and belongsto the normal alkaline group, and is related to the geology of the area [26].

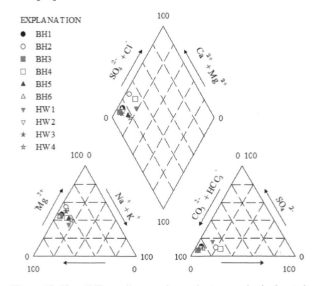

**Figure 18.** Piper Trilinear diagram of groundwater samples in the study area

## 4. Conclusion

Result of analysis of twelve water samples from the different sources reveal that the water in the study area is highly polluted due to presence of high coliform. High

concentration of nitrate and iron occur in few samples. Hence, the water is unfit for human consumption without treatment. The analysis further revealed that the water is generally good for agricultural and industrial uses. The multivariate statistical analysis using PCA and HCA and rock-water interaction diagrams identified diffused form of contamination, leaching of bed rock geochemistry, salinity, natural mineralization, anthropgenic contamination, silicate weathering and oxidation as the major processes controlling the groundwater chemistry. The water is classified as Ca-Mg-HCO₃ facies which belong to the normal alkaline group.

## References

[1]  Adamawa State Rural Water and Sanitation Agency (1998). Lithologic log of boreholes in Toungo area.

[2]  Artimes, D., Ramin, S. and Nagaraju, D. (2011). Hydrogeochemical and rock-water interaction studies in East of Kurdistan, N-W of Iran. International Journal of Environmental Sciences, vol. 1, No 1, pp 16-22.

[3]  Avdullahi, S., Fejza, I. and Timava, A. (2013). Evaluation of groundwater pollution usingmultivariate statistical analysis: A case study from Burimi area, Kosovo. J. Biodiversity and environ. Sci. 3(1): 95-102.

[4]  Belkhiri, L., Boudoukha, A., mouni, L. And Baouz, T. (2010). Multivariate Statistical Characterization of Groundwater Quality in Ain Azel Plan, Algeria. African Journal of Environmental Science and Technology, vol. 4 (8), pp 526-534.

[5]  Bercelona, M.J., Gibb, J.P., Helfrich, J.A. and Garske, E.E. (1985). Practical guide for groundwater sampling. ISWS contract report 374. Illinois State Water Survey Campaign, Illinois, 94p.

[6]  Califonia Fertilizer Committee (1975). Western fertilizer hand-book. The Interstate Printers and Publishers, Inc. Danville, IIinois.

[7]  Carter, J.D., Barber, W.D. and Tait, E.A. (1963). The geology of parts of Adamawa, Bauchi and Borno Province in northeastern Nigeria. Geological Survey of Nigeria (GSN). Bull, vol. 30, pp. 108.

[8]  Consultin Nigeria Ltd (1975). Urban Water Supply Yola. Preliminary Sate Report, pp. 23-47.

[9]  Davis, S.N. and De Weist, R.J.M. (1966). Hydrogeology, John Wile and Sons, New York, pp. 463.

[10]  Deutsch, W.J. (1997). Groundwater Geochemistry: Fundamentals and Applications to Contamination Lewis Publishers, New York.

[11]  Du Preeze, J.W. and Barber, W.D. (1965). The distribution and chemical quality of groundwater in Northern Nigeria (GSN), Bull. Vol. 32, pp. 93.

[12]  Fisher, S.R. and Mullican, W.F. (1997). Hydrogeochemical evaluation of sodium-sulphate and sodium-chloride groundwater beneath the northern Chihuahua desert, Trans-Pecos, Texas, USA. Hydrogeology Journal, v 5, no 2, pp. 4-16.

[13]  Gibbs, R.J. (1970). Mechanisms controlling world water chemistry. Science, vol. 170, pp. 1088-1090.

[14]  Ishaku, J.M. (2011). Assessment of groundwater quality index for Jimeta – Yola area, Northeastern Nigeria. Journal of Geology and Mining Research, vol. 3(9): 219-231.

[15]  Ishaku, J.M., Nur, A. and Matazu, H.I. (2012). Interpretation of groundwater quality in Fufore, North eastern Nigeria. International Journal of Earth Sciences and Engineering, volume 05, No. 03, pp 373-382.

[16]  Ishaku, J.M., Obiefuna, G.J. and Al-Farisu, A. (2002). Assessment of water quality for enhanced rural water supply in Adamawa State, North Eastern Nigeria. Global Journal of Geological Sciences vol. 5 Nos 1&2, pp 13-23.

[17]  Kiser, R.T. (1968). Chemical quality of water in Northern Nigeria (GSN). Open file Report.

[18]  Kumar, S.K., Bharani, R., Magesh, N.S., Godson, P.S. and Chandrasekar, N. (2014). Hydrochemistry and groundwater quality appraisal of partof south Chennai coastal aquifers, Tamil Nadu, India using WQIand fuzzy logic method. Applied Water Science, vol. 4, pp. 341-350.

[19]  Kumar, S.K., Logeshkumaran, A., Magesh, N.S., Godson, P.S. and Chandrasekar, N. (2015). Hydro-geochemistry and application

of water quality index (WQI) for groundwater quality assessment, Ann Nagar, part of Chennai City, Tamil Nadu, India. Applied Water Science, vol. 5, pp. 335-343.

[20] Magnus, U.I., Akaninyere, O.A. and Udoinyang, A.E. (2011). Hydrochemical evaluation of groundwater in Micheal Opara University of Agriculture, Umudike and its environs southeastern Nigeria. Journal of Water Resources and Protection, vol. 3, no. 12, pp. 925-929.

[21] Mandel, S. and Shiftan, Z.I. (1981). Groundwater resources investigation and development. Academic press.

[22] Mohapatra, P.K., Vijay, R., Pujari, P.R., Sundaray, S.K. and Mohanty, B.P. (2011). Determination of processes affecting groundwater quality in the coastal aquifer beneath Puri city, India: a multivariate statistical approach. Water Science &Technology, vol. 64. No. 4, pp 809-817.

[23] Nayak, K.M. and Sahoo, H.K. (2011). Assessment of Groundwater Quality in Tangi-Choudwar and Cuttack Blocks of Cuttack District, Orissa, India. International Journal of Earth Sciences and Engineering, vol. 4, no. 6, pp 986-994.

[24] Ngo Boum-Nkot, S., Ketchemen-Tandia, B., Ndje, Y., Emvouttou, H., Ebonji, C.R. and Huneau, F. (2015). Origin of Mineralization of Groundwater in the Tongo Bassa Watershed (Douala-Cameroon). Research Journal of Environmental and Earth Sciences 7 (2): 29-41.

[25] Ntekim, E.E. (2000). Geochemical characterization of groundwater in Adamawa State. Journal of Pure and Applied Sciences, vol. 1, pp. 55-63.

[26] Nton, M.E., Adejumo, S.S. and Elueze, A.A. (2007). Hydrogeochemical Assessment of Surface water and groundwater quality in Agbowo-Orogun area of Ibadan, South western.

[27] Nur, A., Ishaku, J.M. and Yusuf, S.N. (2012). Groundwater flow patterns and HydrogeochemicalFacies Distribution Using Geographical Information System (GIS) in Damaturu, Northeast Nigeria. International Journal of Geosciences, 3, pp 1096-1106.

[28] Ogunribido, T.H.T. and Kehinde-Philips, O.O. (2011). Multivariate Statistical Analysis for the Assessment of Hydrogeochemistry of Groundwater in Agbabu Area, S.W. Nigeria. Proceedings of the Environmental Management Conference, Federal University of Agriculture, Abeokuta, Nigeria, pp 425-435.

[29] Okafor, D.U. and Adamu, E.A. (1994). Aspect of the geology and hydrogeology of the river bakoji catchments areas of Niger State, Nigeria. Journal of mining Geology, vol. 2,pp. 9.

[30] Olasehinde, P.I., Amadi, A.N., Dan-Hassan, M.A., Jimoh, M.O. and okunlola, I.A. (2015). Statistical Assessment of Groundwater Quality in Ogbomosho, Southwest Nigeria. American Journal of Mining and Metallurgy 3.1, pp 21-28.

[31] Rahaman, M.A. (2003). An address delivered by president, NMGS at the Annual International Conference, Itakpe, IN: The Crust, vol. 26, no. 2.

[32] Saleh, A., Al-Ruwait, F. and Shehata, M. (1999). Hydrogeochemical processes operating within the main aquifers of Kuwait. Journal of Arid and Environment, vol. 42, pp. 195-209.

[33] Singh, S.K., Singh, C.K., Kumar, K.S., Gupta, R. and Mukherjee, S. (2009). Spatial-Temporalmonitoring of groundwater using multivariate statistical techniques in Bareilly District of Uttar Pradesh, India. J. Hydrol. Hydromech, 57, 1, pp 45-54.

[34] SPSS® (Statistical Package for Social Sciences) version 15.0, USA. (2006). SPSS 15.0 Comman Syntax Reference. 233, South Wacker Drive, Chicago.

[35] Suleiman,M.T. and Sameer, A.A. (2015). Interpretation of Groundwater Quality Parameters for Springs in Tafileh Area in South of Jordan Using Principal Component Analysis. Environmental Sciences, vol. 3, no. 1, pp 31-44

[36] Suresh, T.S., Nagana, C. and Srinivas, G. (1991). Study of water quality for agricultural use in Hemavathy river (Kamataka) Hydrology. Journal Indian Association Hydrology, vo. 14(4): pp. 247-254.

[37] Todd, D.K. (1980). Groundwater Hydrology, 2nd ed., John Wiley and Sons, Assessment of groundwater quality using chemical indices, New York, pp. 535.

[38] Upper Benue River Basin Development Authority (2010). Recorded rainfall between 2006 – 2010, Yola, Nigeria.

[39] Vasanthavigar, M.K.S, Gantha, R.R., Vijayaraghavan, K. and Sarma, V.S. (2010). Characterization and quality assessment of groundwater with special emphasis on irrigation utility.

[40] Ward, J.H. (1963). Hierarchical grouping to optimize objective function. J. Am. Stat. Assoc., 69: 236-244

[41] World Health Organization (1997). Guidelines for drinking water quality, vol. 3. World Health Organization, Geneva.

[42] World Health Organization (2006). International standards for drinking water and guidelines for water quality. World Health Organization, Geneva.

[43] World Health Organization (2011). Guidelines for drinking water quality, 4th edn. World Health Organization, Geneva.

# Assessment of Fire Risk of Indian Coals Using Artificial Neural Network Techniques

Devidas S. Nimaje[*], Debi P. Tripathy

Department of Mining Engineering, National Institute of Technology, Rourkela, Odisha, India
*Corresponding author: dsnimaje@nitrkl.ac.in

**Abstract** Spontaneous heating of coal is a major problem in the global mining industry. It has been known to pose serious problems on account of coal loss due to fires and affects not only the coal production but also creates environmental pollution over the years. It is well known that the intrinsic properties and susceptibility indices play a vital role to assess the spontaneous heating susceptibility of coal. In this paper, best correlated parameters from the intrinsic properties with the susceptibility indices were used as input to the different Artificial Neural Network (ANN) techniques viz. Multilayer Perceptron (MLP), Functional Link Artificial Neural Network (FLANN), and Radial Basis Function (RBF) to predict in advance the fire risk of Indian coals. This can help the mine management to adopt appropriate strategies and effective action plans to prevent occurrence and spread of fire. From the proposed ANN techniques, it was observed that Szb provides better fire risk prediction with RBF model vis-à-vis MLP and FLANN.

**Keywords:** *coal; spontaneous heating, ANN, MLP, FLANN, RBF*

## 1. Introduction

Coal is the most important and abundant fossil fuel in India. Coal is the source of about 27% of the world's primary energy consumption and it accounts for about 34% of electricity generated in the world. Hence, in recent years, much attention has been focused on coal as an alternative source of energy [45]. Coal is the dominant energy source in India and meets 55% of the country's primary commercial energy supply. Commercial primary energy consumption in India has grown by about 700% in the last four decades [7]. India is the third largest coal producing country in the world after China and USA [27]. Indian mines have a historical record of extensive fire activity for over hundred years. The fire problem in Indian mines is very complex because of involvement of different seams simultaneously. Such conditions do not exist elsewhere in the world [40]. Spontaneous combustion of coal generally causes mine fires in Indian coalfields despite various preventive measures have been extensively practiced. The spontaneous heating susceptibility of different coals varies over a wide range and it is important to predict their degree of proneness in advance for taking preventive measures against the occurrence of fires to avoid loss of lives and property, sterilization of coal reserves and environmental pollution and raise concerns about safety and economic aspects of mining, etc. [43].

Brief overview of the related works carried out by various researchers in India and other countries are summarized in the following subsection:

Pattnaik et al. [31] investigated intrinsic properties and a few susceptibility indices to characterize the Chirimiri coals of the SECL coalfields, India. Karmakar and Banerjee [15] worked on sixteen Indian coal samples using comparative experimental techniques to measure the susceptibility of coal to spontaneous combustion based on statistical regression analysis. Olpinski index being a convenient and rapid method, and can be used as an alternative to CPT method India. Smith et al. [50] designed and developed the sponcom program in the U.S Bureau of Mines for the assessment of the spontaneous combustion risk of an underground mining operation. It used the available information to make decisions based on a series of rules provided by the programmer. Panigrahi and Ray [48] analyzed 78 Indian coals and used MLP ANN model for obtaining optimum results based on the evaluation of the best combination of wet oxidation potential experimental conditions. Zhang et al. [51] employed feed forward 3-layer MLP model to express relationship between temperature and index gases (CO and $C_2H_4$) and forecasting the coal sponcom in the low-temperature range. Xiao and Tian [52] introduced genetic algorithm and back propagation neural network for the purpose of predicting the danger of coal layer spontaneous combustion based on the selection of three key influencing factors, namely, coal spontaneous combustion inclination, geology conditions and occurrence features of coal seam, and ventilation conditions. Panigrahi et al. [49] investigated Indian coals using susceptibility indices such as CPT, Wet oxidation method, Russian U-index, Szb, etc., and categorized and predicted spontaneous fire risk based on regression analysis. Literature work seems that most of the work carried out by researchers, academicians and coal companies in the world are based on experimental investigations, statistical analysis, mathematical models, and to limited extent ANN models etc. to predict the proneness of coal to spontaneous heating.

In this paper, an attempt has been made to carry out the statistical analysis among the different intrinsic properties (Proximate, Ultimate and Petrographic analysis) and the susceptibility indices (Crossing point temperature (CPT), Olpinski index free of ash (Szb), Wet oxidation Potential difference ($\Delta E$), and Flammability temperature (FT)) to obtain the best correlated parameters. The high significant correlated parameters of ultimate analysis were used as an input to different ANN models such as MLP, FLANN, and RBF. The paper also highlights the performance analysis of different ANN models with different susceptibility indices to predict the fire risk of Indian coals.

## 2. Sample Collection and Preparation

Forty-nine non-coking and coking in-situ coal samples were collected from major coalfields of India viz. South Eastern Coalfield Limited (SECL), Singareni Collieries Company Limited (SCCL), Mahanadi Coalfield Limited (MCL), Western Coalfield Limited (WCL), North Eastern Coalfield Limited (NEC), Northern Coalfield Limited (NCL), Indian Iron and Steel Company (IISCO), Bharat Coking Coal Limited (BCCL) and Tata Iron and Steel Company Limited (TISCO) using channel sampling method. The collected coals were crushed and sieved as per the experimental requirements following Indian Standard IS: 436(Part-I/Section-I)–1964 [9].

**Table 1. Classification of liability of coal to sponcom based on Olpinski index [Tripathy and Pal, 2001]**

| Szb($^0$C/min) | Risk Rating |
|---|---|
| <80 | Poorly susceptible |
| 80-120 | Moderately susceptible |
| >120 | Highly susceptible |

**Table 2. Results of the parameters of proximate, ultimate and petrographic analysis of coal samples**

| Sl. No. | Coal samples | Proximate analysis | | | Ultimate analysis | | | | Petrographic analysis | | |
|---|---|---|---|---|---|---|---|---|---|---|---|
| | | M % | A % | VM % | C % | H % | S % | O % | V% | L% | I% |
| 1 | SECL -1 | 7.63 | 14.10 | 32.42 | 83.75 | 4.74 | 0.38 | 9.64 | 18.91 | 5.85 | 53.04 |
| 2 | SECL -2 | 3.16 | 25.60 | 35.21 | 81.53 | 5.15 | 0.42 | 11.34 | 17.00 | 6.12 | 52.97 |
| 3 | SECL -3 | 6.41 | 16.45 | 24.59 | 80.79 | 4.59 | 0.32 | 12.84 | 32.87 | 5.35 | 49.50 |
| 4 | SECL - 4 | 5.95 | 16.24 | 39.79 | 80.00 | 4.92 | 0.40 | 12.94 | 31.31 | 3.97 | 57.38 |
| 5 | SECL -5 | 8.25 | 12.10 | 39.54 | 81.86 | 6.24 | 1.02 | 9.27 | 58.43 | 1.57 | 24.51 |
| 6 | SECL -6 | 7.62 | 22.55 | 20.77 | 77.26 | 5.75 | 1.24 | 13.60 | 49.23 | 6.17 | 33.96 |
| 7 | SECL -7 | 8.15 | 14.99 | 30.91 | 78.38 | 5.24 | 0.79 | 13.54 | 28.79 | 6.2 | 33.11 |
| 8 | SECL -8 | 8.86 | 11.16 | 30.52 | 77.81 | 6.15 | 0.54 | 13.99 | 39. 76 | 4.71 | 29.26 |
| 9 | SECL -9 | 12.57 | 17.11 | 33.66 | 79.94 | 4.48 | 1.04 | 14.18 | 29.14 | 9.03 | 53.19 |
| 10 | SECL -10 | 8.21 | 19.30 | 28.14 | 78.52 | 5.82 | 0.52 | 13.27 | 30.57 | 11.41 | 48.01 |
| 11 | SCCL-1 | 2.43 | 33.07 | 27.96 | 75.71 | 6.39 | 0.71 | 12.21 | 45.88 | 1.76 | 38.83 |
| 12 | SCCL-2 | 2.13 | 25.94 | 33.42 | 82.32 | 5.55 | 0.76 | 9.94 | 45.72 | 1.67 | 38.29 |
| 13 | SCCL-3 | 2.73 | 14.46 | 35.83 | 81.22 | 3.82 | 0.25 | 12.73 | 42.89 | 6.8 | 33.98 |
| 14 | SCCL-4 | 3.76 | 25.68 | 34.13 | 79.67 | 4.11 | 0.45 | 13.54 | 42.3 | 6.32 | 34.39 |
| 15 | SCCL-5 | 3.17 | 15.28 | 35.99 | 79.45 | 5.16 | 0.75 | 13.02 | 41.66 | 7.22 | 33.96 |
| 16 | SCCL-6 | 3.66 | 37.84 | 25.88 | 78.84 | 4.89 | 1.01 | 14.71 | 50.35 | 4.62 | 27.79 |
| 17 | SCCL-7 | 3.77 | 27.15 | 32.84 | 83.50 | 1.97 | 0.94 | 10.52 | 53.79 | 4.89 | 30.23 |
| 18 | SCCL-8 | 3.69 | 17.41 | 40.40 | 81.84 | 5.46 | 0.85 | 10.26 | 52.15 | 4.12 | 25.69 |
| 19 | SCCL-9 | 2.86 | 11.04 | 38.91 | 80.51 | 4.01 | 0.63 | 13.31 | 54.71 | 4.71 | 32.54 |
| 20 | MCL-1 | 7.13 | 37.48 | 23.17 | 72.93 | 7.64 | 0.79 | 17.64 | 19. 24 | 6.7 | 28.99 |
| 21 | MCL-2 | 6.42 | 35.25 | 25.76 | 74.09 | 7.47 | 1.18 | 15.54 | 18. 88 | 7.89 | 31.34 |
| 22 | MCL-3 | 2.81 | 13.46 | 30.19 | 80.17 | 3.83 | 0.30 | 14.21 | 21.11 | 9.84 | 20.55 |
| 23 | MCL-4 | 6.63 | 11.20 | 40.92 | 82.91 | 3.80 | 0.28 | 11.66 | 39.88 | 7.25 | 35.16 |
| 24 | MCL-5 | 3.89 | 16.20 | 35.55 | 81.67 | 3.62 | 0.31 | 11.51 | 33.19 | 7.88 | 16 |
| 25 | MCL-6 | 6.13 | 37.12 | 26.78 | 66.90 | 7.45 | 1.06 | 19.12 | 23.78 | 2.76 | 25.12 |
| 26 | MCL-7 | 7.77 | 14.01 | 26.46 | 78.31 | 5.91 | 1.14 | 10.91 | 28.2 | 5.11 | 25.23 |
| 27 | MCL-8 | 11.71 | 22.74 | 22.48 | 77.64 | 6.97 | 1.16 | 10.47 | 38.67 | 3.35 | 26.11 |
| 28 | WCL-1 | 6.03 | 14.50 | 39.97 | 82.11 | 3.23 | 0.15 | 12.92 | 58.62 | 4.55 | 17.23 |
| 29 | WCL-2 | 4.00 | 22.00 | 37.00 | 82.78 | 3.18 | 0.20 | 12.13 | 66.74 | 3.56 | 16.63 |
| 30 | WCL-3 | 6.50 | 16.00 | 35.50 | 81.27 | 5.48 | 0.27 | 10.78 | 34.85 | 2.77 | 43.96 |
| 31 | WCL-4 | 3.50 | 23.10 | 32.50 | 82.09 | 4.58 | 0.34 | 9.86 | 56.55 | 5.75 | 26.19 |
| 32 | WCL-5 | 5.50 | 16.00 | 34.50 | 77.95 | 4.80 | 0.48 | 12.23 | 42.18 | 5.94 | 40.59 |
| 33 | WCL-6 | 6.00 | 17.50 | 33.50 | 77.40 | 5.36 | 0.86 | 14.93 | 40.07 | 6.03 | 39.88 |
| 34 | WCL-7 | 7.30 | 16.00 | 31.50 | 79.69 | 5.49 | 0.76 | 12.31 | 40.97 | 8.76 | 37.88 |
| 35 | WCL-8 | 11.00 | 13.50 | 30.00 | 82.28 | 5.46 | 0.60 | 10.05 | 29.15 | 8.52 | 50.93 |
| 36 | WCL-9 | 4.13 | 19.50 | 28.97 | 78.90 | 3.78 | 0.84 | 13.31 | 27.44 | 9.16 | 49.74 |
| 37 | WCL-10 | 4.00 | 16.09 | 30.18 | 80.05 | 4.05 | 0.61 | 11.15 | 31.47 | 8.38 | 50.23 |
| 38 | NEC-1 | 1.32 | 6.20 | 43.26 | 72.72 | 4.54 | 2.27 | 17.41 | 86.87 | 4.32 | 5.10 |
| 39 | NEC-2 | 1.90 | 6.90 | 44.12 | 70.06 | 6.14 | 2.63 | 18.36 | 85.21 | 4.45 | 5.83 |
| 40 | NEC-3 | 4.06 | 11.63 | 54.12 | 72.66 | 4.61 | 1.16 | 18.22 | 85.94 | 4.73 | 5.76 |
| 41 | NEC-4 | 2.36 | 11.21 | 55.45 | 73.14 | 4.55 | 0.59 | 20.05 | 84.18 | 4.56 | 5.42 |
| 42 | NEC-5 | 2.15 | 13.50 | 54.44 | 71.06 | 4.13 | 0.96 | 22.36 | 86.35 | 4.18 | 5.39 |
| 43 | NEC-6 | 2.53 | 8.31 | 56.30 | 72.07 | 3.65 | 0.65 | 18.80 | 85.81 | 4.37 | 5.94 |
| 44 | NCL-1 | 7.94 | 19.40 | 28.40 | 77.43 | 5.37 | 1.36 | 11.82 | 35.08 | 1.6 | 40.94 |
| 45 | NCL-2 | 8.03 | 19.06 | 31.21 | 74.36 | 6.76 | 0.73 | 16.03 | 36. 87 | 0.67 | 41.32 |
| 46 | IISCO-1 | 0.82 | 31.57 | 14.24 | 79.78 | 5.16 | 1.20 | 12.15 | 59.36 | 2.19 | 30.68 |
| 47 | IISCO-2 | 0.97 | 27.96 | 16.58 | 72.12 | 5.92 | 0.82 | 15.21 | 58.33 | 2.17 | 31.92 |
| 48 | BCCL-1 | 1.39 | 16.30 | 18.48 | 81.66 | 5.50 | 0.39 | 8.50 | 59.94 | 2.79 | 27.32 |
| 49 | TISCO-1 | 1.44 | 15.05 | 17.86 | 82.26 | 5.62 | 0.51 | 9.06 | 62.29 | 3.39 | 28.68 |

NB: Coking coal samples – IISCO-1, IISCO-2, BCCL-1 and TISCO-1, and the rest are non-coking coal samples.

## 3. Experimental Investigations

To assess the liability of coals to spontaneous combustion, it is important to investigate the intrinsic properties by proximate, ultimate, and petrographic analysis as well as determination of susceptibility indices viz., CPT, Szb, $\Delta$E, and FT. Parameters of proximate analysis such as moisture (M), volatile matter (VM), and ash (A); elements of ultimate analysis viz., carbon, hydrogen, sulphur, and oxygen; and macerals of petrographic analysis such as vitrinite (V), liptinite (L), and inertinite (I) can be ascertained using the standard procedure [8,10,11,12,13,34] and the results are summarized in Table 2.

Susceptibility indices play a vital role to assess the spontaneous combustion of coal. In this paper, susceptibility indices such as CPT, Szb, $\Delta$E, and FT was determined using standard procedure [1,3,15,27,28,29,42], and the results are depicted in Table 3.

### Table 3. Results of CPT, Szb, $\Delta$E, and FT

| Sample | CPT $^0$C | Szb $^0$C/min | $\Delta$E mV | FT $^0$C |
|--------|-----------|---------------|--------------|----------|
| SECL -1 | 175 | 68 | 132 | 550 |
| SECL -2 | 182 | 74 | 159 | 555 |
| SECL -3 | 156 | 85 | 135 | 540 |
| SECL -4 | 178 | 70 | 130 | 545 |
| SECL -5 | 158 | 99 | 165 | 540 |
| SECL -6 | 163 | 77.36 | 133 | 535 |
| SECL -7 | 176 | 69.14 | 152 | 575 |
| SECL -8 | 182 | 63.26 | 125 | 580 |
| SECL -9 | 154 | 107.89 | 116 | 525 |
| SECL -10 | 188 | 65.42 | 140 | 580 |
| SCCL-1 | 175 | 73 | 140 | 520 |
| SCCL-2 | 180 | 67 | 155 | 500 |
| SCCL-3 | 153 | 111 | 159 | 530 |
| SCCL-4 | 179 | 55 | 151 | 510 |
| SCCL-5 | 164 | 77 | 136 | 510 |
| SCCL-6 | 168 | 78 | 150 | 500 |
| SCCL-7 | 166 | 59 | 143 | 510 |
| SCCL-8 | 157 | 105.74 | 125 | 530 |
| SCCL-9 | 172 | 74.28 | 144 | 525 |
| MCL-1 | 154 | 98 | 73 | 500 |
| MCL-2 | 158 | 104 | 82 | 515 |
| MCL-3 | 151 | 109 | 59 | 525 |
| MCL-4 | 142 | 105 | 104 | 500 |
| MCL-5 | 152 | 110 | 92 | 540 |
| MCL-6 | 168 | 78 | 81 | 535 |
| MCL-7 | 148 | 117.61 | 75 | 480 |
| MCL-8 | 164 | 58.23 | 95 | 540 |
| WCL-1 | 155 | 98 | 131 | 550 |
| WCL-2 | 149 | 112 | 141 | 540 |
| WCL-3 | 142 | 107 | 139 | 535 |
| WCL-4 | 147 | 89 | 145 | 540 |
| WCL-5 | 157 | 98 | 72 | 560 |
| WCL-6 | 148 | 93 | 68 | 540 |
| WCL-7 | 165 | 59.08 | 114 | 540 |
| WCL-8 | 155 | 108.95 | 94 | 530 |
| WCL-9 | 153 | 116.48 | 178 | 520 |
| WCL-10 | 143 | 118.73 | 144 | 475 |
| NEC-1 | 150 | 109 | 56 | 520 |
| NEC-2 | 153 | 118 | 65 | 545 |
| NEC-3 | 152 | 111.88 | 68 | 515 |
| NEC-4 | 151 | 115.58 | 72 | 500 |
| NEC-5 | 154 | 113.16 | 69 | 520 |
| NEC-6 | 176 | 68.69 | 87 | 570 |
| NCL-1 | 141 | 151 | 182 | 490 |
| NCL-2 | 146 | 170 | 148 | 530 |
| IISCO-1 | 180 | 45 | 46 | 570 |
| IISCO-2 | 165 | 70 | 48 | 560 |
| BCCL-1 | 178 | 64.55 | 55 | 575 |
| TISCO-1 | 192 | 58.06 | 67 | 590 |

## 4. Statistical Analysis

Statistical correlation analysis (univariate) reveals that the parameters of ultimate (C, H, and O) analysis show significant correlation with all investigated susceptibility indices (CPT, Szb, $\Delta$E, and FT) as compared to other independent variables (Table 4).

### Table 4. Correlation analysis between intrinsic properties and the susceptibility indices

| Sl. No. | Susceptibility indices / Intrinsic properties | CPT | Sz$_b$ | $\Delta$E | FT |
|---------|-----------------------------------------------|-----|--------|-----------|-----|
| 1 | M | 0.98 | 0.85 | 0.91 | 0.98 |
| 2 | VM | 0.95 | 0.95 | 0.91 | 0.98 |
| 3 | A | 0.92 | 0.85 | 0.88 | 0.91 |
| 4 | C | 0.99 | 0.95 | 0.95 | 0.99 |
| 5 | H | 0.97 | 0.95 | 0.90 | 0.97 |
| 6 | O | 0.96 | 0.95 | 0.90 | 0.97 |
| 7 | V | 0.94 | 0.90 | 0.86 | 0.94 |
| 8 | L | 0.54 | 0.55 | 0.93 | 0.54 |
| 9 | I | 0.92 | 0.88 | 0.93 | 0.92 |

Multivariate analysis has also been carried out on combined parameters of the intrinsic properties such as parameters of proximate analysis (M, VM, and A), elements of ultimate analysis (C, H, and O), and the macerals of petrographic analysis (V, L, and I), and all investigated susceptibility indices. From Table 5, it can be inferred that CPT, Szb, $\Delta$E, and FT show significant correlation with the elements of ultimate analysis(C, H, and O) based on correlation coefficient (r), standard error (SE), and variance ($\sigma$) as compared to the other parameters (Proximate and Petrographic analysis). Hence, these parameters (C, H, and O) can be used as input to ANN models to predict the proneness of coal to spontaneous combustion.

### Table 5. Multivariate analysis between intrinsic properties and the susceptibility indices

| Sl. No. | Independent variable | Multivariate analysis | CPT | Szb | $\Delta$E | FT |
|---------|---------------------|----------------------|-----|-----|-----------|-----|
| 1. | M, VM, and A | r | 0.98 | 0.95 | 0.94 | 0.98 |
| | | SE | 0.28 | 0.27 | 0.39 | 0.89 |
| | | Mean | 1.61 | 0.9 | 1.13 | 5.32 |
| | | Variance | 0.01 | 0.06 | 0.14 | 0.07 |
| 2. | C, H, and O | r | 0.99 | 0.96 | 0.95 | 0.99 |
| | | SE | 0.13 | 0.25 | 0.38 | 0.27 |
| | | Mean | 1.61 | 0.90 | 1.13 | 5.32 |
| | | Variance | 0.01 | 0.06 | 0.14 | 0.07 |
| 3. | V, L, and I | r | 0.97 | 0.93 | 0.91 | 0.97 |
| | | SE | 0.36 | 0.32 | 0.48 | 1.13 |
| | | Mean | 1.61 | 0.9 | 1.13 | 5.32 |
| | | Variance | 0.01 | 0.06 | 0.14 | 0.07 |

## 5. Artificial Neural Network Models

An ANN is an efficient information processing system and performs various tasks such as pattern matching and classification, optimization function, approximation, vector quantization, and data clustering [41,47]. In the model specification, ANN requires no knowledge of the data source but, since they often contain many weights that must be estimated [3]. In this paper, three ANN models such as MLP, FLANN, and RBF were applied to predict fire risk of Indian coals.

## 5.1. Cross-validation Method

Cross-validation is a statistical learning method to evaluate and compare the models by partitioning the data into two portions. One portion of the set is used to train or learn the model, and the rest of the data is used to validate the model. K-fold cross-validation is the basic form of cross validation [21,32]. In K-fold cross-validation, the data are first partitioned into K equal (or nearly equally) sized portions or folds. For each of the K model, K-1 folds are used for training and the remaining one fold is used for testing purpose. In this paper, 5-folds cross-validation was used for designing and comparing the models.

## 5.2. Performance Evaluation Parameters

To assess the performance of prediction models, the most widely used evaluation criterion is the Mean Magnitude of Relative Error (MMRE) [18,44]. Further, determination of software accuracy for a designed model by using performance evaluation parameters [37,46] such as: Mean Absolute Error (MAE), MMRE, Root Mean Square Error (RMSE), and Standard Error of the Mean (SEM). This is usually computed following standard evaluation processes such as cross-validation [19,20].

• **Mean Absolute Error (MAE)**

It determines how close the values of predicted and actual differ.

$$MAE = \frac{1}{n}\sum_{i=1}^{n}(|\,y_i' - y_i\,|) \qquad (1)$$

Where, $n$ is the number of samples, $y_i$ is the actual value, and $y_i'$ is predicted value.

• **Magnitude of Relative Error (MRE)**

The MRE for each observation i can be obtained as:

$$MRE_i = \frac{|\,Actual\ Effort_i - Predicted\ Effort_i\,|}{Actual\ Effort_i} \qquad (2)$$

• **Mean Magnitude of Relative Error (MMRE)**

The mean magnitude of relative error (MMRE) can be achieved through the summation of MRE over N observations

$$MMRE = \sum_{1}^{N} MRE_i \qquad (3)$$

• **Root Mean Square Error (RMSE)**

It determines the differences in the values of predicted and actual differ.

$$RMSE = \sqrt{\frac{1}{n}\sum_{i=1}^{n}\left(y_i - y_i'\right)^2} \qquad (4)$$

• **Standard Error of the Mean (SEM)**

It is the deviation of predicted value from the actual.

$$SEM = \frac{SD}{\sqrt{n}} \qquad (5)$$

Where, SD is the sample standard deviation, and n is the number of samples.

## 5.3. Multilayer Perceptron (MLP)

The multilayer perceptron propagates the input signal through the network in a forward direction, layer-by-layer basis. This system has been applied successfully to solve some difficult and diverse problems by training in a supervised manner with a highly popular algorithm known as the error back-propagation algorithm [4,5,6,16]. The structure of MLP is shown in Figure 1. MLP is widely used for pattern classification, recognition, prediction and approximation.

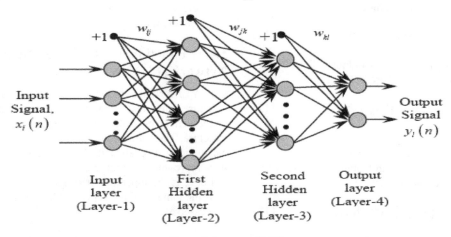

**Figure 1.** Structure of MLP

If $P_1$ is the number of neurons in the first hidden layer, each element of the output vector of first hidden layer can be calculated as:

$$f_j = \varphi_j\left[\sum_{i=1}^{N} w_{ij}x_i(n) + b_j\right], \qquad (6)$$
$$i = 1,2,3,......,N, j = 1,2,3,....,P_1$$

Where, bj - the bias to the neurons of the first hidden layer; N - the number of inputs;

φ - the nonlinear activation function in the first hidden layer.

The time index, n has been dropped to make the equations simpler.

Let $P_2$ be the number of neurons in the second hidden layer. The output ($f_k$) of this layer can be expressed as:

$$f_k = \varphi_k\left[\sum_{j=1}^{P_1} w_{jk}f_j + b_k\right], k = 1,2,3,......,P_2 \qquad (7)$$

Where, $b_k$ - the bias to the neurons of the second hidden layer.

The output of the final output layer can be calculated as:

$$y_1(n) = \varphi_1\left[\sum_{k=1}^{P_2} w_{kl}f_k + b_l\right], l = 1, 2, 3, \ldots\ldots, P_3 \quad (8)$$

Where, $b_l$ - the bias to the neuron of the final layer;
$P_3$ - the number of neurons in the output layer.

So, the output of the MLP neural network can be expressed as:

$$y_1(n)$$
$$= \varphi_n[\sum_{k=1}^{P_2} w_{kl}\varphi_k(\sum_{j=1}^{P_1} w_{jk}\varphi_j \quad (9)$$
$$\{\sum_{i=1}^{N} w_{ij}x_i(n) + b_j\} + b_k) + b_l]$$

### 5.3.1. Back-Propagation (BP) Algorithm

It is the most popular MLP network learning the algorithm. The parameters of the neural network can be updated in both sequential and batch mode operation and the least mean square (LMS) technique is used for the minimization of error [22,35,38].

### 5.3.1.1. Algorithm for Training MLP Based Fire Risk Model

The algorithm for training MLP [5,14] based fire risk model has been represented as follows:

Step 1: Select the total number of layers as m and the number $n_i$ (i=1, 2, . . . , m-1) of the neurons in each hidden layer.
Step 2: Randomly select the initial values of the weight vectors $w^m_{i,j}$ for i=1,2, . . . $n_i$ and m=2 (number of layers).

$$w^m_{i,j} \leftarrow \text{Rand}\left(w^m_{i,j}(0)\right) \quad (10)$$

Step 3: Randomly select the initial values of the bias vectors $b^m_{i,j}$ for i=1,2, . . . $n_i$ and m=2 .

$$b^m_{i,j} \leftarrow \text{Rand}\left(b^m_{i,j}(0)\right) \quad (11)$$

Step 4: Calculation of the neural outputs of the hidden layer and the equation can be represented as:

$$a^m_{i,j} = \varphi\left(\left(w^1_{i,j}\right) * x_k + \text{bias}\right) \quad (12)$$

Where, $\varphi$ - the transfer function;
$w^1_{i,j}$ - weight associated with the neuron
Step 5: Calculation of the neural outputs of the output layer and the equation obtained as:

$$Y_j = \varphi\left(\left(W^2_{i,j}\right) * a^m_{j+} \text{bias}\right) \quad (13)$$

Where, $W^2_{i,j}$ - weight associated with the neuron
Step 6: The final output $y_1(n)$ at the output neuron was compared with the desired output $d(n)$ and the resulting error signal $e_1(n)$ was obtained as:

$$e_1(n) = d(n) - y_1(n) \quad (14)$$

Step 7: Total error obtained by addition of error signals of all neurons in the output layer

$$\xi(n) = \frac{1}{n}\sum_{i=1}^{n} e^2(n) \quad (15)$$

Step 8: The Sensitivity calculation for the output layer is the derivative of activation function of output layer and can be represented as:

$$S_1 = f^{\cdot 2}\left(n^2\right) = \frac{d}{dn}(n) = 1 \quad (16)$$

Step 9: The Sensitivity of hidden layer is the derivative of activation function of hidden layer and can be represented as:

$$S_2 = f^{\cdot 1}(n) = \frac{d}{dn}\left[\frac{1}{1 + \exp^{-n}}\right]$$
$$= \left[1 - \frac{1}{1 + \exp^{-n}}\right] * \left[\frac{1}{1 + \exp^{-n}}\right] \quad (17)$$
$$= \left(1 - a^m_{i,j}\right) * a^m_{i,j}$$

Step 10: The weights of the respective layers are adjusted using the following relationship:
a) Updating the weight for output layer:

$$W^1_{i,j}(\text{new}) = W^1_{i,j}(\text{old}) + \eta S^1_J \quad (18)$$

b) Updating the weight for hidden layer:

$$W^2_{i,j}(\text{new}) = W^2_{i,j}(\text{old}) + \eta S^2_J(a^1_{i,j}) \quad (19)$$

Where, $\eta$ is the momentum parameter of the system.

Step 11: The above process was repeated for steps 4 - 10. The weights and the bias were updated using the iterative method until the error signal reaches minimum. For measuring the degree of matching, the mean square error (MSE) was taken as a performance measurement.
Step 12: After the completion of training of input data, the weights were fixed and the network can be used for future prediction.

### 5.4. Functional Link Artificial Neural Network (FLANN)

In FLANN, the hidden layers are removed, and the structure offers less computational complexity and higher convergence speed than MLP because of its single-layer structure. The mathematical expression and computational calculation was evaluated [22,30,36], and the structure has been represented in Figure 2.
Let X is the input vector of size N×1 which represents N as the number of elements; the $k^{th}$ element, and has been expressed as:

$$X(K) = x_k, 1 \leq K \leq N \quad (20)$$

Each element undergoes trigonometric expansion to form M elements such that the resultant matrix [30] has the dimension of N×M and can be represented as:

$$s_i = \begin{cases} x_k & \text{for } i = 1 \\ \sin(l\pi x_k) & \text{for } i = 2, 4, \ldots, M \\ \cos(l\pi x_k) & \text{for } i = 3, 5, \ldots, M+1 \end{cases} \quad (21)$$

The bias input is unity and the enhanced pattern can be obtained by the trigonometric function $X = [x_1 \cos(\pi x_1)$ $\sin(\pi x_1) \cdots x_2 \cos(\pi x_2) \sin(\pi x_2) \cdots x_1 x_2]^T$ for the prediction purpose. The back propagation algorithm, which is used to train the network, becomes very simple

because of the absence of hidden layers.

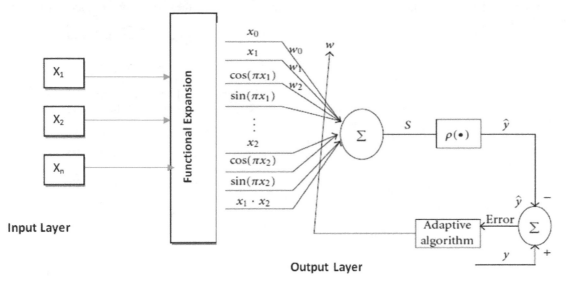

**Figure 2.** Structure of FLANN [25]

### 5.4.1. Algorithm for Training FLANN Fire Risk Model

The algorithm for training FLANN [24] based fire risk model has been represented as follows:

Step 1: Initialize the inputs $x_i$, $(i = 1, \ldots, n)$.

Step 2: Randomly select the initial values of the weight vectors $w_i$, for $i = 1, 2, \ldots l$, where i is the number of functional elements.

Step 3: All the weights $w_i$ were initialized to random number and given as

$$w_i \leftarrow \text{Rand}(w_i(0))$$ (22)

Step 4: The functional block can be represented as:

$$Xi = \begin{bmatrix} 1, x_1, \sin(\pi x_1), \cos(\pi x_1), \\ x_2, \sin(\pi x_2), \cos(\pi x_2)\ldots \end{bmatrix}$$ (23)

Step 5: The output was calculated as follows:

$$O_i = \sum_{i=1}^{N} w_i * X_i$$ (24)

Step 6: The error was calculated as $e_i = d_i - O_i$. It may be seen that the network produces a scalar output.

Step 7: The weight matrix was updated using the following relationship:

$$w_i(k + 1) = w_i(k) + \alpha e_i(k) X_i(k)$$ (25)

Where, k - the time index;

$\alpha$ - the momentum parameter.

Step 8: If error $\leq \varepsilon$ (error limit), then go to Step 9 otherwise, go to Step 3.

Step 9: After the completion of learning, the weights were fixed, and the network can be used for testing.

### 5.5. Radial Basis Function (RBF)

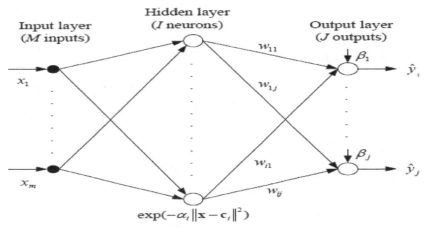

**Figure 3.** Network architecture of RBF [17]

The idea of RBF network derives from the theory of function approximation. RBF networks are very popular curve fitting, time series prediction, and control and classification problems. The architecture of the RBF network is quite simple. An input layer consisting of sources node; a hidden layer in which each neuron computes its output using a radial basis function, that

being in general a Gaussian function, and an output layer that builds a linear weighted sum of hidden neuron outputs and supplies the response of the network (effort) [37]. An RBF network has only output neuron. The structure of RBF has been depicted in Figure 3.

Gradient Descent (GD) [39] is a first-order derivative-based optimization algorithm used for finding a local minimum of a function. The algorithm takes steps proportional to the negative of the gradient of the function at the current point. The output of an RBF network [39] has been written as:

$$\hat{y} = \begin{bmatrix} w_{11} & w_{11} & . & . & w_{1J} \\ w_{21} & w_{21} & . & . & w_{2J} \\ . & . & . & . & . \\ . & . & . & . & . \\ w_{I1} & . & . & . & w_{IJ} \end{bmatrix} \begin{bmatrix} 1 \\ \varphi(\|x - c_1\|^2) \\ . \\ . \\ \varphi(\|x - c_I\|^2) \end{bmatrix} \quad (26)$$

and

$$\hat{Y} = W.H \quad (27)$$

Where, the weight matrix is represented as $W$, and $\phi$ matrix is represented as $H$.

GD algorithm can be implemented to minimize the error after defining the error function:

$$E = \sum (Y - \hat{Y})^2 \quad (28)$$

Where, $Y$ is the desired output.

RBF can be optimized by adjusting the weights and center vectors by iteratively computing the partials and performing the following updates:

$$w_{ij} = w_{ij} - \eta \frac{\partial E}{\partial w_{ij}} \quad (29)$$

$$c_i = c_i - \eta \frac{\partial E}{\partial c_i} \quad (30)$$

Where, $\eta$ is the step size [39].

### 5.5.1. Algorithm for Training RBF Network Based Fire Risk Model

The algorithm for training RBF network [41] based fire risk model has been represented as:

Step 1: Set the weights to small random values.

Step 2: Perform steps 3-9 when the stopping condition is false.

Step 3: Perform steps 4-8 for each input.

Step 4: Each data unit ($x_i$ for all i = 1 to n) receives input signals and transmits to the next hidden layer unit.

Step 5: Calculate the radial basis function.

Step 6: Select the centers for the radial basis function. The centers are selected from the set of input vectors. It should be noted that a sufficient number of centers have to be chosen to ensure adequate sampling of the input vector space.

Step 7: Calculate the output from the hidden layer unit:

$$v_i(x_i) = \frac{\exp\left[-\sum_{j=1}^{r}(x_{ji} - \hat{x}_{ji})^2\right]}{\sigma_i^2} \quad (31)$$

Where, $x_{ji}$ - the center of the RBF unit for input variables; $\sigma_i$ - the width of the $i^{th}$ RBF unit; $x_{ji}$ - the $j^{th}$ variable of an input pattern.

Step 8: Calculate the output of the neural network:

$$y_{net} = \sum_{i=1}^{k} w_{im} v_i(x_i) + w_0 \quad (32)$$

Where, k is the number of hidden layer nodes (RBF function);

$y_{net}$ is the output value of $m^{th}$ node in output layer for the $n^{th}$ incoming pattern;

$w_{im}$ is the weight between $i^{th}$ RBF unit and $m^{th}$ output node;

$w_0$ is the biasing term at the $n^{th}$ output node.

Step 9: Calculate the error and test for the stopping condition. The stopping condition may be the number of epochs or to a particular extent weight change.

## 6. Simulation Results and Discussion

To validate the performance of ANN models for prediction of fire risk of Indian coal seams, three ANN models i.e. MLP, FLANN and RBF were used. Best correlated parameters of ultimate analysis (C, H, and O) were chosen as an input for the simulation process. Simulation studies were carried out using MATLAB. The developed models were designed as per the proposed ANN algorithms. The entire system was a MISO (Multi Input and Single Output) model. To develop these models, initially the real-time data was processed experimentally. Cross validation was adopted to validate the samples after divided 49 samples into five folds.

Initially the Input and the output data were normalized and then it was processed in the system. In MLP and RBF, 3-3-1 structure (3 inputs, 3 hidden layers and 1 output) was used, while in FLANN, due to the non-availability of hidden layers, 3 inputs, and 1 output architecture was adopted. The Mean Square Error (MSE) vs. Epochs plot of all the applied ANN models are represented in Figures 4-6. They indicate that MLP, FLANN and RBF network models provide better results with Szb as compared to CPT, FT, and $\Delta$E and require 10.01, 3.13, and 6.24 secs computation time respectively with 2000 epochs.

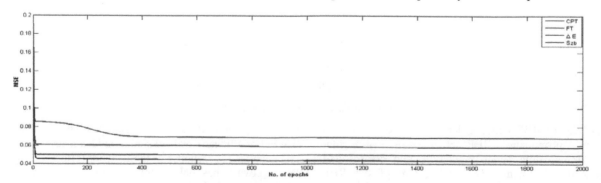

**Figure 4.** Performance curve of MLP

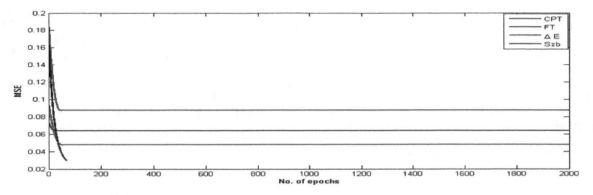

**Figure 5.** Performance curve of FLANN

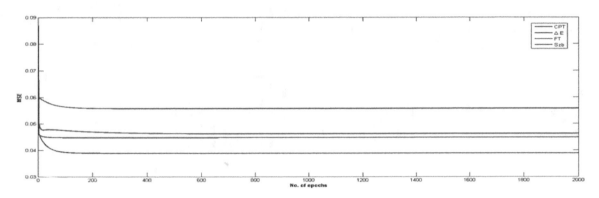

**Figure 6.** Performance curve of RBF network

## 7. Results and Discussion

The results of the investigated intrinsic properties viz., parameters of proximate analysis, elements of ultimate analysis and the macerals of petrographic analysis are summarized in Table 2. Proximate analysis results show that for three non-coking coal samples (SECL-9, MCL-8 and WCL-8), the moisture content was very high i.e. ≥ 11% and it matches with the field investigation; whereas coking coals showed very less (0.82%-1.44%) moisture content. Ash content and volatile matter in the collected non-coking coals varied in the range of 11.04% - 37.84% and 20.77% - 40.92% but in North-Eastern Coalfield (NEC-1 – NEC-6), it ranges from 6.2% - 13.5% and 43.26% - 56.30% . For the coking coals, these values were between 15.05% - 31.57% and 14.24% - 18.48% respectively. High inherent moisture and volatile matter coals have a higher tendency to spontaneous heating [1]. Therefore, only three parameters of proximate analysis viz., Moisture (M), Ash (A), and Volatile matter (VM) have been considered to ascertain the tendency of coal to spontaneous heating.

In the ultimate analysis, the carbon content is an indicator of the rank of coal. Coals containing higher oxygen are more prone to spontaneous combustion [34]. Indian coals have low sulphur content except in North-Eastern Coalfield. The result shows that NEC coals have sulphur content less than 3%, which should not reflect on spontaneous combustion of coal. Additionally, nitrogen content in the collected coal (~ 4%) does not relate to the rank of coal, and therefore it would not have any effect on spontaneous combustion. The classification of the coal was done following the percentage of carbon, hydrogen

and oxygen in coal [2]. Hence, only carbon (C), hydrogen (H) and oxygen (O) have been considered and they play a vital role as compared to other elements Nitrogen (N) and Sulphur (S) of ultimate analysis. The results of the petrographic analysis are summarized in Table 2. The degree of proneness to spontaneous combustion increases with the increase of vitrinite and liptinite, but decreases with the increase of inertinite content [23,31].

The results of the susceptibility indices i.e. CPT, Szb, ΔE, and FT are summarized in Table 3. Usually, CPT decreases with increase in percentages of volatile matter, oxygen, and moisture but more than 35% Volatile Matter; 4-6% Moisture, and 9% Oxygen do not have much effect on CPT [26]. If Szb increases, it implies increases in the susceptibility of coal to spontaneous heating and the fire risk rating can be ascertained using Table 1. The tendency of coal to spontaneous combustion increases with higher wet oxidation potential difference [1]. In FT, coal that is more susceptible towards aerial oxidation burns at low temperature as compared to less susceptible coals.

From the multivariate analysis shown in Table 5, it can be inferred that CPT and Szb show significant correlation results at 5% level of significance with the combined parameters of proximate (M, VM, and A) and ultimate (C, H, and O) analysis based on correlation coefficient, (r= 0.98 and 0.95), standard error, (SE = 0.28 and 0.27), and variance (σ = 0.01 and 0.06) as compared to the other parameters. So these parameters can be used to predict the susceptibility of coal to spontaneous combustion. Macerals of the petrographic analysis and other susceptibility indices show no significant correlation due to less r, high SE and high σ. But the univariate statistical analysis shows that parameters of ultimate (C, H, and O) analysis shows significant correlation with all investigated susceptibility indices (CPT, FT, Szb, and ΔE) and hence can be used as input parameters to ANN models.

The performance analysis of ANN models reveals that Szb provides better results as compared to CPT, ΔE, and FT and can be used for the prediction of fire risk of Indian coals (Figure 4-Figure 6). Further, Figure 7 shows that average MMRE is less in Szb after cross validation viz., MLP (0.56), FLANN (0.72), and RBF (0.49). It implies that RBF network model shows less average MMRE as compared to MLP and FLANN and can provide better prediction of fire risk of Indian coals with Szb.

The collected coals are categorized into three categories based on fire risk rating of Olpinski index (Table 1). Table 6 shows that NCL-1 and -2 are classified into high fire risk, whereas SECL-3,5,9, SCCL-3,8, MCL-1,2,3,4,5,7, WCL-1,2,3,4,5,6,8,9,10, and NEC-1,2,3,4,5 are categorized into medium fire risk, and the rest of the coal samples are low fire risk category. The results of the experimental studies were observed to match closely with the field observations for coal categorization based on the Olpinski index.

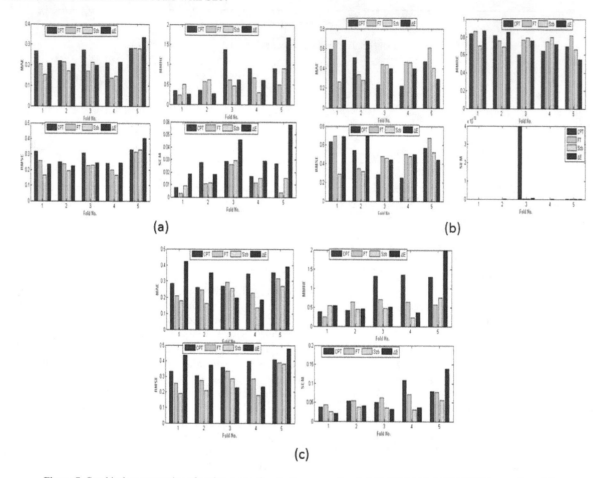

**Figure 7.** Graphical representation of performance of evaluation parameters in (a) MLP (b) FLANN (c) RBF network models

**Table 6. Results of categorization of coals**

| Susceptibility Index | Fire risk rating | | |
|---|---|---|---|
| | Poorly susceptible | Moderately susceptible | Highly susceptible |
| Szb (⁰C/min) | SECL-1,2,4,6,7,8,10, SCCL-1,2,4,5,6,7,9, MCL-6,8, WCL-7, NEC-6, IISCO-1,2, BCCL-1 and TISCO-1 | SECL-3,5,9, SCCL-3,8, MCL-1,2,3,4,5,7, WCL-1,2,3,4,5,6,8,9,10 and NEC-1,2,3,4,5 | NCL-1,2 |

# 8. Conclusions

The following conclusions are drawn from the present investigations:

1. From the statistical analysis (univariate and multivariate), it could be interpreted that parameters of ultimate (C, H, and O) analysis shows better significant correlation with Szb as compared to other susceptibility indices (CPT, ΔE, and FT) and can be used as input parameters to ANN models.

2. The performance analysis of ANN models (MLP, FLANN, and RBF network) revealed that Szb provides better results as compared to CPT, ΔE, and FT and can be used for the prediction of fire risk of Indian coals.

3. The performance evaluation of cross validation implies that RBF network model can provide better prediction of fire risk of Indian coals with Szb than MLP and FLANN based on least MMRE.

4. The simulation study showed that RBF provides appropriate fire risk prediction with Szb as compared

to MLP and FLANN, and, hence can be implemented in hardware.

5. The results of the experimental investigations were matched closely with the field observations based on Olpinski index. NCL-1 and -2 are categorized into high fire risk, whereas rest of the samples were classified into medium and low fire risk category Hence, Olpinski index can be used as a reliable index to predict proneness of Indian coals to spontaneous combustion.

## Acknowledgements

The authors are grateful to Department of Science and Technology, Government of India, New Delhi for partial funding of our research work. The authors are also thankful to the officials and staff of SCCL, SECL, NEC, NCL, MCL, WCL, BCCL, TISCO and IISCO for their assistance in collection of coal samples and field investigation.

## References

[1] S.C. Banerjee, Prevention and combating mine fires, Special Indian ed., Oxford & IBH Publishing Co. Pvt. Ltd., New Delhi, 2000.

[2] W. Francis, Coal-Its formation and composition, Edward Arnold, London, 1961.

[3] S.S. Gultekin, K. Guney, S. Sagiroglu, Neural networks for the calculation of bandwidth of rectangular microstrip antenna, ACES J. Special issue on Neural Network Applications in Electromagnetics, 18, 2(2003) 110-120.

[4] M.M. Gupta, L. Jin, N. Homma, Static and dynamic neural networks: From fundamental to advanced theory, John Wiley & Sons Ltd., USA, 2003.

[5] M.T. Hagan, H.B. Demuth, M.H. Beale, Neural network design, Thomson Leaning, Singapore, 2002.

[6] S. Haykin, Neural networks: A comprehensive foundation, Prentice-Hall, Reading, MA, 1994.

[7] http://www.coal.nic.in/content/coal-indian-energy-choice (accessed on June 9, 2015).

[8] Methods of test for coal and coke: Proximate analysis, IS Stand. 1350 (Part-I), 1984.

[9] Method for sampling of coal and coke: Sampling of coal, manual sampling, IS Stand. 436 (Part I/Section I), 1964.

[10] Methods for petrographic analysis of coal, IS Stand. 9127 (Part- I), 1979.

[11] Methods for petrographic analysis of coal: Preparation of coal samples for petrographic analysis, IS Stand. 9127 (Part- II), 1979.

[12] ICCP (International Committee for Coal and Organic Petrology), International handbook of coal petrology, Second ed., CNRS, Paris, 1971.

[13] ICCP (International Committee for Coal and Organic Petrology), Vitrinite classification, CNRS, Paris, 1994.

[14] Narendra K., Parthasarathy K., Identification and control of dynamical systems using neural networks, IEEE transactions on Neural Networks, 1(1990) 4-27.

[15] N.C. Karmakar, S.P. Banerjee, A comparative study on CPT index, Polish Sz index and Russian U-index of susceptibility of coal to spontaneous combustion, J. of MGMI, 86(1989)109-129.

[16] S.V. Kartalopoulos, Understanding neural networks and fuzzy logic: Basic concepts and applications, IEEE press, New York, 1996.

[17] T. Kurban, E. Beşdok, A comparison of RBF neural network training algorithms for inertial sensor based terrain classification, Sensors, 9(2009) 6312-6329.

[18] L. Briand, I. Wieczorek, Resource modeling in software engineering, Second ed. of the Encyclopedia of Software Engineering, Wiley, Editor: J. Marciniak, 2002.

[19] L.C. Briand, K. El-Emam, I. Wieczorek, Explaining the cost of european space and military projects, in: Proc. of the 21st Int. conference on Software Engineering (ICSE 21), ACM, 1999, 303-312.

[20] H. Li, M. Gupta, Fuzzy logic and intelligent system, Kluwer academic publisher, USA, 1995.

[21] L. Kumar, S.K. Rath, Predicting object-oriented software maintainability using hybrid neural network with parallel computing concept, in: Proc. of the 8th India Software Engineering Conference ISEC '15, ACM, New York, USA, 2014, 100-109.

[22] J. Moscinski, Z. Ogonowski, Advanced control with MATLAB and SIMULINK, Prentice-Hall, Inc., UK, 1995.

[23] H. Munzer, Textbook of coal petrology, in: E. Stach et al. editors, Second ed., Berlin, Gebruder Borntraeger, 1975, 387-388.

[24] S.K. Nanda, S. Panda, P.R.S. Subudhi, R.K. Das, A novel application of artificial neural network for the solution of inverse kinematics controls of robotic manipulators, Int. J. of Intelligent Systems and Applications, 9(2012) 81-91.

[25] S.K. Nanda, D. P. Tripathy, S. S. Mahapatra, Application of legendre neural network for air quality prediction, The fifth PSU-UNS Int. conference on Engineering and Technology (ICET-2011), Phuket, 2011.

[26] D.K. Nandy, D.D. Banerjee, R.N. Chakravorty, Application of crossing point temperature for determining the spontaneous heating characteristics of coal, J. of Mines, Metals and Fuels, Feb., 41(1972).

[27] D.S. Nimaje, D.P. Tripathy, S.K. Nanda, Development of regression models for assessing fire risk of some Indian coals, Int. J. of Intelligent Systems and Application, 2(2013) 52-58.

[28] W. Olpinski, Spontaneous ignition of bituminous coal, in: Proc. Glownego Institute, Gornictwa, 1953, 139.

[29] D. C. Panigrahi, G. Udaybhanu, A. Ojha, A comparative study of wet oxidation method and crossing point temperature method for determining the susceptibility of Indian coals to spontaneous heating, in: Proc. seminar on Prevention and Control of Mine and Industrial Fires- Trends and challenges, Calcutta, India, Dec., 1996, 101-107.

[30] J.C. Patra, R.N. Pal, Functional link artificial neural network-based adaptive channel equalization of nonlinear channels with QAM signal, in: Proc. of IEEE Int. conference on Systems, Man and Cybernetics, 3(1995) 2081-2086.

[31] D.S. Pattanaik, P. Behera, B. Singh, Spontaneous combustibility characterization of the Chirimiri coals, Koriya district, Chhatisgarh, India, Int. J. of Geosciences, 2(2011) 336-347.

[32] R. Kohavi, A study of cross-validation and bootstrap for accuracy estimation and model selection, in: Proc. of the fourteenth Int. joint conference on Artificial Intelligence, San Mateo, 1995, 1137-1143.

[33] G.S.N. Raju, Auto-oxidation in Indian coal mines – An investigation, J. of Mine, Metals and Fuels, Sept., 1998, 437-441.

[34] N.S. Rao, M. Lalitha, D.S. Sastry, Research project on studies of advance detection of fires in coal mines with special references to SCCL, Coal S&T, CMPDIL, Ranchi, 2011.

[35] J. Rogers, Simulating structural analysis with neural network, J. of Computing in Civil Engineering, ASCE, 8,2(1994) 252-265.

[36] D.E. Rumelhart, G.E. Hilton, R.J. Williams, Learning internal representations by error propagation in parallel distributed processing: Explorations in the microstructure of cognition, Editors: D.E. Rumelhart, J.L. McClelland, MIT press, Cambridge, MA, 1986, 318-362.

[37] S.M. Satapathy, Mukesh K., S.K. Rath, Fuzzy-class point approach for software effort estimation using various adaptive regression methods, CSIT, 1,4(2013) 367-380.

[38] F. Shih, J. Moh, H. Bourne, A neural architecture applied to the enhancement of noisy binary images, Engineering Application of Artificial Intelligence, Elsevier, 5,3(1992) 215-222.

[39] D. Simon, Training radial basis neural networks with the extended Kalman filter, Neurocomputing, 48(2002) 455-75.

[40] R.V.K. Singh, Spontaneous heating and fire in coal mines, in: 9th Asia-Oceania symposium on Fire Science and Technology, Procedia Engineering, 62(2013) 78-90.

[41] S.N. Sivanandam, S.N. Deepa, Principles of soft computing, Second ed., Wiley India Pvt. Ltd., New Delhi, 2011.

[42] M.N. Tarafdar, D. Guha, Application of wet oxidation processes for the assessment of the spontaneous heating of coal, Fuel, 68(1989) 315-317.

[43] D.P. Tripathy, B.K. Pal, Spontaneous heating susceptibility of coals-Evaluation based on experimental techniques, J. of Mines, Metals and Fuels, 49(2001) 236-243.

[44] F. Tron, S. Erik, K. Barbara, M. Ingunn, A simulation study of the model evaluation criterion MMRE, IEEE transactions on Software Engineering, 29,11(2003) 985-995.

[45] R.I. Williams, R.H. Backreedy, J.M. Jones, M. Pourkashanian, Modelling coal combustion: The current position, Fuel, 81(2002) 605-618.

[46] Y. Suresh, L. Kumar, S.K. Rath, Statistical and machine learning methods for software fault prediction using CK metric suite: A comparative analysis, ISRN Software Engineering, 2014, Article ID. 251083.

[47] http://iasir.net/IJETCASpapers/IJETCAS13-590.pdf (accessed on June 9, 2015).

[48] D.C. Panigrahi, S.K. Ray, Assessment of self-heating susceptibility of Indian coal seams – A neural network approach, Arch. Min. Sci., 59,4(2014)1061-1076.

[49] D.C. Panigrahi, V.K. Saxena, G. Udaybhanu, Research project report on Development of handy method of coal categorization and prediction of spontaneous fire risk in mines, S&T Ministry of Coal, India, 1, 1999.

[50] A.C. Smith, W.P. Ramancik, C.P. Lazzara, Sponcom - A computer program for the prediction of the spontaneous combustion potential of an underground coal mine, Proc. of the fifth conf. on the use of computers in the coal industry, Editors: S.D. Thompson, R.L. Grayson, Y.J. Wang, Morgantown, West Virginia University, Jan., 1996, 134-143.

[51] X. Zhang, H. Wen, J. Deng, X. Zhang, J.C. Tien, Forecast of coal spontaneous combustion with artificial neural network model based on testing and monitoring gas indices, J. of Coal Science & Engineering (China), 17,3(2011)336-339.

[52] H. Xiao, Y. Tian, Prediction of mine coal layer spontaneous combustion danger based on genetic algorithm and BP neural networks, First Int. symposium on Mine Safety Science and Engineering, Procedia Engineering, 26(2011) 139-146.

# Magnetic Basement Depth Re-Evaluation of Naraguta and Environs North Central Nigeria, Using 3-D Euler Deconvolution

**Opara A.I.[1], Emberga T.T.[2,*], Oparaku O.I.[1], Essien A.G.[1], Onyewuchi R.A.[3], Echetama H.N.[1], Muze N.E.[1], Onwe R.M[4]**

[1]Department of Geosciences, Federal University of Technology, PMB 1526 Owerri
[2]Department of Physics and Industrial Physics, Federal Polytechnic Nekede, Owerri
[3]Department of Geology, University of Portharcourt, Choba, Rivers State
[4]Department of Gelogy/Geophysics, Federal University Ndufu-Alike Ikwo, Abakaliki
*Corresponding author: terhemba4sure@yahoo.com

**Abstract** This paper presents a detailed geological interpretation of the aeromagnetic data over Naraguta and environs using 2-D spectral inversion and 3-D Euler deconvolution techniques. The objectives of the study are to re-evaluate the depth to the magnetic basement and to delineate associated structural features in the study area. Data enhancement was carried out to separate residual features relative to the strong regional gradients and the more intense anomalies due to basement features and igneous intrusives. Results of the 2-D spectral analysis revealed a two layer depth source model. The deeper magnetic source bodies ($d_2$) varies from 1.672km to 2.3km with an average depth of 1.999km while the depth to the shallower magnetic source bodies ($d_1$) ranges from 0.55km to 0.897km with an average depth of 0.711km. The shallower magnetic anomalies are believed to be as the result of basement rocks which intruded into the sedimentary rocks while the deeper layer may be attributed to magnetic basement surface and intra-basement discontinuities like faults, fractures and lineations. Structural analysis of these shallow anomalies using 3-D Euler deconvolution with structural index values ranging from 0-3, revealed three main structural models which include spheres, horizontal pipes/cylinders and sills/ dikes. Similarly, magnetic basement depth estimates using 3-D Euler deconvolution revealed a magnetic basement depth range of 0.5 - 2.5km. Finally, the average sedimentary thickness of 1.999km estimated from the study area is unfavorable for hydrocarbon generation and accumulation. However, the area is more viable for solid mineral exploitation based on the presence of several intrusive and linear features which may have acted as conduits for mineralization.

**Keywords:** *3-D Euler deconvolution, spectral inversion, magnetic basement depth, linear features, Naraguta, Nigeria*

## 1. Introduction

The importance of airborne survey in the delineation of structural discontinuities within sedimentary basins and the underlying basement has been tremendous over the past few decades. This past decade has seen a departure from the usual interpretation of basement structures only to detailed studies with respect to both lithologic and morphologic variations [6]. Basement structures and depths can be accurately delineated and mapped using magnetic data. Definition of various basin and sub-basin geometries can enable the mapping of the regional hydrocarbon and mineral fetch areas. Trends in magnetic features often have related trend in the overlying sediments: systematic offset of magnetic anomalies may indicate strike-slip faults which may have displaced basement rocks and possibly affect the sediment section. Infact magnetic basement interpretation can to a certain extent lead to a better understanding of the structures of the overlying sedimentary rocks.

A typical contoured aeromagnetic map presents several anomalies whose sources are traceable to the basement. It now becomes imperative that depth estimation from those anomalies provide useful clues to approximate estimation of sedimentary thickness. Aeromagnetic survey data are routinely interpreted by estimating source depths or locations; consequently, many processing algorithms have been proposed to assist the estimation [14]. The application of 2-D spectral inversion to the interpretation of potential field data is one method that has been used to determine the basement depth, and is now sufficiently well established [18]. Several authors have applied spectral inversion techniques in the determination of sedimentary thickness in various basins of the world [8,10,12,13,15,22] The magnetic anomaly signature characteristics are results of one or more physical parameters such as the configuration of the anomalous zone, magnetic susceptibility contrasts, as well as the depth to the anomalous body. Similarly, airborne magnetic survey data in grid form may be interpreted rapidly for source positions and depths by deconvolution using Euler's homogeneity relationships [14]. The method employs gradients, either measured or calculated with the details of this method sufficiently discussed by previous scholars [14]. A robust magnetic data set is likely to contain anomalies from sources with various structural indices. It is, therefore, necessary to solve for a range of indices (say 0.0, 0.5, and 1.0) and to plot the results for each index. The maps are then examined feature by feature and the index which gives the best solution clustering is chosen for each feature. This procedure

which was adopted for this study also gives some clue as to the nature of the feature. Thus, a sill edge, dike, or fault with limited throw is best displayed at an index of 1.0, while a fault with large throw may be best displayed at a zero index. Intermediate cases are best shown by an index of 0.5 [14]. However, it should be noted that field data need not be pole-reduced, so that remanence is not an interfering factor. Geologic constraints are imposed by the use of a structural index with the method known to locate or outline confined sources, vertical pipes, dikes, and contacts with remarkable accuracy [14].

This study, therefore attempts to estimate depth to anomalous magnetic sources within the study area using 2-D spectral analysis and 3-D Euler deconvolution techniques. Structural interpretation was also carried out on the contoured aeromagnetic data with a view of locating structures that may possibly accommodate minerals.

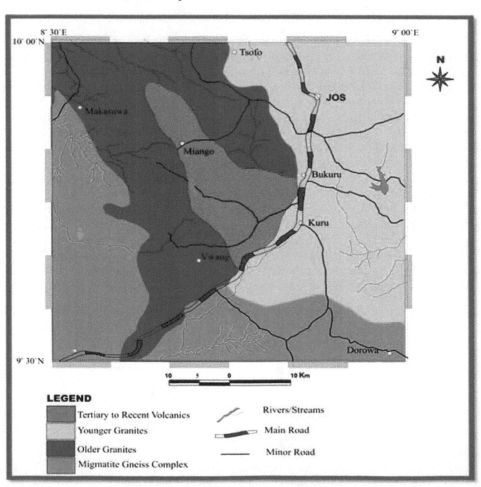

**Figure 1.** Geological map of Naraguta and environs

## 1.1. Background Geology

The area studied is Naraguta and environs which is located within the Jos-Bukuru Younger Granite Ring Complex of North Central Nigeria in the mining district of Jos. It is located at about latitudes $9°30^1 - 10°00^1$N and longitudes $8°30^1 - 9°00^1$E (Figure 1). The Nigerian Younger Granite Ring complexes are a series of petrologically distinctive crystalline igneous rocks of granitic composition [11]. Several individual complexes

have been identified with varying sizes and named after their localities with individual massifs ranging from 640km² to less than 1.68km² [11]. The Younger Granites are believed to be discordant high level intrusions emplaced by means of piecemeal stopping through the collapsed central block [11]. Their outlines are either circular or elliptical. Several cycles of intrusions occur within one complex. The sizes of many of the structures are due to overlapping and superposition of separate intrusive cycles [2,4,17,19]. The Younger Granites were preceded by extensive acid volcanism and emplaced by ring faulting and block subsidence. Granites and rhyolites underlay more than 90 percent of the total area of the Younger Granite province. Intermediate and basic rocks occur in many complexes. Their emplacement is completely unrelated to orogenic activity associated with the study area [2,20]. Their age is Jurassic, around 160 to 170 million years; the Older Granites and the accompanying metamorphism of the basement are dated at about 500 to 600 million years, and represent the Pan African Orogeny in Nigeria [2,4]. However, it seem likely that emplacement of the Younger Granites was associated with epeirogenic uplift [20]. Indirect evidence for this is lack of sediments associated with the volcanic rocks of the younger granite age, which are apparently erupted on to a land surface undergoing erosion, not deposition [20]. The major components of most complexes are granitic ring dykes that range from 5km or less to over 30 km in diameter.

Generally, the Younger Granites are characterized as anorogenic, peralkaline rocks that intruded the basement discordantly to form highly steeped hills. They occur generally as ring dykes and cone sheets. The province contains approximately 50 Jurassic Ring Complexes (165MA), which intruded into the Precambrian basement complex (600MA) within the Central Plateau Region of the Jos Plateau of Nigeria [9]. The geology map (Figure 1) of the study area revealed that the area is underlain by the Precambrian crystalline basement of migmatite-gneiss complex which is directly overlain by the older granites, subsequently by younger granites and then the newer basalts of quaternary-tertiary age that occurs as lava flows and volcanic cone.

## 2. Theory and Method

The data used in this study are part of the airborne aeromagnetic map project of Nigeria obtained from the Geological Survey Agency of Nigeria and was subjected to low pass filtering operation. The generated filtered total field magnetic intensity values are presented in Figure 2 as a contour map. The nature of filtering applied in this study in the fourier domain was chosen to eliminate certain wavelengths and to pass longer wavelengths. Regional - residual separation was also carried out using polynomial fitting. This is a purely analytical method in which matching of the regional by a polynomial surface of low order exposes the residual features as random errors. For the magnetic data, the regional gradients were removed by fitting a plane surface to the data by using multi-regression least squares analysis. The expression obtained for the regional field T(R) is given as:

$$T(R) = 7612.158 + 0.371x - 0.248y \qquad (1)$$

Furthermore, estimation of the source depths for the various anomaly depths were carried out using slope methods which includes Peter's, Half width and Thannel's analytical rules. The original aeromagnetic field maps over the study area are characterized by a series of local anomalies which are not apparent on the residual maps due to the digital filtering and convolution algorithm used. These local anomalies were modeled in terms of intrusions which are known to occur in the area using non linear

optimization techniques. In summary, the method seek to minimize a non linear objective function which represents the difference between the observed and calculated fields through an iterative change of the non linear parameters (location, thickness and depth) by non-linear optimization while at the same time obtaining optimum values for the linear parameters (magnetization components, quadratic regional and composite magnetization angle) by least-square analysis [12]. Many of these prominent local anomalies are associated with intrusive igneous bodies and mineral veins found in the area. Detailed Interpretations of these anomalies in the study area are also presented in this study in Table 2.

In addition, 2-D spectral analysis which allows an estimate of the depth of an ensemble of magnetized blocks of varying depth, width, thickness and magnetization was also carried out in this study. The digitized aeromagnetic data was transformed in the fourier domain to compute the energy (or amplitude) spectrum. This was then plotted on a logarithmic scale against frequency. The slopes of the segments yielded estimates of average depths to magnetic sources of anomalies in the study area [7,12,18]. Given a residual magnetic anomaly map of dimensions l x l, digitized at equal intervals, the residual total intensity anomaly values can be expressed in terms of a double fourier series expression given as [7,18]:

$$T(x,y) = \sum_{n=1}^{N} \sum_{n=1}^{N} \left\{ \begin{array}{l} P_m^n \cos\left\{ \left(\frac{2\pi}{l}\right)(nx+my) \right\} \\ + Q_m^n \sin\left\{ \left(\frac{2\pi}{l}\right)(nx-my) \right\} \end{array} \right\} \qquad (2)$$

where, l = dimensions of the block, and is the Fourier amplitude and N and M are the number of grid points along the x and y directions respectively. Similarly, using the complex form, the two dimensional Fourier transform pair may be written as [7,18]:

$$G(u,v) = \int\int_{-\infty}^{\infty} g(x,y) e^{-j(u_x+v_y)} dxdy \qquad (3)$$

and

$$g(x,y) = \int\int_{-\infty}^{\infty} G(u,v) e^{-j(u_x+v_y)} dudv \qquad (4)$$

where, u and v are the angular frequencies in the x and y directions respectively.

The use of this method involves some practical problems, most of which are inherent in the application of the discrete fourier transform (DFT). They include the problems of aliasing, truncation effect or Gibb's phenomenon, and the problems associated with even and odd symmetries of the real and imaginary parts of the fourier transform [12]. However, in this study, these problems were taken care of by the softwares used in the analysis.

Finally, the objective of the 3-D Euler deconvolution process was to produce maps showing the locations and the corresponding depth estimates of geologic sources of magnetic anomalies in a two-dimensional grid [14,23]. The Standard 3-D Euler method is based on Euler's homogeneity equation, which relates the potential Field (magnetic) and its gradient components to the location of the sources, by the degree of homogeneity N, which can be interpreted as a structural index [19,23]. The method makes use of a structural index in addition to producing depth estimates. In combination, the structural index and the depth estimates have the potentials to identify and calculate depth estimates for a variety of geologic structures such as faults, magnetic contacts, dykes, sills, etc. The algorithm uses a least squares method to solve Euler's equation simultaneously for each grid position within a sub-grid (window). A square window of pre-defined dimensions (number of grid cells) is moved over the grid along each row. At each grid point, a system of equations is solved from which the four unknowns (x, y as location in the grid, z as depth estimation and the background value) and their uncertainties (standard deviation) are obtained for a given structural index. A solution is only recorded if the depth uncertainty of the calculated depth estimate is less than a specified threshold and the location of the solution is within a limiting distance from the center of the data window [14,23]. Thompson [19] showed that for any homogenous, three-dimensional function f(x; y; z) of degree n:

$$f(tx; ty; tz) = t^n f(x; y; z) \qquad (5)$$

It can be shown that, the following equation, which is known as Euler's homogeneity relation can be satisfied [23]:

$$x \frac{\delta f}{\delta x} = y \frac{\delta f}{\delta y} + z \frac{\delta f}{\delta z} = nf \qquad (6)$$

In geophysics, the function f(x,y,z) can have the general functional form [14,19,23]:

$$f(x.y,z) = \frac{G}{r^N} \qquad (7)$$

Where $r^2 = (x - x_0)^2 + (y - y_0)^2 + (z - z_0)^2$, N is a real number (1,2,3...) and G a constant (independent of x,y,z). Many simple point magnetic sources can be described by equation 7 above, with $(x_0; y_0; z_0)$ the position of the source whose field F is measured. The parameter N is dependent on the source geometry, a measure of the fall-off rate of the field and may be interpreted as the structural index (SI). Considering potential field data, Euler's equation can be written as [19,23]:

$$(x - x_0) \frac{\delta T}{\delta x} + (y - y_0) \frac{\delta T}{\delta y} + (z - z_0) \frac{\delta T}{\delta z} = N(B - T) \quad (8)$$

With B the regional value of the total magnetic field and $(x_0; y_0; z_0)$ the position of the magnetic source, which produces the total field T measured at (x; y; z).

Thompson [19] showed that simple magnetic models are consistent with Euler's homogeneity equation. Thus Euler deconvolution provides an excellent tool for providing good depth estimations and locations of various sources in a given area, assuming that appropriate parameter selections are made. Applied to aeromagnetic surveys, the 3-D Euler process is a fast method for obtaining depth and boundary solutions of magnetic sources for large areas. Though it is a general advantage of the Euler deconvolution method, that it is applicable to all geologic models and that it is insensitive to magnetic remanence and geomagnetic inclination and declination, an initial assumption of the source type has to be made.

Dependent upon the potential source type, a structural index is chosen. Table 1 summarizes the structural indices (SI) for given geologic models. The number of infinite dimensions describes the extension of the geologic model in space.

Table 1. Structural Indices for Simple Magnetic Models Used For Depth Estimations by 3-D Euler Deconvolution [14,19,23]

| Geologic Model | Number of Infinite Dimensions | Magnetic Structural Index |
|---|---|---|
| Sphere | 0 | 3 |
| Pipe | 1 (z) | 2 |
| Horizontal cylinder | 1 (x-y) | 2 |
| dyke | 2 (z and x-y) | 1 |
| sill | 2 (x and y) | 1 |
| contact | 3 (x,y,z) | 0 |

The results of the Euler method are displayed in ordinary maps as point solutions combining the location (position of solution) and the depth (colour range). Given the choice of an appropriate structural index, 3-D Euler deconvolution will lead to a clustering of solutions, which can be interpreted. A vertical pipe structure will for example be shown as a cluster of solutions around a specific point, whereas an elongated dyke structure will be recognized as a linear trend of solutions.

# 3. Results and Interpretation

The aeromagnetic data used in this work was obtained from the Geological Survey Agency of Nigeria. It was part of the nationwide survey completed in 1976 by Fairey Survey Ltd. Flight line direction was NNW-SSE at station spacing of 2km with flight line spacing of 20km.It was flown at an altitude of about 150m with tie lines in an ENE-WSN direction. Regional correction of the magnetic data was based on the IGRF (epoch date1 of January, 1974). For this study, aeromagnetic sheet 168 (Naraguta sheet) was used.

The total magnetic field intensity map derived from the digitization is presented as a total field intensity map and 3-D map respectively (Figure 2 & Figure 3). The total field of the aeromagnetic data revealed that the underlying basement within Bukuru, Kuru and Jos have an estimated magnetic intensity range of 7740 to 7780 gammas. This area has a broad relief with low intensity surrounded by fairly high to high intensity area which may be interpreted as trough. The 3-D map revealed two distinctive relief patterns: low and high relief. Areas with low relief are observed in the central part of the study area: Bukuru, Kuru and Jos whereas, the northern and southern fringes of the study area around Tsofo and Makasuwa showed a relatively high relief (Figure 3). On the northern fringe, from Tsofo, an extensive linear anomaly trending in the E-W direction was observed. Along this linear anomaly, there exists a chain of circular magnetic closures suspected to be granitic intrusions which is believed to be part of the volcanic ring complex. The high relief areas are also believed to be more tectonically active than the low relief area owing to the presence of chains of granitic intrusive surfaces present in the area. Another linear anomaly was observed in the southeastern portion of the study area. This anomaly with a sharp gradient was believed to be a linear feature of regional extent. Other magnetic closures suspected to be intrusions were          observed around Makasuwa and environs.

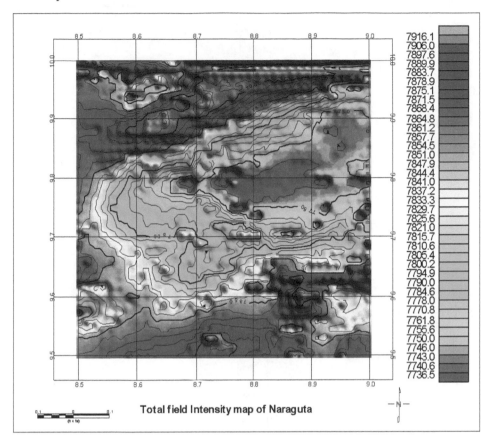

**Figure 2.** Total Magnetic Field Intensity Contour Map of the Study area

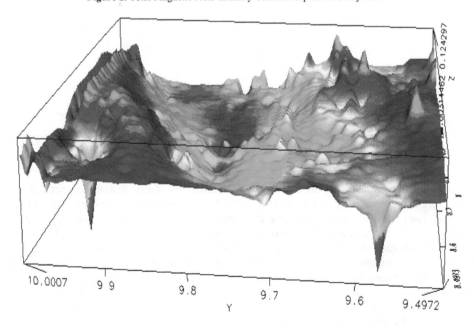

**Figure 3.** 3-D Surface map of the magnetic basement surface of the study area

The Structural trends interpreted from the total field intensity map, polynomial surfaces (Figure 4) and the residual field map (Figure 5) reveled trend directions of NW-SE,N-S, E-W and NE-SW. The NW-SE trend is the dominant trend orientation in the study area and thus reflects the trend of the youngest tectonic episode, the Post pan African orogenic activities in the study area which nearly obliterated the older E-W and NE-SW tectonic

trends of Pan African orogenic phase. The regional fields established the major tectonic elements of the deeper and regional field which affected the structural framework of the study area. These structural trends are correlatable with paleo-structures in the study area which are believed to have affected major geologic events in the study area. Buser [5] suggested the existence of these paleo-structures within the study area and its surroundings.

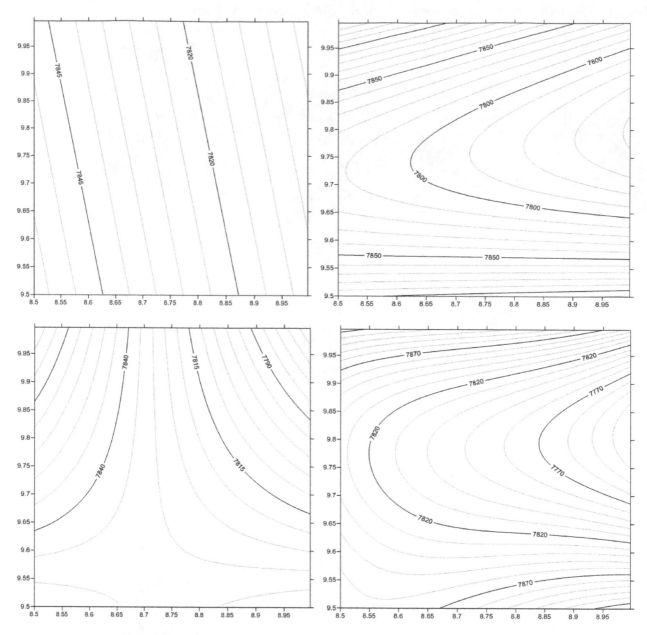

**Figure 4.** First to fourth degree(polynomial) surfaces of the regional fields of the aeromagnetic data

The residual magnetic intensity values (Figure 5) of the study area was discovered to range from -89.9 to 91.9 gammas. The extensive linear feature earlier interpreted from the total field intensity map around Tsofo has a positive residual value of 80 gammas. Other areas with positive residual values are Makasuwa on the central portion and Dorowa on the southern extreme of the study area. Similarly, Bukuru and Kuru on the central portion of the study area was observed to be dominated by low residual magnetic intensity. The residual intensity range in these areas was shown to be between -89.9 to -85.2. The areas with negative magnetic values reflects a typical zone of low magnetization while the positive residual anomalous areas are indicative of an area of high magnetization. This implies that there is an existence of shallow to near surface magnetized bodies in areas having positive residual values. It also indicated that the linear anomalous body seen at Tsofo could be interpreted as a near surface magnetized body. It further implies that Bukuru and Kuru with negative residual values is

underlain by deep seated magnetized bodies. Another probable reason for the predominance of negative residual anomaly signatures in this area may be due to its nearness to the magnetic equator. Apart from the linear anomalies, there are circular closures seen on the extreme northwestern and southern portions of the map. These circular anomaly patterns were interpreted as the Younger Granite Ring Complexes which are believed to be associated with the presence of ore bodies; granitic as well as basic ore bodies.

The local anomalies in the original aeromagnetic field map were modeled in terms of intrusions using non linear optimization techniques. The method seek to minimize a non linear objective function which represents the difference between the observed and calculated fields through an iterative change of the non linear parameters (location, thickness and depth) by non-linear optimization while at the same time obtaining optimum values for the linear parameters (magnetization components, quadratic regional and composite magnetization angle) by least-square analysis. The analytical slope methods mainly Peter's, slope, Thannel and Half-width graphical methods were also used in calculating depth estimates to the anomalous bodies (Figure 6). The estimated depths and linear parameters are presented in Table 2 and Figure 6. The depth to the anomalous magnetic features from the three depth slope methods varies from 0.1601km to 2.478km.

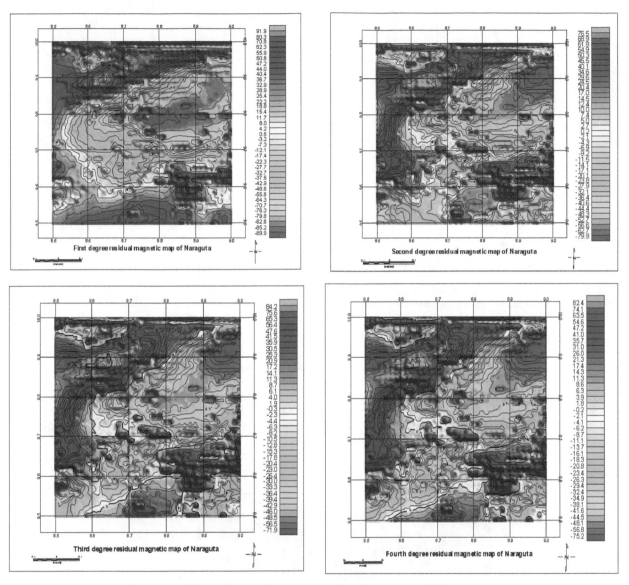

**Figure 5.** First to fourth degree residual fields of the aeromagnetic data of the study area

**Table 2. Calculated Depth to the Magnetic Source of the aeromagnetic map of Naraguta Area**

| SHEET 168 | LOCATION | DEPTH ESTIMATION IN KM | | | WIDTH (KM) | AMPLITUDE IN GAMMA | MAGNETIZATION (A/M) | 1% RADIANCE | TYPE OF ANOMALY |
|---|---|---|---|---|---|---|---|---|---|
| PROFILE | TOWNS | PETER'S SLOPE | THANNEL'S RULE | HALF WIDTH RULE | | | | | |
| AA[1] | 7km NW of Manchox | 0.4726 | 1.5976 | 0.8476 | 4.02 | 7808 | 0.90 | -1.15 | LOW |
| BB[1] | 9km East of Barkin Darowa | 1.0976 | 1.5976 | 1.8476 | 8.0 | 7900 | 0.55 | 1.38 | HIGH |
| CC[1] | 7.5km North of Damakasuwa | 0.1601 | 1.6476 | 0.5476 | 3.75 | 7645 | 1.65 | 1.46 | LOW |
| DD[1] | 8km NW of Bukuru | 0.1601 | 2.6476 | 1.6476 | 4.6 | 7650 | 1.85 | 1.23 | LOW |
| EE[1] | 7km NE of Miango | 0.7876 | 2.2976 | 0.3476 | 6.2 | 7859 | 2.09 | 1.75 | LOW |
| FF[1] | 1km West of JOS | 0.3476 | 1.2476 | 0.4476 | 4.0 | 7854 | 1.52 | 1.83 | HIGH |

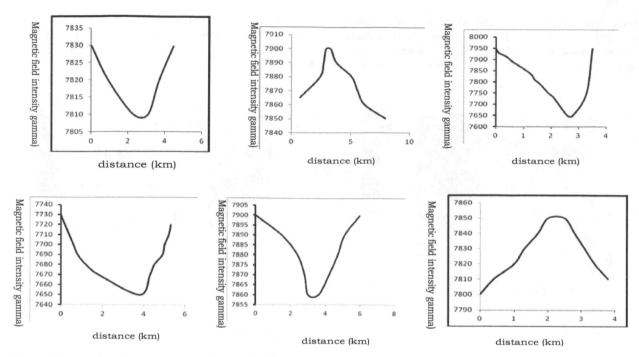

**Figure 6.** Interpretation of some linear magnetic anomalies from Naraguta Sheet. Profiles A→F were taken as follows: Profile[A], 7Km NW of Manchox; Profile[B], 9Km East of Barkin Dorowa; Profile[C], 7.5Km North of Damakasuwa; Profile [D], 8Km NW of Bukuru; Profile [E], 7Km NE of Miango; Profile [F], 1Km West of Jos

Similarly, the second vertical derivative contour map of the study area is presented in Figure 7a below. The zero contours of the second vertical derivative map indicated the lithologic boundaries between the the different lithologies (Figure 7a). A good correlation was also established between the second vertical derivative map and the mineral map of the study area as shown in Figure 7b. In addition, the distribution of mafic and felsic rock

forming minerals were correlated to the positive and negative second vertical derivative anomalies around the study area close to Ifinkpa, Agoi Ibani and Iko Ikperem areas (Figure 7a). The mafic and felsic rock minerals are believed to be the by-product of the re-activation of the trans-oceanic fracture zones that also acted as conduits for primary mineralization. In the study area around Ifinkpa, Agoi.

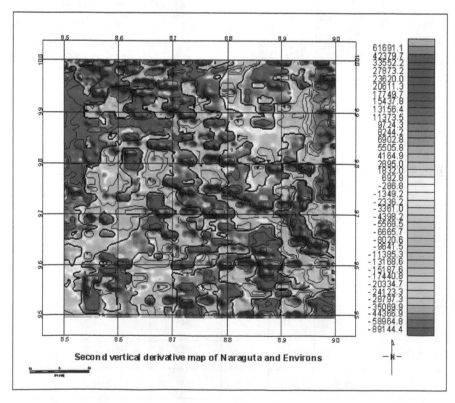

**Figure 7a.** Contour map of Second vertical derivatives of the aeromagnetic map of the study area

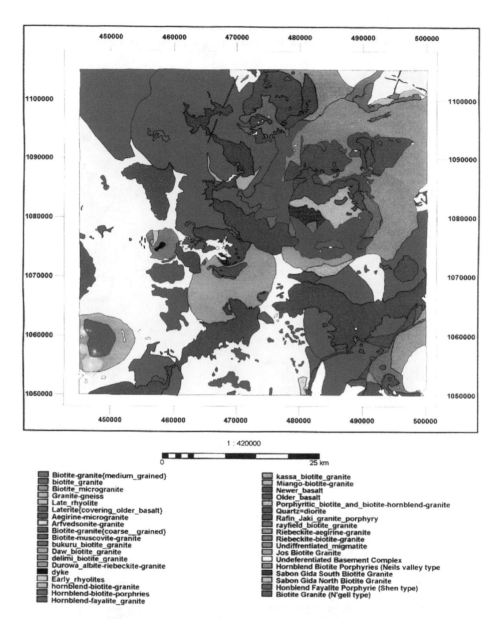

1 : 420000

| Biotite-granite{medium_grained} | kassa_biotite_granite |
| biotite_granite | Miango-biotite-granite |
| Biotite_micrograniite | Newer_basalt |
| Granite-gneiss | Older_basalt |
| Late_rhyolite | Porphyritic_biotite_and_biotite-hornblend-granite |
| Laterite{covering_older_basalt} | Quartz=diorite |
| Aegirine-microgranite | Rafin_Jaki_granite_porphyry |
| Arfvedsonite-granite | rayfield_biotite_granite |
| Biotite-granite{coarse__grained} | Riebeckite-aegirine-granite |
| Biotite-muscovite-granite | Riebeckite-biotite-granite |
| bukuru_biotite_granite | Undiffrentiated_migmatite |
| Daw_biotite_granite | Jos Biotite Granite |
| delimi_biotite_granite | Undeferentiated Basement Complex |
| Durowa_albite-riebeckite-granite | Hornblend Biotite Porphyries (Neils valley type) |
| dyke | Sabon Gida South Biotite Granite |
| Early_rhyolites | Sabon Gida North Biotite Granite |
| hornblend-biotite-granite | Honblend Fayalite Porphyrie (Shen type) |
| Hornblend-biotite-porphries | Biotite Granite (N'gell type) |
| Hornblend-fayalite_granite | |

**Figure 7b.** Geology map of the study area showing mineral distribution (NGSA)

The spectral analysis plot of the power spectrum of the residual field of the study area was used to estimate the average depth to buried magnetic rocks and is presented as Figure 8 below. The spectral analysis presents a two layer ($D_1$ and $D_2$) depth source model. These depths were established from the slope of the log- power spectrum at the lower end of the total wave number or spatial frequency band. The method allows an estimate of the depth of an ensemble of magnetized blocks of varying depth, width, thickness and magnetization. The estimated depths to magnetic basement are shown as $D_1$ and $D_2$ (Table 3).

**Table 3. Location and magnitude of first and second layer spectral depths**

| SPECTRAL BLOCK | LONGITUDE | | LATITUDE | | DEPTH (KM) | |
|---|---|---|---|---|---|---|
| | $X_1$ | $X_2$ | $Y_1$ | $Y_2$ | $D_1$ | $D_2$ |
| A | 8.5 | 8.75 | 9.75 | 10 | 0.55 | 1.672 |
| B | 8.75 | 9 | 9.75 | 10 | 0.63 | 2.3 |
| C | 8.5 | 8.75 | 9.5 | 9.75 | 0.765 | 1.863 |
| D | 8.75 | 9 | 9.5 | 9.75 | 0.897 | 2.161 |

The first layer depth ($D_1$), is the depth to the shallower source represented by the second segment of the spectrum (Figure 8). This layer ($D_1$) varies from 0.55km to 0.897km, with an average of 0.711km. The second layer depth ($D_2$) varies from 1.672km to 2.3km, with an average of 1.999km (Table 3). This layer may be attributed to magnetized rocks of the basement surface. Another probable origin of the magnetic anomalies contributing to this layer are the lateral variations in basement susceptibilities, and intra-basement features like faults and fractures [8,12]. It can be deduced from the spectral plots that the $D_2$ values obtained from the spectral plots represent the average depths to the basement complex in the blocks considered. The depth of the shallower magnetic source and that of the magnetic basement increases from west to east (Figure 9). This revealed that the basement outcropped around Makasuwa suggesting epirogenic uplift. Similarly, the sediment thickness is believed to be more within Kuru, Bukuru and Jos areas indicating a thicker overburden.

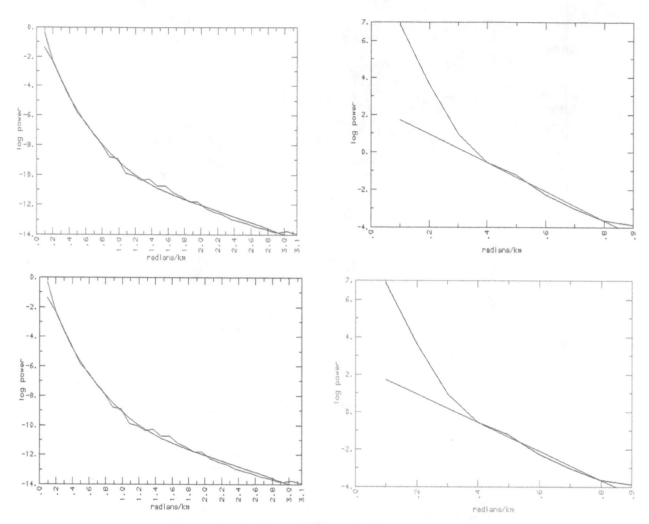

**Figure 8.** Energy Spectra for Blocks A, B, C, and D for the determination of basement depths

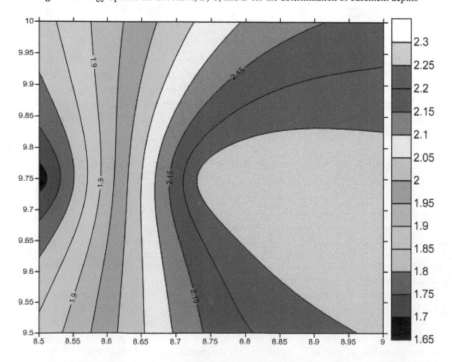

**Figure 9.** Depth to basement map estimated from spectral inversion contoured in metres

3-D Standard Euler deconvolution of the aeromagnetic data of the study area using different geological scenarios based on several vintages of structural indices (structural index 0 for contacts, structural index 1 for sills and dikes, structural index 2 for horizontal cylinders and pipes, and structural index 3 for spheres), revealed standard Euler solution clusters as shown in Figure 10a – Figure 10d. For a better comparison of the obtained solutions, only solutions in the depth range between 0m to 3000 m were used, which entailed an elimination of a few spurious solutions. The Euler solutions used for the plots were gotten under the assumption of a wide range of geologic models determined represented by contacts, dike/ sill, horizontal cylinder/pipes and spheres; thus structural indices of 0 to 3 were used. While solutions were gotten for the standard Euler deconvolution, using the different structural indices, there is however a similarity in all the solutions. The Standard Euler solutions for four geological models are presented in this model. Figure 10a is the Standard Euler solution for contacts (structural index =0). This figure revealed a depth range of 250 – 3000m with isolated cluster of solutions around Makusuwa, Jos and Vwang areas. This indicates that magnetic contacts are not a dominant geological feature of the study area. Figure 10b on the other hand represents the Standard Euler

Solutions for sills/dikes. The depth estimate of this geological model in the study area ranges from 0m to a little above 3000m. It is therefore revealed that sills/dikes are dominant geological features in the study area as the cluster of solutions all over the map. However, there is a complete absence of this geological structure around Tsofo. Similarly, for structural index 2(horizontal cylinders/pipes), the estimated depth ranges from 1000 to 3000m as shown in Figure 10c. Several clusters of solutions were observed all over the map of the study area with Jos however having no noticeable cluster of solution. Horizontal cylinders/pipes are therefore observed to be dominant structural/geological feature in the study area. Finally, Figure 10d (spheres) revealed a concentration of clusters of solution at the central and southern parts. There is also a concentration of the cluster of solutions representing spheres around Tsofo area in the northern part. The depth of the interpreted cluster of solutions varies between 1000- 3000m. Generally, the Euler deconvolution suggests a depth to the magnetic source of between 0m to a little above 3000m. Comparing these depth estimates with the information gotten from the spectral analysis, there is a positive correlation between the results from both depth estimation methods.

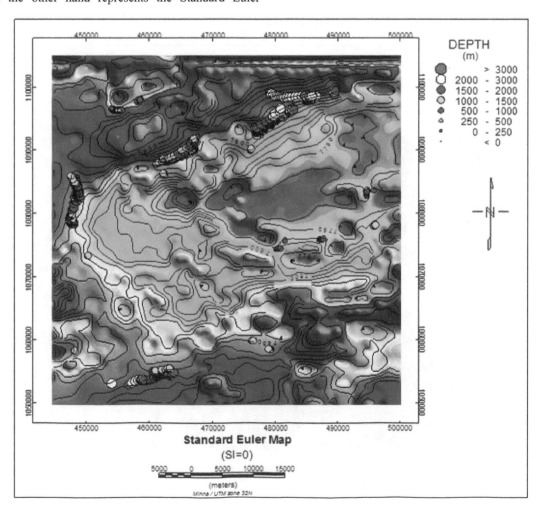

**Figure 10a.** Standard Euler Deconvolution Depth Solution Plot Draped on the Colour Shaded Grid of the Total Magnetic Intensity of the Study area (Structural Index=0)

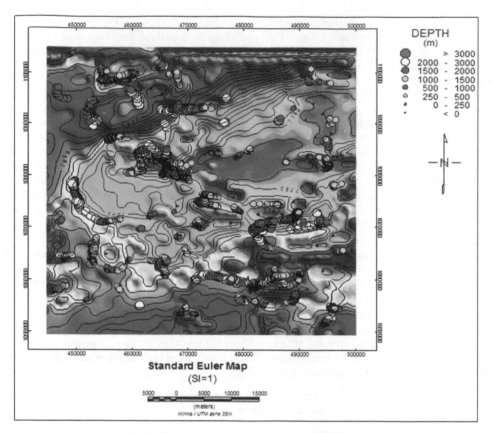

**Figure 10b.** Standard Euler Deconvolution Depth Solution Plot Draped on the Colour Shaded Grid of the Total Magnetic Intensity of the Study area (Structural Index=1)

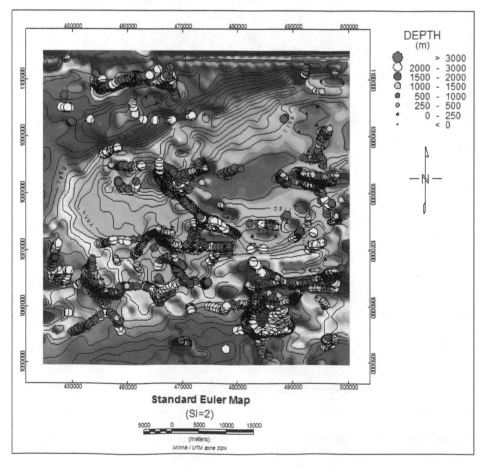

**Figure 10c.** Standard Euler Deconvolution Depth Solution Plot Draped on the Colour Shaded Grid of the Total Magnetic Intensity of the Study area. (Structural Index=2)

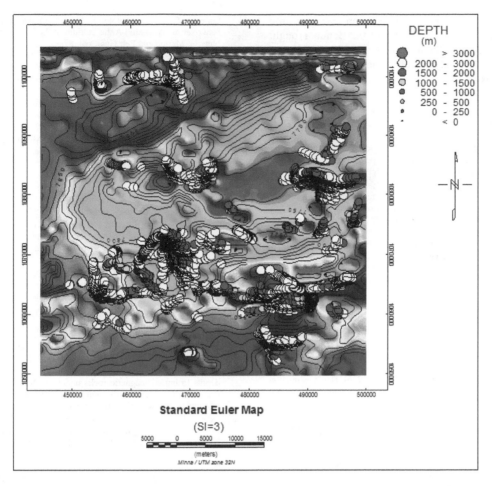

**Figure 10d.** Standard Euler Deconvolution Depth Solution Plot Draped on the contour map of the total map intensity field of the study area (Structural Index=3)

## 4. Discussion and Conclusion

A detailed structural interpretation and re-evaluation of the magnetic basement depth in the study area was carried out using 3-D Standard Euler deconvolution and spectral inversion techniques. The presence of intrusive bodies in some parts of the study area suggests that some parts of the area are more tectonically active than others. The extensive linear anomaly observed at Tsofo is suspected to be part of the ring faulting prevalent in the Younger Granite Complex. Structural analysis of the interpreted anomalies(intrusive bodies) using 3-D Standard Euler deconvolution with structural index values ranging from 0-3 revealed the dominant presence of three geological structural features which include spheres, horizontal pipes/cylinders and sills/ dikes. Results of the 2-D spectral analysis revealed a two layer depth source model. The spectral depths obtained in this study revealed the depth to the shallow magnetic layer to be in the range of 0.55km to 0.897km with an average depth of 0.711km, while the depth to the magnetic basement varies from 1.672km to 2.3km with an average depth of 1.999km.It can be deduced that the $D_2$ values obtained from the spectral depth represent the average depth to the magnetic basement in the blocks considered. The shallower magnetic anomalies are however believed to be as the result of basement rocks which intruded into the sedimentary rocks while the deeper layer may be attributed to magnetic basement surface and intra-basement discontinuities like faults. Similarly, the depth to the anomalous magnetic features from the three depth slope methods revealed depth values that range from 0.1601km to 2.478km.Finally, magnetic basement depth estimates using 3-D Euler deconvolution revealed a magnetic basement depth range of 0.5 - 3.0km. This average depth of 1.999km from this study is fairly close to that of the neighbouring Nupe Basin which has a basement depth of 3.3km [1,21]. Elsewhere in the Benue Trough, a sedimentary thickness of 3.3km was estimated [9]. Depth to source interpretation of aeromagnetic field data provides important information on basin architecture for mapping areas where the basement is shallow enough for mineral exploration.

Economically, the area has low hydrocarbon exploration potential owing to the shallow sedimentary thickness and numerous tectonic activities that affected the study area which are seen on the aeromagnetic map as lineations (fractures, joints, faults, etc). The average sedimentary thickness of 1.999km estimated from this study may be unfavorable for hydrocarbon generation and accumulation. Another reason that may have affected the hydrocarbon bearing potentials of the area are the high thermal gradients associated with igneous activities believed to have taken place in the region. However, the area is more viable for solid mineral exploitation based on the presence of linear features which may have acted as conduits for mineralization. Finally, the NE-SW trend of the study area suggests a relationship between the Younger Granite Complex of Jos and the Benue Trough. This trend suggests that the tectonic events that took place in the Benue trough may have extended to the Younger Granite Complex of Jos.

# References

[1] Adeniyi, J.O., (1986). Polynomial Regional Surfaces and Two dimensional models in parts of Nupe Basin and Adjacent Basement Complex, Niger State, Nigeria. Nigerian Jour. of Applied Sciences. Vol. 4, pp. 25-34.

[2] Akanbi, E.S., Ugodulunwa, F.X.O. and Gyang, B.N., (2012). Mapping Potential Cassiterite Deposits of Naraguta Area, North Central, Nigeria using Geophysics and Geographic Information System (GIS). Journal of Natural Sciences Research. Vol.2, (8),

[3] Blakely, R.J., (1995). Potential theory in gravity and magnetic applications: Cambridge Univ. Press.

[4] Bowden, P, and Kinnaird, J.A. (1984). Geology and mineralization of the Nigerian anorogenic ring complexes. Geologisches Jahrb (Hannover) B56, 3–65.

[5] Buser, H. (1966). Paleostructures of Nigeria and adjacent Countries. Geotectonic Res. E. Schweizerbart Sci. Publi., Johannesstr. 3A D-70176 Stuttgart, Germany. Vol. 24.

[6] Gunn, P. J. (1997). Application of Aeromagnetic Survey to Sedimentary Basin Studies. AGSO Journal of Australian Geology and Geophysics 17, 2, 133-144.

[7] Hahn, A., Kind, E. G. and Mishra, D. C. (1976): Depth estimation of Magnetic sources by means of Fourier amplitude spectra, Geophysical. Prospecting, 24, 287-308.

[8] Kangoko, R., Ojo, S.B. and Umego, M.N., (1997). Estimation of Basement depths in the Middle Cross River basin by Spectral analysis of the Aeromagnetic field. Nig. Journ. of Phys. Vol. 9, pp.30-36.

[9] Kogbe, C.A., (1989). The Cretaceous and Paleogene Sediments of Southern Nigeria. Geology of Nigeria 2nd Edition. Rock View (Nig) Ltd, pp. 325-334.

[10] Liu, X., (2003). On the use of different methods for estimating magnetic depth: The Leading Edge, November, pp. 1090-1099.

[11] Obaje, N.G., (2009). Geology and Mineral Resources of Nigeria. Springer-Verlag Berlin Heidelberg, P. 219.

[12] Ofoegbu, C. O., and Onuoha, K. M., (1991). Analysis of magnetic data over the Abakaliki Anticlinorium of the Lower Benue Trough, Nigeria. Marine and Petr.Geol. vol. 8.

[13] Opara, A.I., (2011). Estimation of the Depth to Magnetic Basement in Part of the Dahomey Basin, Southwestern Nigeria. Australian Journal of Basic and Applied Sciences, Vol. 5(9), 335-343.

[14] Reid, B., Allsop, J.M., Granser, H. Millet, A.J., and Somerton., I. W., (1990). Magnetic interpretation in three dimensions using Euler deconvolution., Geophysics. Vol. 55, No. I; P. 80-91.

[15] Salem, A., Ushijima, K., Elsirafi, A. and Mizunaga, H., (2000). Spectral analysis of aeromagnetic data for Geothermal reconnaissance of Uaseir area, Northern Red Sea, Egypt. Proceedings of World Geothermal Congress, Kyushu-Tohokie, Japan, 28: 10.

[16] Sharma, P. V. (1987). Magnetic Method Applied to Mineral Exploration; Ore Geology, Review. 2, 323-357.

[17] Samaila, C. A and Solomon, N. Y., 2011. Groundwater Potential on the Jos-Bukuru Plateau, North Central Nigeria, using lineaments from gravity measurements; Journal of Water Resource and Protection, vol. 3; pp. 628-633.

[18] Spector, A. and Grant, F.S., (1970). Statistical models for interpreting aeromagnetic data. Geophysics, Vol.35, pp. 293-302.

[19] Thompson, D. T. (1982). EULDPH: A new technique for making computer-assisted depth estimates from magnetic data, Geophysics, 47, 31-37.

[20] Turner., D. C.1989. Structure and Petrology of the Younger Granite Ring Complexes," In: C.A Ogbe, Ed., Geology of Nigeria, Rock View International, France, 1989, pp. 175-190.

[21] Udensi, E.E. and Osazuwa, I.B., (2000). Spectral determination of depths to buried magnetic rocks under the Nupe Basin, Nigeria. Proceedings of the 23rd annual conference of the Nigerian Institute of Physics "The Millenium Conference". pp. 170-176.

[22] Udensi, E.E. and Osazuwa, I.B., (2004). Spectral determination of depth to buried magnetic rocks under the Nupe Basin, Nigeria. Nigerian Association of Petroleum Explorationists Bulletin, Vol. 17(1), pp. 22-27.

[23] Whitehead, N. and Musselman, C. (2008). Montaj Gravity/Magnetic Interpretation: Processing, Analysis and Visualization System for 3D Inversion of Potential Field Data for Oasis montaj v6.3, Geosoft Incorporated, 85 Richmond St. W., Toronto, Ontario, M5H 2C9, Canada.

# 15

# Cut-off Grade and Hauling Cost Varying with Benches in Open Pit Mining

**Siwei He[1,*], Xianli Xiang[2], Gun Huang[2]**

[1]CAMCE Mining and Tunneling, Lougheed Hwy, Burnaby, Canada
[2]Guizhou University of Engineering Science, Bijie, Guizhou, China
[3]Resource and Environmental Sciences, Chongqing University, Shapingba Zhengjie, Chongqing, China
*Corresponding author: jhe@procongroup.net, swjhe118@gmail.com

**Abstract** One of the most challenging problems geologist and engineer encountering in open pit mining is how much ore at benches can be extracted and sent to mill at the same time. Cutoff grade is playing an important role in solving this problem. The author, in this paper, presented a hauling cost-bench model, which directly impact cutoff grade. The cost model indicates that hauling cost, increasing with bench depth, largely depends upon the dynamic stripping ratio, representing the fundamental nature of resource and economic requirements of mineable ore at benches. The overall ore grade at benches is calculated according to operating cost model provided, while cutoff grade is found using trial-and-error technique based on resource model of a deposit. An example representing cut-off grade-bench curve is given, illustrating what relationship among cutoff grade, average grade and tonnage at benches and how much ore can be extracted and sent from these bench to mill. The example also shows how a way is found to mine out low-grade ore from one bench in a desired cut-off grade for an expected profit of the company. The cutoff grade vs. bench curve is a simple but useful tool for engineers to facilitate sound decision on how much ore tonnage at benches can be extracted and sent to mill at a time.

*Keywords:* cutoff grade, bench, average grade, expected profit, ore tonnage, stripping ratio

## 1. Introduction

Mineral resource is naturally occurring, but mineable reserve is created by human effort, which is varying with times and places (R.V. RAMANI at al., 1995) [1]. In mine investment or project assessment, analysts focus more on ore reserve varying with times for a long-term planning and financial model purpose. But in mining operation, mine geologist and engineer are much interested in mineable ore varying from bench to bench for creating a better short-term plan to generate an expected profit (such as CF)in open pit(Peter Darling, 2011; Hustrulid at al., 2001; William A at al., 2013;Xinming Tang, at al., 2007; G. Matheron, 1987) [2,3,4,5,6].

In mining operation, we need to extract ore from one to six benches at a time. The main problem that mine geologist and engineer are encountering is how much ore is sent from these benches to mill. Cut-off grade is the key to success in solving this problem. The cut-off grade could be varying greatly from bench to bench due to two reasons below:

- Mineral nature occurrence such as resource grade distribution, thickness, dip and strike and weathering of a deposit varying from bench to bench (A.E. Annels, 2012) [7]
- Ore hauling cost increasing with depth.

Therefore, delineating a cutoff-depth chart at benches will be significant in mining operation. It should be, at least, a tool for engineers to facilitate sound decision on which portion of ore can be immediately sent to millfor an expected profit. To create this chart, first, we need to builda hauling cost model at benches. Using the model, we can calculate the hauling cost varyingwith bench depth and tonnage-grade distribution in any bench of the pit. Second, we need to look at the inter-relationship between operating or hauling cost, average grade and cut-off grade

## 2. Hauling Cost Model (Varying with Benches)

Here the hauling cost includes two parts: ore hauling cost and waste dumping cost.

**(1) Ore hauling cost**

An open pit is divided into benches (i) from the top (surface) to the bottom. The hauling cost (Coi) at each bench can be expressed as:

For i=1 (first bench), $Co1 = L*r + (\frac{h}{2})/1000/\beta*r$

For i= 2 (second one) $Co2 = L*r + (h + \frac{h}{2})/1000/\beta*r$

Or in general,

$$Coi = L*r+[(i-1)*h+\frac{h}{2}]/1000/\beta*r \qquad (1)$$

Then the cost increases in benches: $\Delta C = Ci - Ci-1$

$$\Delta Co = \frac{h \times r}{1000\beta} \qquad (2)$$

Where i: bench number counting from top to bottom (1, 2 ....n), h: bench height (m), $\beta$: hauling road slope (%), r: hauling cost rate ($/t.km), L: the hauling distance from first bench to mill (km),G&A and mill costin operation.

Normally, ore is extracted from one to six benches at a time in mining operation. The average hauling cost/t can be determined as follow:

$$C(h) = \frac{1}{n}\sum_{i=1}^{6}\frac{[(i-1)h+\frac{h}{2}]r}{1000\beta}+L*r \qquad (3)$$

Obviously, the hauling cost changes with depth, similar to what happen in underground mining(S.M. Rupprecht, 2012) [8]

**(2) Waste hauling cost**

It is supposed that waste dump is located beside pit, and then for any bench, the waste hauling is simply expressed as:

$$Cwi=SR*[(i-1)*h+\frac{h}{2}]/1000/\beta*r \qquad (4)$$

Where SR: stripping ratio (waste tonnage /ore tonnage)

It is important to note that the SR may vary from bench to bench due to both the fundamental nature of resource and economic requirement of reserve. This is an important variable in determining ore-waste hauling cost.The SR is given below:

$$SR = \frac{t0-tj}{tj} \qquad (5)$$

Where $t0$ material tonnage above 0 cut-off, $tj$: ore tonnage at any cut-off grade in a bench

The SR depends, mainly, on ore tonnage-grade distribution; it also depends, to some extent, upon thickness, dip and strike of a deposit and different mining methods employed.

Therefore Cwi is solved by using eq. (4) and (5):

$$Cwi = \left(\frac{t0-tj}{tj}\right)*\left[(i-1)*h+\frac{h}{2}\right]*\frac{r}{1000\beta} \qquad (6)$$

**(3) Hauling Cost and Operating Cost**

The hauling cost ($/t) for ore and waste is yielded using Eq.1 and 6:

$$C = L*r+\left(1+\frac{t0-tj}{tj}\right)*\left[(1-i)*h+\frac{h}{2}\right]*\frac{r}{1000\beta} \qquad (7)$$

And total operating cost is expressed as:

$$C = L*r+Cf+\left(1+\frac{t0-tj}{tj}\right)*\left[(i-1)*h+\frac{h}{2}\right]*\frac{r}{1000\beta} \qquad (8)$$

Where $Cf$: fixed cost including drilling, blasting, mucking cost, G&A and mill cost.

## 3. Cutoff Grade and Average Grade Varying with Benches

On each bench the ore tonnage above the break-even cutoff grade are measured and the average grade of the ore is calculated (*Bruce A. Kennedy, 1990*) [9]. It is important to note that profit generated in operation such as net smelter return or NSR depends upon average grade, while the average grade solvedis theoretically based on cut-off grade, which is geology and economy constraint of ore reserve. Normally, for an assigned cut-off, the average grade can be quickly calculated based on resource model provided by resource geologist; on the other hands, if the average grade is known, we can also easily find the cut-off grade using this resource model.

(1) NSR model and average grade

NSR can be given below:

$$NSR >= G*D*R*S$$

Where G: average grade (recovered), D: overall geology and mining dilution in the bench, R: mill recovery (%), S: metal price after deducting the market cost,

If the profit NSR is greater than or equal to the operating cost previously mentioned (C), then the average grade required at benches will be given below:

$$G >= \frac{L*r+Cf+\left(1+\frac{t0-tj}{tj}\right)*\left[(i-1)*h+\frac{h}{2}\right]*(\frac{r}{1000\beta})}{D*R*S} \qquad (9)$$

The average grade is an expected mean value of ore grade for reserve reporting purpose. It is also expressed as:

$$G = \frac{1}{n}\sum_{k=1}^{n}gk \qquad (10)$$

The grade value, $gk$, is assigned to any block in a bench, which is sorted from lowest to highest, and then Eq. 10 can be expressed as:

$$G = \frac{1}{n}\times(g1+g2+...gn) \qquad (11)$$

These grade values from g1 to gn are clearly displayed from plan to plan or from section to section based on resource model provided.

(2) Relationship between Cut-off grade Gc and average grade G

The cut-off grade in a bench can be quickly selected if an expected average grade is determined using Eq.9. and then four relationships between cut-off and average grade are summarized below.

- Gc = Cifall value in blocks are: g1= g2 =...gn
- Gc = 0 if any grade value in blocks, $gk$, is at a profit
- Gc = g1 if g1< g2 <...gn
- Gc could be any selection between g1 and gn for achieving the company's expected profit.

## 4. Delineating a Cutoff (and the Overall Grade)-depth Chart at Benches

An Example given for Cutoff Grade (and the overall grade) vs. Bench Curve at Benches

An operating mine located in China is going to change cutoff grade at benches because it was found that a 1% Zn cut-off grade given on the basis of FS study was not suitable for generating an expected profit. This example is used to demonstrate how a cutoff-bench curve is created using the method proposed in this paper. The procedure can be generally divided into four steps below:

**Step 1: Bench Ore info prepared**

The basic ore information from bench to bench, which is based on resource model (blockmodel), is illustrated in Table 1 below:

**Table 1. bench ore info**

| Cut-off Grade %Zn | 0 | 0.5 | 1 | 1.5 | 2 | 2.5 |
|---|---|---|---|---|---|---|
| Bench1-tonnage t | 4720275 | 4046125 | 3811563 | 3434575 | 2873650 | 2113263 |
| Bench2-tonnage t | 4567725 | 3920912.5 | 3693775 | 3270550 | 2697475 | 1976575 |
| Bench3-tonnage t | 3786750 | 3215875 | 3071763 | 2600175 | 2181000 | 1536713 |
| Bench4-tonnage t | 338850 | 234463 | 211850 | 163925 | 139800 | 119425 |
| Bench5-tonnage t | 1508625 | 1091750 | 445850 | 666300 | 389788 | 221475 |
| Bench6-tonnage t | 90450 | 79775 | 58425 | 34713 | 13913 | 9400 |
| Bench1-grade %Zn | 2.71 | 2.89 | 3 | 3.28 | 3.6 | 3.83 |
| Bench2-grade %Zn | 2.59 | 2.68 | 2.78 | 3.46 | 3.95 | 4.54 |
| Bench3-grade %Zn | 2.83 | 2.88 | 2.96 | 3.16 | 3.37 | 3.71 |
| Bench4-grade %Zn | 2.33 | 2.35 | 2.46 | 2.68 | 2.97 | 3.26 |
| Bench5-grade %Zn | 2.33 | 2.56 | 2.68 | 3.4 | 3.11 | 3.39 |
| Bench6-grade %Zn | 3.19 | 3.21 | 3.24 | 3.33 | 3.56 | 3.86 |

In Table 1, red and black colors represent tonnage and grade respectively.

**Step 2: SR calculation**

The SR is calculated based on the data from Table 1 and Eq. 5, summarized in Table 2 below:

**Table 2. SR varying with average grade at benches**

| Cutoff grade g/t | 0.5 | 1 | 1.5 | 2 | 2.5 |
|---|---|---|---|---|---|
| Bench1_SR | 0.17 | 0.24 | 0.37 | 0.64 | 1.23 |
| Bench2_SR | 0.16 | 0.24 | 0.40 | 0.69 | 1.31 |
| Bench3_SR | 0.18 | 0.23 | 0.46 | 0.74 | 1.46 |
| Bench4_SR | 0.45 | 0.60 | 1.07 | 1.42 | 1.84 |
| Bench5_SR | 0.38 | 2.38 | 1.26 | 2.87 | 5.81 |
| Bench6_SR | 0.13 | 0.55 | 1.61 | 5.50 | 8.62 |

**Step 3: The parameters used for cutoff (break even and optimum) calculation at benches**

The parameters needed for the calculation of cut-off at benches is summarized in Table 3 below:

**Table 3.**

| Item | Value | Unit |
|---|---|---|
| Fixed Mining cost (not including hauling cost) | 5 | US$/ mined t |
| Processing cost | 15.00 | US$/ milled t |
| G&A | 5.00 | US$/ milled t |
| Ore recovery (inside the pit) | 100.00 | % |
| Ore dilution | 15.00 | % |
| Mill recovery | 85.00 | % |
| Zinc Price | 0.95 | US$/lb |
| refining | 95.00 | % |
| Sales Tax | 12.00 | % |
| road slope | 8.00 | % |
| ore mining bench height | 10.00 | m |
| distance from first bench to mill | 4.00 | km |
| Waste dump from pit first bench | 3.0 | km |

**Step 4: Average grade calculation and cutoff grade determined**

The Eq. 9 above is used for calculating average grades and determining cut-off grade. The calculation is carried out using iterative techniques in excel.

**Step 5 Cut-off –Bench curve**

Data from the tables above is used to generate the cut-off grade (also ave. grade) vs. bench grapics as shown in Figure 1.

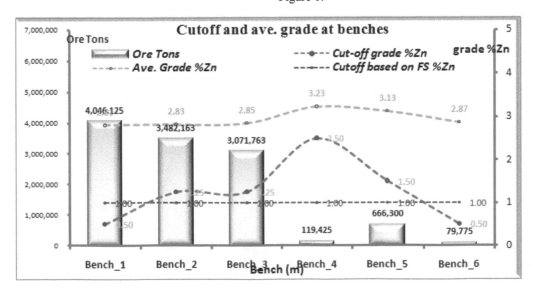

**Figure 1.** Cut-off grade moving at benches

Figure 1 illustrates that the cut-off including break-even and the overall grade varies from bench to bench, and the previously given 1% cut-off grade is too low to generate the lowest profit at bench 2-5. The example also shows the low-grade material with bench 4 and there is no way to extract ore from the bench in a desired cut-off grade. The best solution would be to remain the low-grade material in its original place or transport it to some place beside the pit after blasting. This is a way reducing waste hauling cost so that we may have a desired cut-off and average grade for a higher production with the bench.

## 5. Summary

The author presented a hauling cost model to facilitate the dynamic hauling cost calculation. The model indicates that hauling cost increases with depth, but largely depends upon the dynamic stripping ratio, representing the fundamental nature of resource and economic requirements of mineable ore at benches.

The overall grade (and average grade) at a profit can be calculated using the formula presented in this paper, then cut-off grade quickly selected using trial-and-error technique.

The author believe that, by following a five-step procedure as shown in the given example, the cutoff gradevs. bench curve is easily delineated, which will be a simple but useful tool for engineers to facilitate sound decision on how much ore at benches can be extracted and sent to mill.

## References

[1]    R.V. Ramani, K.V.K. Prasad, and R.L. Frantz, (March 1995), "Ore Reserve estimation: Myth and Reality of Computer Models", *Min. Res. Eng.*4 (2), 15-28

[2]    Peter Darling, (2011), *Mining Engineering Handbook, Third Edition,*SME, USA, 401-403.

[3]    William A. Hustrulid, (2001), *Underground Mining Methods: Engineering Fundamentals and International Case Studies,*SME, 293-296.

[4]    William A. Hustrulid, Mark Kuchta, Randall K. Martin, (2013), *Open Pit Mine Planning and Design,*CRC Press, 334-496.

[5]    Xinming Tang, Yaolin Liu, Jixian Zhang, Wolfgang Kainz, (2007), *Advances in Spatio-Temporal Analysis,* CRC Press, 840-860.

[6]    G. Matheron, (1987), *Springer Science & Business Media,* Geostatistical Case Studies, 239-241.

[7]    A.E. Annels, (2012), *Mineral Deposit Evaluation: A practical approach,* Springer Science & Business Media, 321-323.

[8]    S.M. Rupprecht, (2012), "MINE DEVELOPMENT – ACCESS TO DEPOSIT", the Southern African Institute of Mining and Metallurgy, 101-120.

[9]    *Bruce A. Kennedy,(1990), Surface Mining, Second Edition,* SME, 466-467.

# 16

# Knowledge-Based Intellectual DSS of Steel Deoxidation in BOF Production Process

Zheldak T.A.[*], Slesarev V.V., Volovenko D.O.

Department of Systems Analysis and Control, National Mining University, Dnipropetrovs'k, Ukraine
*Corresponding author: zheldak@dniprograd.org

**Abstract** This article describes one of possible approaches to deoxidant cost optimization in steel production, based on the expert system. Education of the system is based on successfully completed heats. Bayesian networks and decision trees are suggested as the mechanisms for knowledge extraction.

*Keywords:* Bayesian networks, decision trees, decision-making, deoxidizing, knowledge, management, rules

## 1. Introduction

Liquid or solid iron, scrap, deoxidants, alloying and slag-forming materials are used as basic material in the manufacture of steel-making. Several major problems are being resolved at redistribution of iron and scrap into steel: melting and heating blend to the temperature that provides the following operations (typically 1600-1650°C), refining steel from impurities (typically, these include sulfur, phosphorus, hydrogen and nitrogen), alloying, and finally obtaining steel ingot or continuous casting billet from liquid steel [1].

Heating up to the desired temperature, partial refining and alloying are performed in steelmaking units, particularly in the basic oxygen furnace, final refining and alloying - in steel teeming ladle after the release of the unit using specialized facilities and spill - through molds or continuous casting machines (CCM).

## 2. Problem Description

Released from the converter, non-deoxidized liquid steel contains a significant amount of dissolved oxygen. Lowering of the metal temperature during filling and crystallization is accompanied by oxygen solubility decrease, leading to carbon monoxide formation and separation, bubble castings and leaky bars [2]. The first task of deoxidation is to reduce dissolved oxygen in the steel and linking it into the stable compounds that do not give gaseous emissions during the solidification of the metal. Another problem is the maximum removal of liquid steel deoxidation products (non-metallic inclusions). Non-metallic inclusions are characterized by physical properties different from the base metal, causing the formation of local stress concentration, contact fatigue of metals, intercrystalline fractures, depleting and crashes of moving parts. To acquire high quality steel, the content of nonmetallic inclusions shouldn't be more than 0.005-0.006% [1].

The most common steel deoxidants are silicon, manganese and aluminum. Calcium, chromium, vanadium, cerium, titanium are used in some cases. Meanwhile, ferrosilicon, silicomanganese and ferromanganese are often used in the domestic metallurgy. All of them have different deoxidization ability, extirpation quality (silicon is the best for oxygen binding, although it is the worst for oxides deriving from the melt) and price. The problem of optimization of alloys used in the deoxidation seems to be actual because the price of alloys makes quite a substantial share of the cost of finished steel.

This problem is one of the integrated system control functions of the multistage steel production. The main approaches to solving the problem, principles structure and functioning are set out in [3] and [4]. In particular, the system should implement the optimality principle of each production stage to obtain optimal process plan by generalized economic criterion [5].

Generalized process control at the stage of deoxidation is presented in scheme in Figure 1.

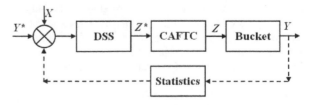

**Figure 1.** Scheme of deoxidation process control

As inputs for decision support system (DSS) of deoxidation process control a number of parameters, which are caused by produced steel brand (vector $Y^*$) and parameters of steel in the converter (vector $X$ which in general is a disturbance) was selected. Depending on the component values of these vectors, DSS generates control action (vector $Z^*$) as the task of different kind deoxidants that served in buckets. The task is mined by closed automatic ferroalloy tract control (CAFTC) which consists of the regulator, dispenser, valves system and sensors [5]. At the initial phase of converter steel production selected ferroalloys (vector $Z$ ) are added to the bucket, that helps to maximize deoxidants absorption by steel melt and reduces their consumption in the slag zone [1]. As a result we have control vector $Y$ that describes the chemical composition of the obtained steel.

Figure 1 shows the main feature of the deoxidation control system: it is open-loop and uses the principle of management by objectives. The usage of any present feedback system in the process is impossible: because when the numerical characteristics that describe the output are obtained, the meaning of the control action will be lost as steel becomes cold ingot, in which deoxidation processes are completed.

Existing techniques used in manufacturing [1,2] provide the usage of empirical formulas for deoxidants calculating depending on their chemical composition, considering grade of steel, carbon, silicon and manganese content before turndown and casting kind, either steel cast is boiling or calm. Target carbon and alloying elements value content in finished steel (vector $Y^{*}$ ) defined as the average value of the range allowed for this brand by standard. These formulas are made with some reserve and do not provide ferroalloys savings.

Research objective: To propose a mechanism of previous positive steel deoxidation experience using knowledge extracted from fusions databases and their inclusion in decision support systems, introducing feedback into a deoxidizing control system.

## 3. Materials and Methods

Decision support systems that use intelligent (including fuzzy) conclusion are used in industry, particularly in the ferroalloys production. A mathematical model of fuzzy optimization of multicomponent mixture and methods and algorithms that improve the predictions consequences control action quality in the face of uncertainty regarding the structure and parameters of the processes and actions of uncontrolled disturbances are proposed in [6]. Thus, neuro-fuzzy model using is optimal regarding the economic criterion of the charge composition.

Similar ideas developed in [7], allow simulating steel alloying process and predicting the mechanical properties of rolled results by chemical analysis of steel in the ladle, using fuzzy production rules, in particular, Mamdani model. Even a small number of rules in such intelligent systems have high resolution in result space.

There is also another approach [8] to modeling the optimal behavior of the BOF production operator based on the use of standards. Under this approach, during the smelting of steel and brand known initial conditions of melting, the example that most accurately describes the current conditions is sought in a database containing information about all previous melting. It applies multidimensional mathematical optimization of a model interpolated to the conditions that are different. The authors propose taking the main charge materials and ferroalloys similar to previously known successful examples from the database.

In this paper, key features of both approaches are proposed to be used in an expert DSS. As a production model, the so-called naive Bayes network [9] is proposed to be used.

This method of processing knowledge has several important features. Firstly, because the model determined the relationship among all the variables, the situation when the values of some changes are unknown is easily handled. Secondly, the approach allows combining patterns derived from the data and background knowledge obtained in an explicit form (e.g., experts) in a natural way. Finally, using the described method avoids the problem of overfitting, i.e. excessive models complexity which many methods are affected (such as decision trees) because of too close distribution of noisy data imitation.

Despite its simplicity, speed and ease of interpretation of results, naive Bayes algorithm has several disadvantages, the key one of which is the basic assumption of mutual uncorrelatedness of all input variables (reason of the "naive" in the title). However, the method does not permit the direct processing of continuous variables - they must be divided into a number of intervals to discretize values.

Considering all the method above, the data from the database of fusion performed in the converter shop of PJSC "Evraz - DMP named after Petrovsky" in 2008 and 2009 (12039 fusions in total) were discretized, and then factor analysis was conducted for them to determine the linearly independent factors. The results are reported in Table 1, where the separation of variables (control action, disturbance, and state variables) is presented in accordance with the scheme of Figure 1.

**Table 1. Research problem variables**

| Variable | Dimension | The value in the database | Range of variation | Number of intervals |
|---|---|---|---|---|
| X1 | °C | The temperature of the metal in the converter | 1559-1698 | 3 |
| X2 | % | Mn content before deoxidation | 0,03-0,8 | 4 |
| X3 | % | S content before deoxidation | 0,012-0,05 | 4 |
| X4 | % | P content before deoxidation | 0,002-0,031 | 3 |
| X5 | % | C content before deoxidation | 0,04-0,86 | 6 |
| X6 | % | Mn content in FeMn | 68,5-79,2 | 3 |
| X7 | % | Mn content in SiMn | 66,2-74,9 | 3 |
| X8 | % | Si content in FeMn | 0,41-2,8 | 3 |
| X9 | % | Si content in SiMn | 16,67-18,2 | 3 |
| Y1 | % | Set Mn content in the finished steel | 0,5-0,8 | 3 |
| Y2 | % | Set S content in the finished steel | 0,013-0,05 | 3 |
| Y3 | % | Set P content in the finished steel | 0,002-0,038 | 3 |
| Y4 | % | Set C content in the finished steel | 0,28-0,37 | 3 |
| Y5 | % | Set Si content in the finished steel | 0,05-0,12 | 3 |
| Z1 | kg. | FeMn deoxidizer weight | 0-1070 | 5 |
| Z2 | kg. | SiMn deoxidizer weight | 0-675 | 4 |
| Z3 | kg. | FeSi deoxidizer weight | 0-360 | 3 |

Separation of continuous variables values was carried out according to the ranges of these recommendations [10]:
- If the value distribution law is uniform or normal - 3 bands ("small", "medium" and "high") at values intervals of the basic scale;
- If the distribution law is negative or it is difficult to establish - the minimum number for quantile with equal (or as close as possible) probability density intervals.

Since the variables $z_k$, $k = \overline{1,3}$ depend not only on the semi-independent $x_i$, $i = \overline{1,9}$ and $y_j$, $j = \overline{1,5}$, but are also interdependent (ferroalloys are complementary and interchangeable), each combination of their linguistic values needs to match a specific class. According to reports such classes made up 34. These included all possible combinations of terms of output variables, which met in the database at least once.

According to the method [9] all variables were quantized and for each interval central values and priori probabilities were calculated. In particular, for the output variables were defined probabilities $P(z = c_r)$, where $c_r$ makes vector combinations of central importances, $r$ interval of output variables, such as $z_1$ = «Little» | $z_2$ = "Many» | $z_3$ = "Nothing".

Putting independent variables into line to output variables, posteriori probability formula can be defined.

$$P(x_i = c_{i,j}, y_k = c_{k,j} \mid z = c_r) =$$

$$= \frac{P(z = c_r) \cdot \prod_i \left( P(x_i = c_{i,j}) \right) \cdot \prod_k P(y_{1k} = c_{1k1j})}{\sum_r \left( P(z = c_r) \cdot \prod_i \left( P(x_i = c_{i,j}) \right) \cdot \prod_k P(y_k = c_{k,j}) \right)}, \quad (1)$$

where $r = \overline{1,34}$ is the current number of output class; $c_{i,j}$, $i = \overline{1,9}$ and $c_{k,j}$, $k = \overline{10,14}$ are central values of input terms, with individual number (3 to 6) for each variable (see Table. 1).

Thus, we calculate probability for complex rules, such as "If $x_1 = c_{1,j}$ and $x_9 = c_{9,j}$ and ... $y_1 = c_{10,j}$ and ... $y_5 = c_{14,j}$ then $z = c_r$". The total number of such rules with certainty $P_\Sigma = 1$ describing the specified subject area makes up 1589.

Defuzzification of fuzzy solutions to precise is performed using the formula

$$z_m = \sum_{r=1}^{34} \mu(c_r) \cdot c_{m,r}, \quad m = \overline{1,3} \quad (2)$$

where $c_{m,r}$ is the central meaning of each term that refers to the necessary mass of $m$ ferroalloys.

As an alternative to the technique of obtaining precise production rules, membership of which is estimated at probability occurrence conditions and result in the database chosen for study, a method for constructing decision trees was proposed. This method, detailed in [10], is widely used in the construction of expert systems as an algorithm for knowledge discovery.

To build a knowledge base based on decision trees partitioning results presented in Table 1 were used. Let us consider in detail the variable selection criterion for which a partition is produced in the next node. If the variable $x_k$ takes $h$ values $c_{k1}, c_{k2}, ..., c_{kh}$ then partitioning $T$ by variable $x_k$ will subset $T_1, T_2, ..., T_h$. Variable chosen is based on information about how classes $c_{m,r}$ of output variables distributed on the set $T$ and its subsets.

Let $freq(c_r, I)$ is the number of objects from subset $I$ belonging to the same class $c_r$. Then the probability that randomly chosen object from subset $I$ belongs to the class $c_r$ is equals

$$P = \frac{freq(c_r, I)}{|I|}, \quad (3)$$

According to information theory, estimation of the average amount of information needed to classify an object from a set $T$, calculated from

$$Info(T) = -\sum_{j=1}^{l} \left( \frac{freq(c_r, T)}{|T|} \right) \log_2 \left( \frac{freq(c_r, T)}{|T|} \right), \quad (4)$$

Because we are using the logarithm to the base 2, this expression provides a quantitative assessment in bits. The same assessment, but after splitting sets $T$ by $x_k$, gives the following expression:

$$Info_{x_k}(T) = \sum_{i=1}^{h} T_i / |T| \, Info(T_i), \quad (5)$$

The criterion for selecting a variable, which hold partitioning at the next node is

$$\max_k \left( Gain(x_k) \right) = Info(T) - Info_{x_k}(T), \quad (6)$$

Variable with the highest *Gain* becomes the key in the current node. Further, tree building continues regarding its value. The principle of choice (3) - (6) is applied to the resulting subsets $T_1, T_2, ..., T_h$ thus, recursive tree construction continues until the node contains only the objects of one class.

In order to limit the depth of the tree to its basic algorithm next rule was added: stop dividing in the next node, if subset that is associated with this node contains objects which belong to no more than 10% of the classes of the total number of the study samples. The restriction above does not materially affect the classifying ability of the tree, but can significantly reduce the total number of rules. After constructing, the tree was truncated, thus reducing the number of rules from 2152 (clear rules identifying at least one object) to 422. Truncation was carried out with a minimal level of support in 0.1% (for training data of 8023 fusion it included 9 fusions).

Levels of 10% of the minimum variation and 0.1% of the minimum support were chosen empirically while retaining ranging capability during the structure maximum simplify.

## 4. Results

Testing of knowledge discovery methods within DSS was performed using fusions data for 2008-2009. The evaluation of the accuracy of approximating productive models of root mean square error is given in Table 2.

**Table 2. The mean square error of approximation**

| Sample | Deoxidants | Bayesian network | | Decision Tree | |
|---|---|---|---|---|---|
| | | Absolute, kg | Relative | Absolute, kg | Relative |
| Training | FeMn | 37.40 | 0.0926 | 45.089 | 0.1116 |
| | SiMn | 40.59 | 0.0551 | 47.138 | 0.0640 |
| | FeSi | 2.83 | 0.0195 | 2.849 | 0.0196 |
| Test | FeMn | 36.86 | 0.0844 | 71.026 | 0.1625 |
| | SiMn | 44.16 | 0.0601 | 58.139 | 0.0791 |
| | FeSi | 4.54 | 0.0325 | 4.1411 | 0.0296 |

You can see that the knowledge base developed on naive Bayesian networks, gives accurate results approximations both in the training sample and the test. The prediction error does not exceed 10%, which is close to the instrumental error of ferroalloys supply RSA and can be considered as acceptable for the rest of this production.

The number of rules in the knowledge base for the following methods differs significantly - for Bayesian network they are 1598, for decision trees – 422, but studies have shown that the speed of knowledge processing bases and calculation of the final result in both systems embedded in real-time limits of this process.

Application of the existing knowledge bases in practice showed that the predicted mass required for melting alloys exceed the value calculated theoretically. That is, in practice, the operator of converter production, consciously or not, often uses excess ferroalloys, which affect the cost of steel.

The study was conducted to determine the boundaries of ferroalloys economy in which the central values of output terms gradually decreased. In this case, the projected steel parameters, consumption of ferroalloys and prediction error was controlled.

With a shift in terms center of the naive Bayes network by 20%, compromise optimum is achieved: total savings of ferroalloys 0.398 kg / t DSS approximation error increases by only 1%. In monetary terms, 4013 meltings made in 8 months corresponds to saving $206 thousands, which in turn provides monthly savings at $26 070 and yearly saving about $312 000. Further shift of the central values of the terms leads to a redistribution of training examples and a sharp increase in errors of approximation.

Unfortunately, while using decision trees, even slight shifting of terms causes significant changes in the structure of the tree, altering hundreds of rules. However, the displacement of some terms down by up to 10% can provide saving ferroalloys in average 0.118 kg / t, which financially give $147 000 per year.

# 5. Conclusions

To form the task at deoxidizing, controls DSS were proposed production model of expert system. As methods of obtaining knowledge probabilistic approach was used regarding buildings Bayesian networks and ID3 algorithm of constructing decision trees. The first knowledge base contains 1589 probabilistic rules with generalizing fuzzificator, the second one – 422 clear rules concluded in a tree of 14 variables.

DSS on Bayesian network shows higher accuracy of approximation at the training and the test sample. Prediction error does not exceed 10%, which is acceptable for this process.

With a shift in terms of the naive Bayes network by 20% is compromise achieved: with total savings of ferroalloys at 0.398 kg / t DSS approximation error increased by 1%. In monetary terms, this corresponds to savings of about $312 thousands per year.

Prospects for the development of this subject are seen at applying the proposed method to other metallurgical industries, including electric one, which also uses ferroalloys. Another possible direction of development involves the use of a neural network with radial basis functions, which is equivalent to the simultaneous classification and approximation of nonlinear dependencies.

To improve the accuracy of forecast models, it necessary pay attention to the factors of production processes, which are usually not included in the passport of melting, for example, the degree of wear of the lining, casting method, and etc.

# References

[1]    Demidov V. *Production of converter steel [Instruction manual]* TI-233 ST-CC-02-2002. DMP. Dnepropetrovsk. 2002.

[2]    Byheev A., Baytman V., "Using thermodynamic deterministic mathematical models in management BOF process", *Proceedings of the Chelyabinsk Scientific Center,* V. 4(30). 73-76. 2005.

[3]    Zheldak T., Garanzha D., "Decision Support System of production planning and control process flow" in *17th International Conference of Automatic Control "Automation - 2010",* Kharkov:KNURE. 1, 212-214. 2010.

[4]    Slesarev V. T. Zheldak, "Integrated control multistage manufacturing steel pipes for example rolling", *System technology. Regional Interuniversity collection of scientific papers.* V.75, 78-85. 2011.

[5]    Boyko V., Smolyak V., *Automated process control systems in the steel industry,* Nauka I osvita. Dneprodzerzhinsk, 1997.

[6]    Mikhalev A., Lisaya N., "The use of neuro-fuzzy algorithms for the analysis and prediction of dependency process of smelting of ferroalloys", *System technology. Regional Interuniversity collection of scientific papers.* V.26, 29-34. 2003.

[7]    Novikova E., Mikhalev A., Bublykov Yu., "Fuzzy identification of micro-alloying process steel with carbonitride hardening", *Modern problems of metallurgy: Proceedings.* System Technology. Dnipropetrovsk. 113-127. 2006.

[8]    Bohushevskyy V., Litvinov L., *Mathematical models and the control system converter process.* NPK "Kiev Institute of Automation". Kiev. 1998.

[9]    Barseghyan A., Kupriyanov M., Stepanenko V., Holod I., *Data mining technology: Data Mining, Visual Mining, Text Mining, OLAP.* - 2nd ed. BHV-Petersburg. St.-Petersburg. 2007.

[10]   Witten I. H., Eibe F., Hall M. A., *Data mining: practical machine learning tools and techniques.* - 3rd ed. Elsevier. 2011.

# Stream Sediment Geochemical Survey of Gouap-Nkollo Prospect, Southern Cameroon: Implications for Gold and LREE Exploration

**Soh Tamehe Landry[1], Ganno Sylvestre[1,*], Kouankap Nono Gus Djibril[1,2], Ngnotue Timoleon[3], Kankeu Boniface[4], Nzenti Jean Paul[1]**

[1]Laboratory of Petrology and Structural Geology, University of Yaoundé I, Cameroon
[2]Department of Geology, HTTC, University of Bamenda, Cameroon
[3]Department of Geology, University of Dschang, Dschang, Cameroon
[4]Institut de Recherches Géologiques et Minières, Yaoundé, Cameroun
*Corresponding author: sganno@uy1.uninet.cm; sganno2000@yahoo.fr

**Abstract** Stream sediments play a significant role in geochemistry exploration by identifying possible sources of anomalous element concentration. This work is the baseline stream sediments geochemical study which brings general information on the geochemical dispersion of the metal elements (especially gold) at Gouap-Nkollo prospect (SW Cameroon) with the aim of providing a useful guide for future exploration strategies. For this study a concentration of 47 elements was measured in 10 stream sediment samples using BLEG and ICP-MS methods, but emphasis was given to the following 21 chemical elements: Al, Ca, Fe, K, Mg, Na, P Ag, Au, B, Co, Cr, Cu, Mn, Ni, Ti, Zn, Ce, La Th, U and Zr. Averaged elemental concentration for each samples obtained by statistical analysis showing patterns of enrichment and depletion which may relate to localized mineralization conditions or local lithological changes. Results showed that the stream sediments have high concentrations of Au, Ce and La with average values of 314.85ppm, 19081ppm and 11808ppm respectively for gold, cerium and lanthanum. Cerium and Lanthanum have considerably high concentrations when compared with other Rare Earth Elements (REE) analyzed. These concentrations represent interesting indices for Au and LREE mineralization's. The geochemical dispersion of the metal elements (especially gold) reveals that high concentrations are recorded in the northern part of the prospect, close to the quartz-tourmaline vein within the quartzite. This result indicates that the Au and other metal elements probably originated from the quartz-tourmaline veins hosted by surrounding rocks. Detailed exploration work including geochemical soil sampling and geophysical survey is highly recommended in the northern part of the Gouap-Nkollo prospect, where anomalous concentrations of Au were observed, for further investigation.

*Keywords:* stream sediments geochemistry, anomalous concentrations, Gouap-Nkollo, Cameroon

## 1. Introduction

One of the most widely used methods in regional geochemical approaches is the stream sediments sampling. Stream sediment geochemistry is extensively used in mineral exploration and environmental studies. Active sediments in the channels of streams and rivers can contain low levels of metals derived from weathering of mineralized rocks within the upstream catchment [1]. Natural concentrations of heavy metals as a result of the weathering processes of mineral deposits can be quite high in stream sediments close to the deposit, but decrease with increasing distance downstream, due to dissipating energy and dilution of sediments from other unpolluting sources [2]. Geochemical maps have been constructed using stream sediment geochemical data over the world to identify possible sources of anomalous element concentrations [3]. Ore deposits form when a useful commodity is sufficiently concentrated in an accessible part of the Earth's crust so that it can be profitably extracted [4]. The stream sediment technique has played a major part in the discovery of many ore bodies around the world. A good example being the discovery of the Panguna porphyry copper/gold deposit on Bougainville Island, Papua New Guinea [1,5].

Gold prospection in Cameroon using stream sediments has been by large receiving little attention. The recent work of Embui et al. [6] investigating the concentrations of gold and associated elements in stream sediment samples from the Vaimba-Lidi drainage system in northern Cameroon represents the only published works on the use of stream sediment in gold exploration in

Cameroon. It's worth noting that alluvial gold exploitation commenced in the early 1940s and continues till date. Most of these small-scale alluvial gold mining operations are located in the eastern part of country, especially in the Betare Oya and Batouri gold district [7]. Also, others but not more active alluvial gold mining operations are located in southern and southeastern Cameroun, respectively around Akom II and Mintom areas, but little or no exploration works have been carried out in these areas. In an attempt to discover new gold potentials in the southern part of the country, we designed a stream sediment survey targeting the Gouap-Nkollo area. The selection of this area is based on two main factors: (i) the lithology which consists of quartzite crosscut by pegmatite tourmaline-quartz vein. Gold ore deposits with tourmaline in wall rocks and sometimes in the ore vein itself have been fully documented around the world [eg. [8-13]]; (ii) the presence of visible gold grain in pan concentrate and the existence of small-scale alluvial gold mining in Nkollo site.

This work was undertaken at a time of renewed interest in the gold potential of Gouap-Nkollo prospect for which G-Stones Resources Ltd has exclusive exploration rights (see www.g-stonesresources.com/php.index). The current paper is the baseline geochemical study which brings general information on the geochemical dispersion of the metal elements (especially Gold), REE, and the relationship of this distribution to the geology of the area to provide a useful guide for future exploration strategies.

## 2. Geographical and Geological Settings

### 2.1. Geographical Location and Drainage Pattern

The Gouap-Nkollo prospect is located between latitudes 3°08'N and 3°15'N and longitudes 10°13'E and 10°18'E and covers an area of 120Km². The area is covered by the tropical rain forest. The local climate is equatorial climate with two dry seasons (mid-November to mid-March; mid-June to mid-August) and two rainy seasons (mid-August to mid-November; mid-March to mid-June). Annual average rainfall is 563 mm and the annual average temperature is 27°C. The relief of the site is quite plain with hills. The detail geomorphology could be divided into two topographical units including a lower unit made up of plains with low altitude (<200 m) and a higher unit formed of hills with elevations ranging from 200 to 400 m. There is no major river in Gouap-Nkollo prospect. However, the drainage system over the areas is a dendritic network with the proliferation of many smaller stream channels.

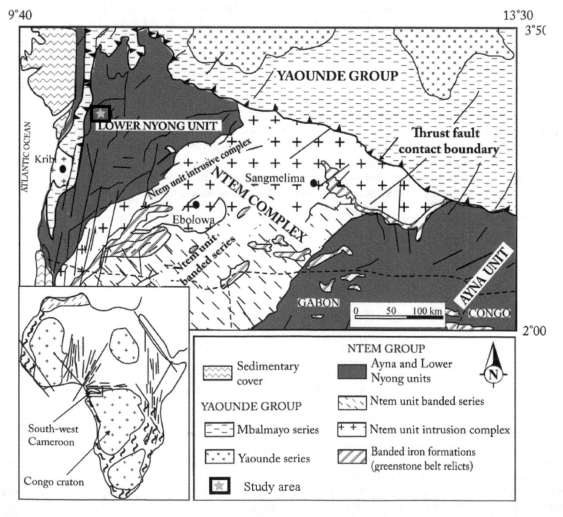

**Figure 1.** Geological map of South-West Cameroon after Maurizot et al. [20] showing the location of Gouap-Nkollo prospect within the Lower Nyong Unit

## 2.2. General on Local Geological Setting

The Gouap-Nkollo prospect belongs to the Lower Nyong unit which corresponds to the NW corner of the Congo Craton in South-Cameroon (Figure 1). This unit is of Palaeoproterozoic age and is a well-preserved granulitic unit of the West Congo craton resting as an Eburnean nappe on the Congo Craton [14,15]. The Lower Nyong unit underwent high-grade tectono-metamorphic event at 2117–2045 Ma [15,16,17]. It is associated with charnockite formation [18]. This led to the assertion that the Nyong Group is a reactivated portion of the Archean

Ntem complex [14,19,20]. It is constituted of both Archean and Paleoproterozoic materials associated with iron formation or BIF (Banded Iron Formation), plutonics (TTG, charnockites, dolerites, alkaline syenites), greenstone (serpentinites, ...) and biotite hornblende gneisses, which locally appear as grey gneisses of TTG composition, orthopyroxene–garnet gneisses (charnockites), garnet–amphibole–pyroxenites, and banded iron formations (BIF). The metamorphic evolution is polycyclic with Paleoproterozoic granulitic assemblages overprinted in the western part of the group by Pan-African high-grade recrystallization [15].

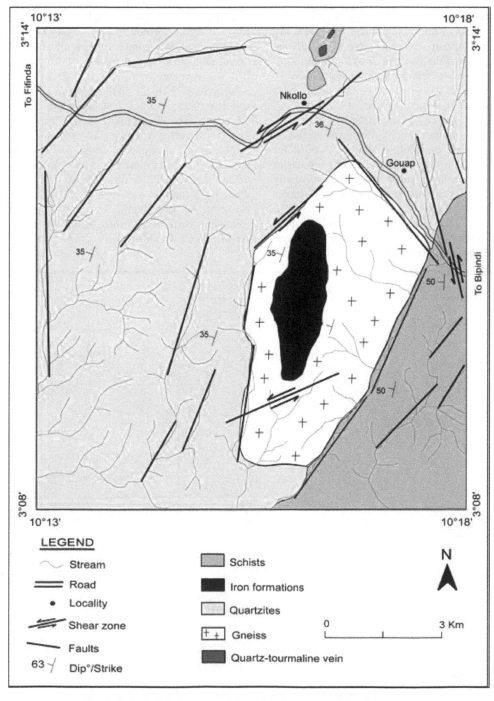

**Figure 2.** Geological sketch map of Gouap-Nkollo prospect, modified from [21]

Previous geological investigations have recognized four rock types in the area, namely muscovite quartzites, gneiss, tourmaline micaschists and tourmalinites [21,22] (Figure 2). All these rocks have heterogranular granoblastic microstructures with mineral assemblages' characteristics of plurifacial prograde metamorphism from greenschist facies to medium grade amphibolite facies [22]. Tourmalinization, sericitization and silicification are the main observed wall rock alteration processes. These processes are probably related to the circulation of hydrothermal fluids in felsic to mafic rocks (sericitization and silicification) and shale, slate and schist (tourmalinization) [23].

# 3. Methodology

## 3.1. Sampling and Preparation

Systematic stream sediment sampling was carried out using topographic maps with 1:5,000 scales. For this study, a total of forty-five stream sediments were sampled. The location coordinates and the position of each sample were quickly recorded with a global positioning system (GPS). Active stream sediments were collected along stream beds of the area using a pick, a panning dish, a stainless steel sieve and collection pan (a –80 mesh sieve size). They were sampled mostly at a depth of 30-45 cm and sieved on site in order to provide a sample of suitable weight for assay (2 kg of <1.2 mm fraction). Since some tributaries of the seasonal streams have had no water flow for many months, their stream bed was dogged and the samples were collected with extreme care. For the technique to work with maximum effectiveness, sieving has been done by washing the sample through the mesh of the sieve. Clean polyethylene bags were used in storing stream

sediment samples before it was transported to the laboratory. The sample bags were carefully labeled with permanent markers to avoid mix up. During the sampling process, simultaneous site surveys carried out in order to provide specific information relating to the geology of the sampling point.

## 3.2. Analytical Methods

Forty-five stream sediments were sampled in the whole study area and ten samples yielded visible gold in pan concentrate were selected for chemical analysis. The analyses were performed at OMAC Laboratories (Alex Stewart Assayers Group), Ireland. The chemical analysis involved the use of BLEG (Bulk Leach Extractable Gold) and ICP/MS (Inductively Coupled Plasma/Mass Spectrometry) following a lithium metaborate/ tetraborate fusion and nitric acid digestion of 0.2g sample. BLEG-ICP/MS involves weighing a 1-2 kg sample into a polyethylene bottle, adding an appropriate cyanide solution (0.25% to 1% NaCN) and bottle rolling for a determined period of time [24]. A pH of 10 or greater is maintained during leaching. The gold is dissolved through formation of its cyanide complex [24]. The resultant cyanide solution is diluted and analyzed by Perkin Elmer Sciex ELAN 6000, 6100 or 9000 ICP/MS [24]. After dilution, solutions were analyzed for a series of 47 elements. In this study, emphasis was given on the following chemical elements: Al, Ca, Fe, K, Mg, Na, P Ag, Au, B, Co, Cr, Cu, Mn, Ni, Ti, Zn, Ce, La Th, U and Zr.

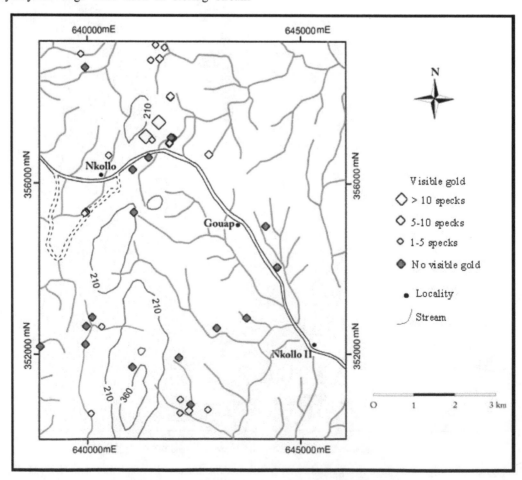

**Figure 3.** Stream sediment sampling map showing gold field survey results

# 4. Results and Discussion

## 4.1. Field Survey

The results of the field investigation are presented in Figure 3. On this map, twenty (20) sampling sites showed visible gold grain in pan concentrate. These sites are predominately located in the Northern part of the study area and few in the southern part. Twenty-five (25) samples didn't show visible gold in the concentrate. The sampling sites with visible gold in pan concentrate are located downstream of quartzite with quartz-tourmaline vein, dominated by silification and sericitization hydrothermal alteration. This results shows that the gold mineralization in Gouap-Nkollo prospect could be related to the quartz-tourmaline vein.

## 4.2. Geochemical Survey

Chemical composition of the studied stream sediment samples is presented in Table 1 and Table 2.

**Table 1. Statistical distribution of major elements (wt%)**

| Samples | Ti | Fe | Al | Mg | Mn | Ca | K | Na | P |
|---|---|---|---|---|---|---|---|---|---|
| BPS002 | 0.5078 | 2 | 2.7 | 0.34 | 0.0286 | 0.38 | 0.15 | 0.08 | 0.059 |
| BPS003 | 0.3511 | 1.72 | 1.7 | 0.05 | 0.0196 | 0.02 | 0.11 | < 0.01 | 0.08 |
| BPS004 | 0.2462 | 1.69 | 1.18 | 0.12 | 0.0178 | 0.03 | 0.17 | < 0.01 | 0.049 |
| BPS004B | 0.131 | 3.59 | 3.51 | 0.96 | 0.0303 | 0.04 | 0.89 | < 0.01 | 0.042 |
| BPS005 | 0.3729 | 1.56 | 0.84 | 0.12 | 0.0312 | 0.05 | 0.05 | < 0.01 | 0.012 |
| BPS005B | 0.3038 | 0.88 | 0.82 | 0.01 | 0.0207 | 0.01 | 0.03 | < 0.01 | 0.034 |
| BPS006 | 1.7941 | 7.74 | 1.62 | 0.08 | 0.096 | 0.1 | 0.04 | 0.01 | 0.039 |
| BPS006B | 0.86 | 2.02 | 0.89 | 0.04 | 0.0544 | 0.09 | 0.06 | 0.02 | 0.009 |
| BPS007 | 0.5498 | 1.33 | 1.02 | 0.05 | 0.0373 | 0.09 | 0.07 | < 0.01 | 0.031 |
| BPS007B | 0.3847 | 1.35 | 3.01 | 0.05 | 0.024 | 0.07 | 0.14 | < 0.01 | 0.029 |
| Minimum | 0.131 | 0.88 | 0.82 | 0.01 | 0.0178 | 0.01 | 0.03 | < 0.01 | 0.012 |
| Maximum | 1.7941 | 7.74 | 3.51 | 0.96 | 0.096 | 0.38 | 0.89 | 0.08 | 0.08 |
| Mean | **0.55014** | **2.39** | **1.73** | **0.18** | **0.03599** | **0.09** | **0.17** | **0.04** | **0.04** |
| Standard deviation | 0.47987 | 2.01 | 0.99 | 0.29 | 0.02365 | 0.11 | 0.257 | 0.04 | 0.021 |

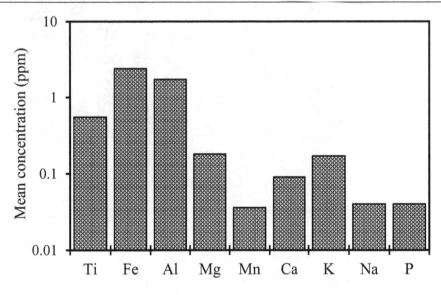

**Figure 4.** Histogram of mean values against major elements (wt%)

Table 1 presents the statistical analysis of major elements occurring in the stream sediments while Figure 4 indicates the histogram of distribution of these elements against their mean values. The result of the analysis of major elements (Table 1) revealed that these samples yielded weak to very strong Na, P, Ca, K, Mg, Fe Al, Ti and Mn contents (Figure 4). Fe and Al are the mean major element with mean values of 2.39% and 1.73% respectively (Table 1). The abundance of iron (0.88-7.74%) in the stream sediments indicates that these sediments are highly ferruginous and should come from a rock rich in ferromagnesian minerals such as biotite, pyroxene, and iron oxide. Fe has its highest concentration in BPS006 and its lowest concentration at BPS005B. The highest content of Fe in stream sediment could be attributed to the leaching of banded iron formation which crops in the central part of the studied prospect. The weathered materials enriched in iron oxide are transported by water towards the stream bed situated downstream. The second major element with high content is Alumimium (Al) which ranges from 0.82% to 3.51%. The abundance of Al in the stream sediments indicates the presence of disintegrated aluminosilicate minerals such as feldspars and muscovite. Al values in the stream sediment are higher in BPS004B and lower in BPS005B (Table 1). The presence of aluminosilicate minerals in the stream sediments might be as a result of solubility of some minerals caused by the action of running water on the surrounding rocks. Titanium (Ti) has average concentration value of 5501.35ppm. Higher Ti content is observed in sample BPS006 and the lowest content in sample BPS004B (Table 1). These strong concentrations in Ti indicate that the sediments come from a rock rich in titanohematite such as ilmenite. Due to more resistivity of Ti to weathering, this accessory mineral can be concentrated in stream sediments. Mn concentration ranges from 178-960ppm with higher concentration in BPS006 and lowest in sample BPS004 (Table 2).Calcium (Ca), Potassium (K), Magnesium (Mg), Sodium (Na) and Phosphor (P) are low in concentration. Their values range as follows: Ca (0.01-0.38)%, K (0.03-0. 89)%, Mg (0.01-0.96)%, Na (<0.001-0.08)%, P (0.012-0.080)% and their averages are 0.09, 0.17, 0.18, 0.04 and 0.04 respectively.

The trace elements and LREE geochemical composition of stream sediment samples at Gouap-Nkollo prospect is presented in Table 2 while Figure 5 presents the histogram of distribution of these elements against their mean values. Stream sediment geochemical sampling anomalous values

ranging from 2 to 3278ppm Au, 368 to 719ppm Cr. Gold (Au), and chromium (Cr) are the most abundant trace elements with mean concentration values of 468.90ppm and 493.10ppm respectively (Table 2, Figure 6A). Au has its highest concentration in BPS003 and its lowest in five samples (BPS002, BPS004, BPS005, BPS006 and BPS007). Their values range from 2 to 3278 ppm. Cr records its maximum concentration in BPS004 and its minimum in BPS006, and their value range from 368 to 719ppm (Table 2). These concentrations are indicative for the presence of precious metal such as Au and Cr in the studied area. Cobalt (Co), Copper (Cu), Nickel (Ni), zinc (Zn), thorium (Th), lead (Pb) and Zirconium (Zr) have low concentrations with mean concentration values of 6.39ppm, 9.85ppm, 19.52ppm and 50.33ppm respectively. Their values range from 2-14ppm, 2-29ppm, 5-48ppm, 17-210ppm, 8-263ppm, 6-43ppm and 11-88ppm respectively for Co, Cu, Ni, Zn, Th, Pb and Zr.

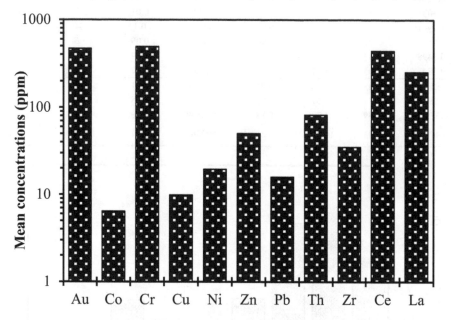

**Figure 5.** Histogram of mean values against the trace elements (ppm)

**Table 2. Statistical distribution of trace and REE elements (ppm)**

| Samples | Au | Co | Cr | Cu | Ni | Zn | Pb | Th | Zr | Ce | La |
|---|---|---|---|---|---|---|---|---|---|---|---|
| BPS002 | 2 | 8 | 400 | 9 | 26 | 51 | 29 | 171 | 88 | 875 | 497 |
| BPS003 | 3278 | 3 | 560 | 5 | 8 | 23 | 43 | 263 | 64 | 1307 | 771 |
| BPS004 | 2 | 5 | 719 | 9 | 10 | 44 | 17 | 84 | 51 | 618 | 308 |
| BPS004B | 22 | 10 | 407 | 29 | 25 | 210 | 19 | 68 | 16 | 548 | 276 |
| BPS005 | 2 | 7 | 494 | 4 | 48 | 22 | 7 | 11 | 14 | 47 | 34 |
| BPS005B | 86 | 2 | 538 | 2 | 5 | 19 | 16 | 100 | 35 | 472 | 236 |
| BPS006 | 2 | 14 | 368 | 11 | 27 | 65 | 16 | 77 | 22 | 224 | 169 |
| BPS006B | 1290 | 5 | 659 | 7 | 9 | 23 | 6 | 8 | 29 | 30 | 25 |
| BPS007 | 2 | 3 | 376 | 4 | 8 | 17 | 11 | 31 | 21 | 257 | 183 |
| BPS007B | 3 | 7 | 410 | 17 | 31 | 30 | 14 | 9 | 11 | 54 | 36 |
| Minimum | 2 | 1.71 | 368 | 2 | 5 | 17 | 6 | 8 | 11 | 30 | 25 |
| Maximum | 3278 | 14 | 719 | 29 | 48 | 210 | 43 | 263 | 88 | 1307 | 771 |
| **Mean** | **468.90** | **6.39** | **493.10** | **9.85** | **19.52** | **50.33** | **15.93** | **82** | **35.19** | **443** | **253** |
| Standard deviation | 1065.55 | 3.65 | 123.61 | 8.03 | 14.016 | 58.29 | 17.62 | 81.48 | 25.01 | 413 | 233 |

The REE content of studied stream sediment samples is ranging from 30 to 1307ppm Ce, 25 to 771 ppm La (Table

2). Cerium (Ce) and Lanthanum (La) have considerably high concentrations when compared with other Rare Earth

Elements (REE) analyzed (Table 2). Ce, La and Th    concentrations might indicate the presence of monazite.

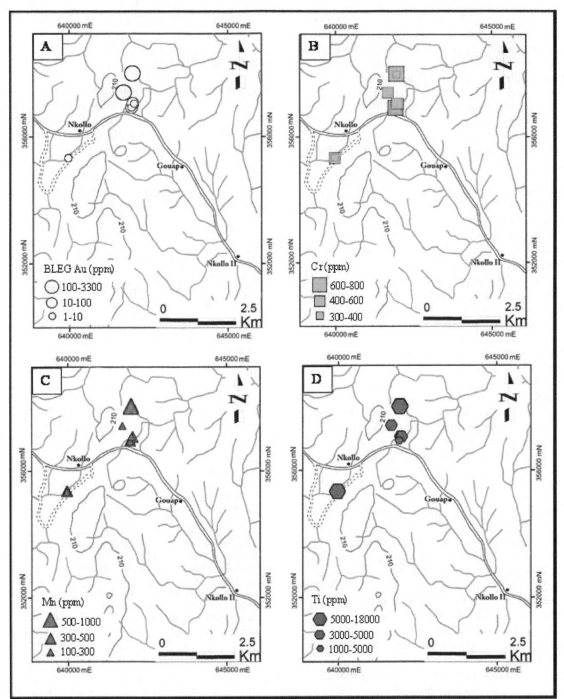

**Figure 6.** Graduated symbol plot for (a) Au, (b) Cr, (c) Mn and (d) Ti scores (ppm) superimposed on the drainage map of the Gouap-Nkollo prospect. High positive scores (large symbols) indicate locations with anomalous concentrations

## 4.3. Source of the Stream Sediment and Mineralization Potential

To enhance the data presentation and interpretation, we generated point symbol maps for Au, Cr, Mn, and Ti (Figure 6). The data for each element were then superimposed on a georeferenced drainage map of the study area. These plots show four elements (Au, Cr, Mn and Ti) with high values in the same sampling site, located at the northern of Nkollo village (Figures 6A, 6B, 6C and 6D). The sample with high Au values corresponds to that collected just some few meters down of the tourmaline-

quartz vein crosscutting the quartzite which is the main lithology of the area. Overall, the collected stream sediment samples present ferromagnesian, aluminosilicate minerals, precious and ferrous metals reflecting the lithological composition of the area comprising quartzite, gneiss, tourmaline bearing micaschist and iron formations [21]. So, the element distribution patterns and chemical composition of stream sediments of Gouap-Nkollo are greatly influenced by the local geology of the area. The geochemistry of the stream sediments has revealed that the source of sediments collected in the stream bed is the surrounding rocks. The recorded high concentrations of titanium probably indicative of ilmenite mineralization while the presence of manganese and chromium might indicate the potential of ferrous metals mineralization. Cerium, Lanthanum and Thorium concentrations are indices of monazite mineralizations. Fe is associated with Co, Mn, and Ti in BPS006 while Au is in relation with Ce, La and Th in BPS003. Ce concentration might indicate the presence of an unusual distinctive form of monazite [25], the cerium and lanthanum phosphate is the major commercial source of cerium.

## 4.4. Recommendation for Future Exploration Strategies

Owing to the high concentrations of Fe and Al as the dominant major elements, as well as high concentrations of Mn, P and Ce as the dominant trace and rare earth elements in the stream sediments analyzed from Gouap-Nkollo prospect, it is recommended that a more comprehensive survey of stream sediments should be carried out. However, the present result provides baseline geochemical information needed to carry out a more detailed investigation on the occurrence of heavy minerals in stream sediments of the region. Due to high concentrations of elements (major, trace and LREE) recorded in the northern part of the prospect, especially at the vicinity of Nkollo area, it is recommended that more detailed geophysical and geochemical surveys are needed in this area for possible mineralization of gold. We also recommend detailed geological mapping of the study prospect to evaluate the possible mineralization zones and potential mineable areas as this will help in determining the level, quantity and quality (tonnage) of mineralization in place.

## 5. Conclusion

A stream sediment survey was undertaken in the Gouap-Nkollo area. Considering the geochemical analysis of the stream sediments as well as its interpretation, the following conclusions can be drawn.

1. Fe and Al are the dominant major elements in the stream sediments. Fe indicates the possible occurrence of ferromagnesian minerals while Al indicates that the stream sediments are enriched in aluminosilicate minerals such as feldspar and mica.

2. High concentrations of titanium probably indicative of ilmenite mineralization while the presence of manganese and chrome might indicate the potential of ferrous metals mineralization. Cerium, Lanthanum and Thorium concentrations are indicators of monazite mineralization's.

3. Occurrences and indicators of Fe, Co, Mn, Ti, Au, Ce, La, Pb and Th mineralization's were discovered. Gold has some interesting concentrations which merit more detailed investigation.

4. The element distribution patterns and chemical composition of stream sediments of Gouap-Nkollo is greatly influenced by the local geology of the area and the geochemistry of the stream sediments originated from their surrounding rocks.

5. Future exploration work will be focused on the northern part of the Gouap-Nkollo prospect, where some anomalous concentrations of Au were observed.

## Aknowledgements

The data presented here form a part of the first author's Ph.D thesis supervised by J.P. Nzenti at the University of Yaoundé I, Cameroon. We gratefully acknowledge the mineral exploration company 'G-Stones Resources SA' for their permission to work within their permit area and for financial support for geochemical analysis. We thank the anonymous reviewers for their critical and constructive comments of the manuscript. This is the contribution to ICGP-Y 616 project.

## References

[1] Marjoribanks, R. Geological Methods in Mineral Exploration and Mining. *Second Edition, Springer-Verlag Berlin Heidelberg*, 2010.

[2] Plumlee, G.S. The environmental geology of mineral deposits. *In:* PLUMLEE, G.S. & M.J. LOGSDON (Eds), The Environmental Geochemistry of Mineral Deposits, Part A. Processes,Techniques, and Health Issues, *Society of Economic Geologists, Reviews in Economic Geology*, 6A, 71-116, 1999.

[3] Atsuyuki, O., Noboru, I., Shigeru, T., Yoshiko, T., 2005. Influence of surface geology and mineral deposits on the spatial distributions of element concentrations in the stream sediments of Hokkaido Japan. Journal of Geochemical Exploration, 86, 86-103, 2005.

[4] Robb, L. Introduction to ore-forming processes. *Blackwell Publishing*, ISBN 0-632-06378-5 p373 pp., 2005.

[5] Baume,r A., and Fraser, R.B. Panguna porphyry copper deposit, Papua New Guinea. In: CLKnight (ed) Economic geology of Australia and Papua New Guinea I – Metals. *Australasian Institute of Mining and Metallurgy, Melbourne*, 855-866, 1975.

[6] Embui, V.F., Omang, B. O., Che, V. B., Nforba, M.T. 4, Suh C. E. Gold grade variation and stream sediment geochemistry of the Vaimba-Lidi drainage system, northern Cameroon (West Africa). *Natural Science*, (5)2A, 282-290, 2013.

[7] Suh, C.E., Lehmann, B. and Mafany, G.T. Geology and geochemical aspects of lode gold mineralization at Dimako—Mboscorro, SE Cameroon. *Geochemistry: Exploration, Environment, Analysis*, 6, 295-309, 2006.

[8] King, R.W. Geochemical characteristics of tourmaline from superior province Archaean lode-gold deposits: implications for source regions and processes. In: Bicentennial Gold '88. *Geological Society Australia Abstract Series*, 2, 445-447, 1988.

[9] Dommanget, A., Milési, J.P. and Diallo, M. The Loulo gold and tourmaline-bearing deposit: A polymorph type in the Early Proterozoic of Mali (West Africa). *Mineralium Deposita* 28, 253-263, 1993.

[10] Anglin, C.D., Jonasson, I.R. and Franklin, J.M. 1996. Sm–Nd dating of scheelite and tourmaline: implications for the genesis of Archean gold deposits, Val d'Or, Canada. *Economic Geoogy*, 91, 1372-1382, 1996.

[11] Deksissa, D.J., and Koeberl, C. Geochemistry and petrography of gold-quartz-tourmaline veins of the Okote area, southern Ethiopia: implications for gold exploration. *Mineralogy and Petrology, 75, 101-122*, 2002.

[12] Baksheev, I.A., Prokof'ev, V.Y., Yapaskurt, V.O., Vigasina, M.F., Zorina, L.D. and Solov'ev, V.N. Ferric-iron-rich tourmaline from the Darasun gold deposit, Transbaikalia, Russia. *Canadian Mineralogist*, 49, 263-276, 2011.

[13] Tornos, F., Wiedenbeck M., and Velasco, F. The boron isotope geochemistry of tourmaline-rich alteration in the IOCG systems of northern Chile: implications for a magmatic-hydrothermal origin. *Mineralium Deposita 47, 483-499*, 2012.

[14] Feybesse, J.L., Johan, V., Maurizot, P. and Abessolo, A. Evolution tectono métamorphique libérienne et éburnéenne de la partie NW du Craton Zaïrois (SW Cameroun). *In G. Matheis and H. Schandelmeier (Editors), Current research in African Earth Sciences. Balkema, Rotterdam*, 9-12, 1987.

[15] Toteu, S.M., Van Schmus, W.R., Penaye, J., and Nyobé, J.B. U-Pb and Sm-Nd evidence of eburnean and pan African high grade metamorphism in Cratonic rock of southern Cameroon. *Precambrian Research, 67, 321-347*, 1994.

[16] Van Schmus, W.R., Toteu, S.F. Were the Congo craton and the Sào Francisco craton joined during the fusion of Gondwanaland? *Eostrans AGU, 73(14),* Spring Meeting, Supplement p. 365, 1992.

[17] Penaye, J., Toteu, S.F., Michard, A., Van Schmus, W.R. and Nzenti, J.P. U/Pb and Sm/Nd preliminary geochronologic data on the Yaoundé series, Cameroon: reinterpretation of granulitic rock as the suture of the collision in the « Centrafricain » belt. *Comptes Rendus de l'Académie des Sciences, Paris,* 317, 789-794, 1993.

[18] Lerouge, C., Cocherie, A., Toteu, S.F., Penaye, J., Milesi, J.P., Tchameni, R., Nsifa, N.E., Fanning, C.M. and Deloule, E. SHRIMP U/Pb zircon age evidence for paleoproterozoic sedimentation and 2.05Ga syntectonic plutonism in the Nyong Group, South-western Cameroon: consequences for the eburnean-transamazonian belt of NE Brasil and central Africa. *Journal of African Earth Sciences*, 44, 413-427, 2006.

[19] Lasserre, M., Soba, D. Age Libérien des granodiorites et des gneiss à pyroxènes du Cameroun Méridional. *Bulletin BRGM 2(4)*, 17-32, 1976.

[20] Maurizot, P., Abessolo, A, Feybesse, J.L., Johan V. and Lecomte P. Etude et prospection minière du Sud-Ouest Cameroun. Synthèse des travaux de 1978 à 1985. *Rapport BRGM*, Orléans 85, CMR 066, 274 pp, 1986.

[21] Sikaping, S. Métamorphisme et minéralisations associées dans le secteur de Gouap-Nkollo (Région du Sud). Unpublished Master thesis, University of Yaoundé 1, 77p., 2012.

[22] Soh Tamehe, L. Roches à tourmaline et prospection alluvionnaire de l'or à Nkollo (Région du Sud Cameroun). Unpublished Master thesis, University of Yaoundé 1, 82p., 2013.

[23] Ganno, S. Gouap prospect: iron and gold mineralization potentials. *Technical report*, pp15, 2012.

[24] Hoffman, E.L., Clark, J.R. and Yeager, J.R. Gold Analysis – Fire Assaying and Alternative Methods. *Exploration and Mining Geology*, 7(1, 2), 155-160, 1998.

[25] Basham, I.R. and T.K. Smith. On the occurrence of an unusual form of monazite in panned stream sediments in Wales. *Geology Journal*, 18, 121-127, 1983.

# Economic Feasibility Study of Hard rock Extraction Using Quarry Mining Method at Companiganj Upazila in Sylhet District, Bangladesh

**Mohammad Kashem Hossen Chowdhury***, **Md. Ashraful Islam Khan, Mir Raisul Islam**

Department of Petroleum and Mining Engineering, Shahjalal University of Science and Technology, Sylhet, Bangladesh
*Corresponding author: kashem.hossen@gmail.com

**Abstract** Companiganj is one of the resourceful upazila of Sylhet district, though no gas or oil field have not been found yet, renowned for its hard rock. Hard rock, one of the main geo-resources of Bangladesh after gas and coal, is very useful in construction sector and cement industry. The source of this rock is near Meghalaya of India. This part of India is relatively in higher than the ground level of Bangladesh. That's why the rock comes down by gravity especially in the rainy season. The water is the main transportation media which brings these rocks. People normally use manual hand tools for rock extraction and use two type of crusher to crush these rocks. In this research paper, we tried to find the economic feasibility of extracting these rocks. Though the extraction method of this quarry is not technologically advanced, but after doing this research, we can say that this quarry is economically feasible.

*Keywords: companigonj, Sylhet, Geo-resources, hard rock, crusher, economic feasibility*

## 1. Introduction

The feasibility Study is an exercise that involves documenting each of the potential solutions to a particular business problem or opportunity. Feasibility studies aim to objectively and rationally uncover the strengths and weaknesses of the existing business or proposed venture, opportunities and threats as presented by the environment, the resources required to carry through, and ultimately the prospects for success [1]. In this paper we will discuss about the economic feasibility of hard rock quarry at Companiganj Upazilla in Sylhet District. The purpose of the economic feasibility assessment is to determine the positive economic benefits to the quarry mining. This assessment typically involves a cost benefits analysis.

### 1.1. Geology and Geography of Companiganj

Geologically, this region is complex having diverse geomorphology; high topography of Plio-Miocene age such as Khasi and Jaintia hills and small hillocks along the border. At the centre there is a vast low laying flood plain, locally called Haors. Available limestone deposits in different parts of the region suggest that the whole area was under the ocean in the Oligo-Miocene [2].

Companiganj is located at 25°04′45″N91°45′15″E-25.0791°N 91.7542°E. It has 13620 units of house hold and total area 278.55 km² is bounded by Meghalaya (state of India) on the north, Sylhet sadar on the south, Gowainghat upazila on the east, Chatak upazila on the west. Main rivers are Surma, Piyain. Notable Haors: Baors, Pokohair. Notable Beels: Panichapara, Nagar, Rauti and Kalenga [3].

### 1.2. Some Features of Companiganj Upazila

Companiganj (Town) consists only of one mouza. It has an area of 9.21 km². It has a population of 6365; male 54.12%, female 45.88%; density of population is 691 per km². Literacy rate among the town people is 15.2%.

**Administration:** A police outpost was established at Companiganj in 1976 which was upgraded to an upazila in 1983. The upazila consists of three union parishads, 74 mouzas and 131 villages.

**Religious institutions:** Mosque 130, temple 4, tomb 3.

**Population:** 85169; male 51.86%, female 48.14%; Muslim 91.16%, Hindu 8.73% and ethnic people 0.1% (Monipuri).

**Literacy and educational institutions:** Average literacy 43.05%; male 50.6%, female 35.5%. Educational Institutions: College 1, Secondary School 7, Junior High School 3, Primary School (Government) 32, Primary School (Non-government) 27, Community School 10, Satellite School 6, Madrasa 11. Vatrai High School (1957), Parua Noagaon Dakhil Madrasa (1965) are notable educational institutions.

**Cultural organisations:** Rural club 14, playground 5, Women's organization 12.

**Main occupations:** Agriculture 45.87%, Agricultural Laborer 16.54%, wage laborer 9.58%, Fishing 4.32%, Commerce 7.01%, Service 1.72%, others 14.96%.

**Land use:** Arable land 29676.40 hectares, Khas land 999.84 hectares.

**Land control:** Among the peasants 51% are landless, 16% marginal, 25% intermediate and 8% rich.

**Value of land:** Market value of the first grade agricultural land is about Tk. 2500 per 0.01 hectare.

**Main crops:** Paddy, mustard, cassia leaf, betel leaf and betel nut.

**Extinct and nearly extinct crops:** Tobacco, ganja (hemp) is entirely extinct.

**Main fruits:** Orange, jackfruit, mango, lemon, pineapple, satkora.

**Fisheries, dairies, poultries:** Fishery 15, dairy 20, poultry 35.

**Communication facilities:** Roads: pucca 41.15 km and mud road 160 km.

**Traditional transport:** palanquin and bullock cart. These means of transport are extinct of nearly extinct.

**Cottage Industries:** Weaving 20, Goldsmith 40, Blacksmith 30, Potteries 80, Welding 3.

**Hats, bazaars and fairs:** Hats and bazaars are 14, fair 3; Companiganj Bazaar, Islamganj Bazaar, Tuker Bazaar, Parua Bazaar are notable.

**Main Exports:** Orange, fish, stone (boulder and pebble), limestone.

**NGO Activities:** CARE, BRAC, Grameen Bank and Simantik.

**Health Centre:** Upazila health complex 1, satellite clinic 1, family planning center 3 [4].

## 1.3. Basic information of Rocks in Companiganj

Hard rock a term used loosely for igneous and metamorphic rock, as distinguished from sedimentary rock. Hardrocks in Bangladesh are of four types. (i) Maddhyapara subsurface hard rock (ii) Bholaganj-Jaflong hard rock concretions (Companiganj) (iii) Tetulia-Patgram-Panchagar hard rock concretions and (iv) Chittagong hilly track sedimentary concretions. The terms (ii), (iii) and (iv) are usually considered as gravel deposits. The Bholaganj (under Companiganj Upazilla) hard rock project is approximately 850 km$^2$. The hard rock is mined following the open pit technique. The worker extracts hard rock by using their hand operating tools. In so far as the Dupitila formation, this immediately overlies the hard rocks in the region. The hard rocks are to be extracted from a depth of 2.5 meter to 10meter below the surface [5].

The Sona Tila gravel bed is equivalent to the lower Pleistocene series and belongs to the Madhupur clay formation while the Bholaganj gravel bed is equivalent to the upper Pleistocene to Holocene series. Similarly, the former is weathered and the latter is fresh, hard and high quality derived from the Khasi-Jaintia hill ranges. The gravels of both beds are of igneous and metamorphic origins. They have high sphericity and roundness values and as such suggest long transportation and long time abrasion of the gravel sediment. They are made of river borne deposit [6].

## 1.4. Quarry Mining

A quarry mine is a type of open-pit mine from which rock or minerals are extracted. Quarries are generally used for extracting building materials, such as dimension stone, construction aggregate, riprap, sand, and gravel. They are often collocated with concrete and asphalt plants due to the requirement for large amounts of aggregate in those materials. The word quarry can include underground quarrying for stone, such as Bath stone.

Quarries in level areas with shallow groundwater or which are located close to surface water often have engineering problems with drainage. Generally the water is removed by pumping while the quarry is operational, but for high inflows more complex approaches may be required. For example, the Coquina quarry is excavated to more than 60 ft (18 meter) below sea level. To reduce surface leakage, a moat lined with clay was constructed around the entire quarry. Ground water entering the pit is pumped up into the moat. As a quarry becomes deeper water inflows generally increase and it also becomes more expensive to lift the water higher during removal - this can become the limiting factor in quarry depth. Some water-filled quarries are worked from beneath the water, by dredging [6].

## 1.5. Crusher

A crusher is a machine designed to reduce large rocks into smaller rocks, gravel, or rock dust. Crushers may be used to reduce the size, or change the form, of waste materials so they can be more easily disposed of or recycled, or to reduce the size of a solid mix of raw materials (as in rock ore), so that pieces of different composition can be differentiated. Crushing is the process of transferring a force amplified by mechanical advantage through a material made of molecules that bond together more strongly, and resist deformation more, than those in the material being crushed do. Crushing devices hold material between two parallel or tangent solid surfaces, and apply sufficient force to bring the surfaces together to generate enough energy within the material being crushed so that its molecules separate from (fracturing), or change alignment in relation to (deformation), each other. The earliest crushers were hand- held stones, where the weight of the stone provided a boost to muscle power, used against a stone anvil. Querns and mortars are types of these crushing devices [7].

There are two types of crushers (small and large) are found in Bholaganj area. Small size crusher are named "Tom tom" by local people.

## 2. Materials and Methods

In order to conduct this study, steps like field investigation, data collection, analysis, economic consideration are done. Field investigation includes site visiting, conversation with Upazila Nirbahi Officer, visiting different quarry, visiting different type of crusher and data collection from small and large crusher. Data collected based on survey and questionnaire by visiting the location several time.

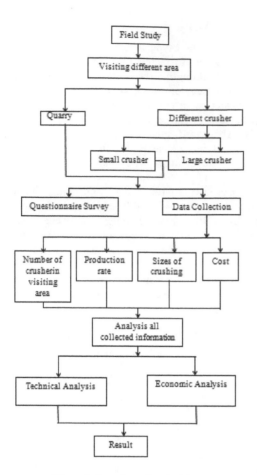

**Figure 1.** Flow chart of methodology

## 3. Data Collection

Data mainly collected from three locations-Companigonj Bajar, Volagonj and Konabari.

**Companigonj Bajar:** There were 40 small crushers and 75 large crushers in the Companiganj area. We observed most of the area and collect data from 2 small crushers and 2 large crushers.

**Bholaganj:** There were 80 small crushers and 110 large crushers in the Bholaganj area. We observed and collected data from 3 small crushers and 2 large crushers.

**Kolabari:** There were 40 small crushers and 60 large crushers in the Kolabari area. We observed and collected data from 2 small crushers and 2 large crushers.

There are approximately 350 large size crushers in the project area. Data are collected from six large crushers at different place of the project area. They are P&S Stone Crusher, Borak Stone Crusher, Akash Stone Crusher, S.M Stone Crusher and Surma Trading and Kashem Stone Crusher.

There were approximately 160 small crushers in the project area. Data was collected from local businessmen or owner of small crusher. Few data were collected from the workers. There were some business firm from which we collected data and they are Gazi Trading, Bangla Trading, S.Alam Trading, Zia Trading and Protik Stone Crushing.

Local businessmen gather hard rock by using day laborer or worker. Then they crush different size rock and sell them to local customer. The local customers collect

crushing rock and send them to different part of the country.

**Table 1. Collected data from the location**

| Data Collecting Point | Total Number of Crusher | | Number of visited crusher | |
|---|---|---|---|---|
| | No. of small crusher | No. of large crusher | small crusher | large crusher |
| Companiganj | 40 | 75 | 3 | 2 |
| Bholaganj | 80 | 110 | 3 | 2 |
| Kolabari | 40 | 60 | 2 | 2 |

**Production of Small Size Crusher:**

The production of a small size crusher is 700-800 $ft^3$/day. The number of total small size crushers is 160, so according to this the total production by small size crushers are 45000 $ft^3$/day and the annual production of crushed rock is $1.65 \times 10^7$ $ft^3$.

**Production of Large Size Crusher:**

The production of a large size crusher is 2500-3000 $ft^3$/day. The total number of large size crushers is 350, so according to this the total production by large size crusher is 962500 $ft^3$/day.

**Selling price of crushed rock using small crusher:**

Hard rock is crushed into different required size. Their prices are also varying with their size and also with the size of crusher. The most required size is 3/4. The size 3/4(boulder) are most costly than others. The cost of 3/4 (boulder) is 0.84-0.85 USD/$ft^3$ and the cost of 3/4 (bhuto) is 0.70-0.72 USD/$ft^3$. The size 1/2 and 5/10 are residual and their price are less than 3/4 sizes. The cost of 1/2 size is 0.57-0.58 USD/$ft^3$. The price of 5/10 size are 0.28-0.30 USD/$ft^3$ which are less costly than other size. The prices are also varying with cost of raw material.

**Selling price of crushed rock using large crusher:**

The cost of 3/4 (boulder) is 0.86-0.90 USD/$ft^3$ and the cost of 3/4 (bhuto) is 0.73-0.80 USD/$ft^3$. The size 1/2 and 5/10 are residual and their prices are less than 3/4 size. The cost of 1/2 size is 61-0.64 USD/$ft^3$. The price of 5/10 size is 0.28-0.30 USD/$ft^3$ which is less costly than other size.

**Investment cost of crusher:**

Investment cost is the cost to set up the machinery or corresponding material during the early life of an industry. 20616 USD is required to set up the small size crusher and 47674.5 USD is required to set up the large size crusher. All the crushers are imported from the China.

**Table 2. Total Production of crusher**

| Crusher type | Total Production | |
|---|---|---|
| | Per day ($ft^3$) | Per year ($ft^3$) |
| Small crusher | 45,000 | $1.64 \times 10^7$ |
| Large crusher | 9,62,500 | $3.51 \times 10^8$ |

**Table 3. Labor cost per crusher**

| Crusher | Labor consumption/crusher | Labor cost per day/crusher (USD) |
|---|---|---|
| Small crusher | 8 | 30.92 |
| Large crusher | 22 | 85.04 |

**Table 4. Selling price of crushed rock from small crusher**

| Crushed type | Size of crushed rock | Selling price (USD) |
|---|---|---|
| Small size crusher | 3/4 (Bhuto) | 0.70-0.72 |
| | 3/4 (Boulder) | 0.84-0.85 |
| | ½ | 0.57-0.58 |
| | 5/10 | 0.28-0.30 |

**Table 5. Selling price of crushed rock from large crusher**

| Crushed type | Size of crushed rock | Selling price (USD) |
|---|---|---|
| Large size crusher | 3/4 (Bhuto) | 0.73-0.80 |
| | 3/4(Boulder) | 0.86-0.90 |
| | ½ | 0.61-0.64 |
| | 5/10 | 0.28-0.30 |

**Table 6. Cost of raw material of a crusher**

| Crushed type | Amount of rock bought by a crusher/ft$^3$ | Total cost of raw material* (USD) |
|---|---|---|
| Small crusher | 1325 | 256.09 |
| Large crusher | 3975 | 768.27 |

\* cost of raw material 0.19 USD/ft$^3$

**Table 7. Cost scenario per crusher**

| Crusher type | Investment cost per crusher | | Total income per crusher per year(USD) |
|---|---|---|---|
| | Set up cost (USD) | Operating cost* (USD)/day | |
| Small crusher | 20616 | 38.66 | 21260.25 |
| Large crusher | 59271 | 118.54 | 59271 |

\* Operating cost includes labor cost and fuel consumption by a crusher

**Table 8. Total profit from crusher**

| Crusher type | Profit per year (USD) |
|---|---|
| Small crusher | 7086.75 |
| Large crusher | 15977.4 |

\* Profit = Total income - operating cost

## 4. Result

### For Small crusher
Return on investment [%] = (Income-Cost) / Total cost [7]

= ( 7086.75/14173.5) × 100

= 50% per year

### For Large crusher
Return on investment [%] = (Income-Cost) / Total cost [7]

= (15977.4/43293.6) × 100

= 36.9% per year

## 5. Discussion

From the above calculation we can see that a small crusher is beneficiary with respect to a large crusher. But net income in a large crusher is more than twice of a small crusher.

It can be noticeable that when raw materials are crushed into different size and shape then their selling price increase because it's public demands is much more than it 's original size and shape which is found in quarry and consequently in this regard the efficiency of investment (return on investment) increased.

So, from this study we can say that –

This rock extraction project is economically profitable. And the benefits from selling the rock worth the cost.

## 6. Conclusion

Hard rock is known as the building material which usually used in construction. Bholaganj is one of the main sources of hard rock and are used in construction all over the Bangladesh. But no appropriate engineering technology is used here to extract this hard rock. Local people are extracting this hard rock by using hand operating tools. At least 9,000 people including 3000 women and 1000 children is working as stone laborer, on the bank of the Dholai River, in Bholaganj, Companianj, Sylhet. The average income of the stone laborers is less than 150 taka per day. Stone extraction goes on in the area for about eight months a year, except the rainy season. On an average 300 truck load of stones are sent to Sylhet and other parts of Bangladesh every day. Based on this, local people get involve in rock business and crushing business. Maximum labors in crusher mills and workers who working in quarry for extraction purposes are local people of Sylhet district. Not only local people but also people of other districts involve here. An unemployed people can get involve here easily. So it is clearly visible that a great working place has been created here. These rocks of Companiganj are assets of local area of Bangladesh. Future study is required to extract these economically valuable rocks by environment friendly way using the modern mining technology.

## Acknowledgements

The authors would like to thank Almighty for empowering and guiding throughout the completion of this research paper. The authors would also like to thank the teachers of the Department of Petroleum and Mining Engineering, Shahjalal University of Science and Technology and the local community of this research area for their support.

## References

[1] Georgakellos, D. A. &Marcis, A. M. (2009). Application of the semantic learning approach in the feasibility studies preparation training process. Information Systems Management 26 (3) 231-240.

[2] Imam, B. (2005) "*Energy resources of Bangladesh*", University Grant commission of Bangladesh Dhaka, pp. 19-37.

[3] <http://www.lged.gov.bd/ViewMap2.aspx?DistrictID=61>(LGED MAP) (Accessed Dec 08, 2012).

[4] The Bangladesh bureau of Statistics website [online], http://www.bbs.gov.bd.2010. (Accessed Dec 08, 2012).

[5] Chowdhury, B.S., 1994, Mining and Quarrying sector poised to take off: *The Daily Star*, Dhaka, April 18, 1994

[6] Encyclopedia of Bangladesh Sylhet region <http://encyclopedia.com> (Accessed Dec 08, 2012).

[7] Barry A. Wills, Tim Napier-Munn, "*Mineral Processing Technology*", (2009) seventh edition, Elsevier Science & Technology Books.

# Empirical Evaluation of Slag Cement Minimum Setting Time (SCMST) by Optimization of Gypsum Addition to Foundry Slag during Production

C. I. Nwoye[1,*], I. Obuekwe[1,2], C. N. Mbah[3], S. E. Ede[3], C. C. Nwangwu[1], D. D. Abubakar[1,4]

[1]Department of Metallurgical and Materials Engineering, Nnamdi Azikiwe University, Awka, Nigeria
[2]Scientific Equipment Development Institute, Enugu, Nigeria
[3]Department of Metallurgical and Materials Engineering, Enugu State University of Science & Technology, Enugu, Nigeria
[4]Ajaokuta Steel Company, Kogi State, Nigeria
*Corresponding author: nwoyennike@yahoo.com

**Abstract** An empirical evaluation of slag cement minimum setting time (SCMST) has been successfully carried out through optimization of gypsum addition to foundry slag in the course of the cement production. A model was derived and used as a tool for predictive analysis of the cement setting time based on gypsum input. The model aided optimization of gypsum addition indicates a minimum setting time of 14.1054 minutes at an optimum gypsum input concentration of 6.4847%. Beyond 6.4847% gypsum addition, the slag cement setting time increases drastically; a situation typifying immiscibility and lack of homogeneity between the cement slurry and the extra gypsum addition. This is because increased gypsum addition (above a specified quantity) is unlikely to forms a coherent mass with a specified and fixed liquid volume, resulting to delayed and differential setting. The derived model expressed as;

$$\gamma = 0.7168\,\alpha^2 - 9.2965\,\alpha + 44.2478$$

is quadratic and single factorial in nature. The slag cement setting time per unit gypsum addition are as obtained from experiment and derived model are 4.75 and 4.996 mins./% respectively. Statistical analysis of the experimental and derived model-predicted results for each value of the gypsum input concentration considered shows standard errors of 1.8974 and 1.5485% respectively. Deviational analysis indicates 10.46% as the maximum deviation of the model-predicted slag cement setting time from the corresponding experimental value. The validity of the model was rooted on the expression $0.0226\,\gamma + 0.2101\,\alpha = 0.0162\,\alpha2 + 1.0001$ where both sides of the expression are correspondingly approximately equal. The validity of the derived model-predicted results also was ascertained using SPSS 17.0. The results indicate check variance = 0.001, standard deviation = 0, and model operational confidence of 95.0% at a significant level: 0.05.

*Keywords:* evaluation, optimization, gypsum addition, slag cement setting time

## 1. Introduction

The growing need to support research and development aimed at improving the quality of cement produced through incorporation of industrial wastes and cheaper locally acquired material stems on the fact that cement plays a very important and unavoidable role in the building and construction sector. There is also need to evaluate other materials to produce materials of similar characteristics to cement.

The need to satisfy specification on application of cement has influenced the addition of materials in the course of cement production that would have the necessary performance characteristics [1]. As it concerns health and environment, it should be inoffensive. Finally, the incorporation of the waste should not impair concrete durability. Traditional assessment methods must therefore be adopted to evaluate these new materials [2]. The high content of soluble salt (Sugar) in ash has identified it as experimental waste [3]. The environmental impact and durability of concrete are closely connected to its transport properties, which control the kinetics of the penetration of water and aggressive agents into concrete [4]. Concrete diffusivity is practically linked to the movement of chemical species within the material and the leaching of certain chemicals [5].

The essential components of blast furnace slag has been discovered [6] to be of same oxide (such as lime, silica, and aluminum) as are present in Portland cement, but differ in proportion. Blast furnace slag is a by-product obtained in the manufacture of pig iron in the blast furnace. It has found application in the road and building industries and in the production of cementing materials, as an aggregate in concrete and in the manufacture of slag wool for thermal insulation.

The aim of this work is to evaluate empirically the slag cement minimum setting time (SCMST) through optimization of gypsum addition to foundry slag in the course of the cement production. A model would be derived and used as a tool for predictive analysis of the cement setting time based on gypsum input.

## 2. Materials and Methods

Materials used and their respective sources and details of the experimental procedure and associated process conditions are as stated in the past report [6].

### 2.1. Model Formulation

Experimental data obtained from research work [6] were used for this work. Computational analysis of the experimental data [6] shown in Table 1, gave rise to Table 2 which indicate that;

$$e1\gamma + e2\alpha = e3\alpha2 + e4 \text{ (approximately)} \qquad (1)$$

Introducing the values of the empirical constants $e_1$, $e_2$, $e_3$ and $e_4$ into equation (1) reduces it to;

$$0.0226 \ \gamma + 0.2101 \ \alpha = 0.0162 \ \alpha2 + 1.0001 \qquad (2)$$

$$0.0226 \gamma = 0.0162 \ \alpha^2 - 0.2101 \ \alpha + 1.0001 \qquad (3)$$

$$\gamma = 0.7168 \ \alpha^2 - 9.2965 \ \alpha + 44.2478 \qquad (4)$$

Where

$(\gamma)$ = Setting time (mins.)

$e_1$ = 0.0226; Empirical constant (determined using C-NIKBRAN [7]

$(\alpha)$ = Concentration of gypsum added (%)

$e_2$ = 0.2101; Empirical constant (determined using C-NIKBRAN [7])

$e_3$ = 0.0162; Empirical constant (determined using C-NIKBRAN [7])

$e_4$= 1.0001; Slag grade (determined using C-NIKBRAN [7])

During the model development, a soft ware; C-NIKBRAN was used to map-link parameters such setting time $\gamma$, and concentration of gypsum addition $\alpha$, with the view to establishing an empirical relationship between $\gamma$ and $\alpha$. Map-linking was carried out using experimental results in Table 1. The result of the data mapping is the core model expression in equation (2), which on evaluation gives Table 2.

Table 1. Variation of setting time with the concentration of gypsum added [6]

| $(\gamma)$ | $(\alpha)$ |
|---|---|
| 35 | 1 |
| 30 | 2 |
| 22 | 3 |
| 18 | 4 |
| 16 | 5 |

## 3. Boundary and Initial Condition

Consider foundry slag paste of required consistency (in a cylindrical container) interacting with quantities of gypsum. The container's atmosphere was not contaminated i.e (free of unwanted gases, dusts and other micro organisms). Range of gypsum addition: 1-5%. Range of setting time considered (based on added gypsum): 16 - 35 mins., mass of wastes used, resident time, treatment temperature, and other process conditions are as stated in the experimental technique [6].

The boundary conditions are: free movement of oxygen across the cylindrical container. At the bottom of the particles, a zero gradient for the gas scalar are assumed and also for the gas phase at the top of the slag paste. The treated slag paste was stationary. The sides of the waste particles are taken to be symmetries.

## 4. Results and Discussion

The derived model is equation (4). The computational analysis of Table 1 gave rise to Table 2.

Table 2. Variation of 0.0226 $\gamma$ + 0.2101$\alpha$ with 0.0162 $\alpha^2$ + 1.0001

| 0.0226 $\gamma$ + 0.2101 $\alpha$ | 0.0162 $\alpha^2$ + 1.0001 |
|---|---|
| 1.0011 | 1.0162 |
| 1.0982 | 1.0648 |
| 1.1275 | 1.1458 |
| 1.2474 | 1.2592 |
| 1.4121 | 1.4050 |

### 4.1. Model Validation

The validity of the model is strongly rooted on equation (2) where both sides of the equation are correspondingly approximately equal to 1. Table 2 also agrees with equation (2) following the values of 0.0226 $\gamma$ + 0.2101$\alpha$ and 0.0162 $\alpha^2$ + 1.0001 evaluated from the experimental results in Table 1. Furthermore, the derived model was validated by comparing the setting time (of the slag cement) predicted by the model and that obtained from the experiment [6]. This was done using the 4th Degree Model Validity Test Techniques (4th DMVTT); computational, graphical, statistical and deviational analysis [8].

The empirical model derived for evaluating the slag cement minimum setting time $\gamma$ relates $\gamma$ with $\alpha$ such that slag cement minimum setting time could be determined (as model out-put) at any point in time without further experimental work by substituting the highlighted input parameter $\alpha$ (concentration of gypsum addition) into the derived model, providing the input parameter is within the range of values generated during the initial experimental work under the initial experimental conditions (Boundary and initial conditions).

Based on the forgoing, substitution of the model input parameter $\alpha$ into the derived model, outside these boundary and initial conditions: gypsum addition: 1-5%, setting time considered: (based on gypsum addition): 16 - 35 mins, makes the predicted slag cement minimum setting time unreliable. The initial condition is rooted on the condition that the foundry slag should be in paste form and the atmosphere inherent in the processing container must not be contaminated i.e (free of unwanted gases, dusts and other micro organisms).

The limitation of the derived model is that it cannot operate outside the highlighted boundary and initial conditions of the initial experimental work. The derived model could be improved on by adding a correction factor to the derived model expression so as to take care of the prevailing experimental conditions, which play vital roles during the experiment, but were not considered during model formulation.

### 4.1.1. Computational Analysis

A critical comparative analysis of the results computed from experimental and model-predicted setting time was carried out to ascertain the degree of validity of the derived model. This was done by comparing setting time per unit gypsum addition obtained by calculations involving experimental results, and model-predicted results obtained directly from the model.

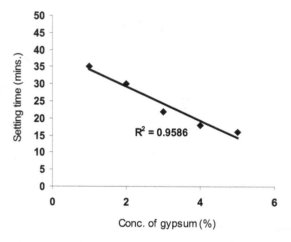

**Figure 1.** Coefficient of determination between setting time and concentration of gypsum added as obtained from the experiment [6]

Setting time of slag cement per unit gypsum addition $\gamma_g$ (mins./%) was calculated from the equation;

$$\gamma_g = \gamma \,/\, \alpha \tag{5}$$

Therefore, a plot of setting time against concentration of gypsum addition as in Figure 1 using experimental results in Table 1, gives a slope, S at points (1, 35) and (5, 16) following their substitution into the mathematical expression;

$$\gamma_g = S = \Delta\gamma \,/\Delta\alpha \tag{6}$$

Equation (6) is detailed as

$$S = \gamma_2 - \gamma_1 /\, \alpha_2 - \alpha_1 \tag{7}$$

Where

$\Delta\gamma$ = Change in the setting time $\gamma_2$, $\gamma_1$ at two gypsum concentrations $\alpha_2$, $\alpha_1$. Considering the points (1, 35) and (5, 16) for ($\alpha_1$, $\gamma_1$) and ($\alpha_2$, $\gamma_2$) respectively, and substituting them into equation (7), gives the slope as 4.75 mins./% which is the slag cement setting time per unit gypsum addition during the actual experimental process. Also similar plot (as in Figure 2) using model-predicted results gives a slope. Considering points (1, 35.6681) and (5, 15.6853) for ($\alpha_1$, $\gamma_1$) and ($\alpha_2$, $\gamma_2$) respectively and substituting them into equation (7) gives the value of slope, S as 4.996 mins./%. This is the model-predicted slag cement setting time per unit gypsum addition. A

comparison of these two values of the slag cement setting time per unit gypsum addition shows proximate agreement.

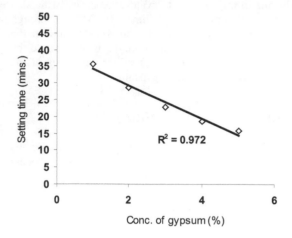

**Figure 2.** Coefficient of determination between setting time and concentration of gypsum input as predicted by derived model

## 4.2. Statistical Analysis

### 4.2.1. Standard Error (STEYX)

The standard errors (STEYX) in predicting the setting time (using results from derived model, and experiment) for each value of the concentration of added gypsum are 1.5485 and 1.8974% respectively. The standard error was evaluated using Microsoft Excel version 2003.

### 4.2.2. Correlation (CORREL)

The correlations between setting time and gypsum input concentration as obtained from derived model and experiment considering the coefficient of determination $R^2$ from Figure 1 and Figure 2 was calculated using Microsoft Excel version 2003.

$$R = \sqrt{R^2} \tag{8}$$

The evaluations correlations are shown in Table 3. These evaluated results indicate that the derived model predictions are significantly reliable and hence valid considering its proximate agreement with results from actual experiment.

**Table 3. Comparison of the correlations evaluated from D-Model predicted and ExD results based on gypsum input concentration**

| Analysis | Based on gypsum input concentration | |
|---|---|---|
| | ExD | D-Model |
| CORREL | 0.9791 | 0.9859 |

## 4.3. Graphical Analysis

Comparative graphical analysis of Figure 3 shows very close alignment of the curves from the experimental (ExD) and model-predicted (MoD) slag cement setting time relative to gypsum addition respectively. Furthermore, the degree of alignment of these curves is indicative of the proximate agreement between both experimental and model-predicted setting time.

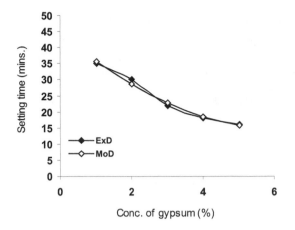

**Figure 3.** Comparison of the setting times relative to concentration of gypsum input as obtained from experiment [6] and derived model.

## 4.4. Comparison of Derived Model with Standard Model

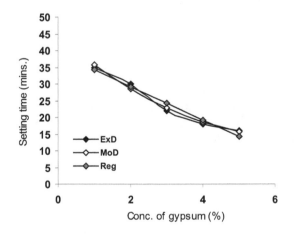

**Figure 4.** Comparison of the setting time relative to concentration of gypsum input as obtained from experiment (ExD) [6], derived model (MoD) and regression model (Reg).

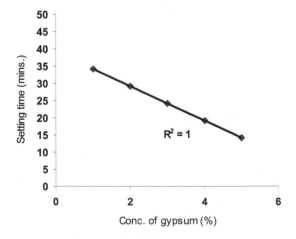

**Figure 5.** Coefficient of determination between setting time and concentration of gypsum added as predicted by regression model

The validity of the derived model was further verified through application of the regression model (Reg) (Least Square Method using Excel version 2003) in predicting the trend of the experimental results. Comparative analysis

of Figure 3 and Figure 4 shows very close alignment of curves and significantly similar trend of data point's distribution for experimental (ExD) and derived model-predicted (MoD) results of setting time. Figure 5 shows that the calculated correlations between setting time and gypsum input concentration for results obtained from regression model gives 1.000. The slag cement setting time per unit gypsum addition as obtained from regression model is 5.0 mins./%. These values are in proximate agreement with both experimental and derived model-predicted results.

Results generated from experiment, derived model and regression model show that the slag cement setting time decreases with increase in the concentration of gypsum addition. The correlations between slag cement setting time and concentration of gypsum addition as obtained from experiment, derived model and regression model were all > 0.97.

## 4.5. Validation of Model-predicted Results Using SPSS 17.0

The validity of the derived model-predicted results also was ascertained using SPSS 17.0. The results indicate check variance = 0.001, standard deviation = 0, and model operational confidence of 95.0% at a significant level: 0.05.

## 4.6. Deviational Analysis

Comparative analysis of setting time from experiment [6] and derived model revealed insignificant deviations on the part of the model-predicted values relative to values obtained from the experiment. This is attributed to the fact that the surface properties of the slag material and the physiochemical interactions between the slag material and added gypsum (under the influence of the treatment temperature) which were found to have played vital roles during the process [6] were not considered during the model formulation. This necessitated the introduction of correction factor, to bring the model-predicted compressive strength to those of the corresponding experimental values.

Deviation (Dv) of model-predicted setting time from that of the experiment [6] is given by

$$Dv = \left(\frac{Ps - Es}{Es}\right) \times 100 \qquad (9)$$

Correction factor (C) is the negative of the deviation i.e

$$Cr = -Dv \qquad (10)$$

Therefore

$$Cr = \left(\frac{Ps - Es}{Es}\right) \times 100 \qquad (11)$$

Where
Ps = Model-predicted setting time ( mins.)
Es = Setting time obtained from experiment (mins.)
Cr = Correction factor (%)
Dv = Deviation (%)
Introduction of the corresponding values of Cr from equation (11) into the model gives exactly the corresponding experimental setting time.

Comparative analysis of Figure 6 shows that the maximum deviation of the model-predicted slag cement setting time from the corresponding experimental values is less than 11% and quite within the acceptable deviation limit of experimental results. This invariably translated into over 89% operational confidence for the derived model as well as over 0.89 dependency coefficients of slag cement setting time on gypsum addition.

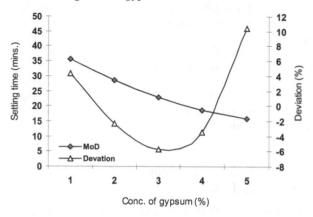

**Figure 6.** Variation of model-predicted setting time with its associated deviation from experimental values

These figures show that least and highest magnitudes of deviation of the model-predicted setting time (from the corresponding experimental values) are + 10.46 and − 2.32% which corresponds to setting times: 15.6853 and 28.522 mins./% as well as concentration of gypsum addition: 5 and 2% respectively.

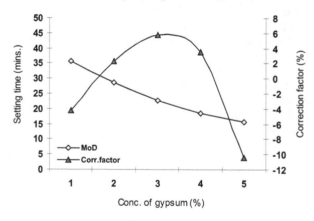

**Figure 7.** Variation of correction factor (to model-predicted values) with setting time

Analysis of Figs. 6 and 7 indicates that the orientation of the curve in Figure 7 is opposite those of the deviation of model-predicted setting time in Figure 6. This is because correction factor is the negative of the deviation as shown in equations (10) and (11). It is believed that the correction factor takes care of the effects of the surface properties of the slag material and the physiochemical interaction between the slag material and the added gypsum which (affected experimental results) were not considered during the model formulation. Figure 7 indicate that the least and highest magnitudes of correction factor to the model-predicted compressive strength are − 10.46 and + 2.32% which corresponds to setting times: 15.6853 and 28.522 mins./% as well as concentration of gypsum addition: 5 and 2% respectively.

## 4.7. Optimization of Gypsum Addition

The setting time predicted by the derived model (equation (6)); $\gamma = 0.7168\ \alpha^2 - 9.2965\ \alpha + 44.2478$ is based on the concentration of gypsum addition. Optimization of gypsum addition was achieved by differentiating the derived models (equations (6)) with respect to $\alpha$ (and equating to zero) in order to determine the value of $\alpha$ at which $\gamma$ is minimum.

$$d\beta/d\alpha = \gamma = 0.7168\ \alpha2 - 9.2965\ \alpha + 44.2478 \quad (12)$$

Differentiation of equations (12) with respect to $\alpha$ reduces them respectively to;

$$d\beta / d\alpha = 1.4336 - 9.2965\ \alpha = 0 \quad (13)$$

$$1.4336 - 9.2965\ \alpha = 0 \quad (14)$$

The value of $\alpha = 6.4847\%$, evaluated from equations (14) is the optimum concentration of gypsum addition which invariably gives the minimum setting time, $\gamma$ as 14.1054 mins. on substituting the value of $\alpha = 6.4847\%$ into the derived model in equation (6).

Figure 8 shows that beyond 6.4847% gypsum addition, the slag cement setting time increases drastically; a situation typifying immiscibility and lack of homogeneity between the cement slurry and the extra gypsum addition. This is because increased gypsum addition (above a specified quantity) is unlikely to forms a coherent mass with a specified and fixed liquid volume, resulting to delayed and differential setting.

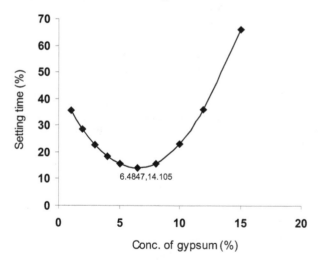

**Figure 8.** Predicted minimum setting time of the slag cement at predicted optimum concentration of gypsum addition

Confirmation of the minimum and optimum values of $\gamma$ and $\alpha$ respectively was carried out by substituting an assumed range of values of the concentration of gypsum addition $\alpha$ = 8, 10, 12 and 15%. Figure 8 strongly shows that addition of gypsum below or above the optimum value 6.4847% would give untargeted and obscured result. This is because gypsum addition below the optimum value only lowers the cement setting time and above this optimum, cement setting time increases at the expense of quality transfer to other productions. For example, increase in the cement setting time might hamper some mechanical properties of concrete made with the cement. This is so because prolonged setting of the concrete could be negatively affect by natural constraints like rain fall which drops after periodic increase in the atmospheric relative humidity.

# 5. Conclusion

An empirical evaluation of slag cement minimum setting time (SCMST) was successfully carried out through optimization of gypsum addition to foundry slag in the course of the cement production. A model was derived and used as a tool for predictive analysis of the cement setting time based on gypsum input. The model aided optimization of gypsum addition indicated a minimum setting time of 14.1054 minutes at an optimum gypsum input concentration of 6.4847%. Beyond 6.4847% gypsum addition, the slag cement setting time increased drastically; a situation typifying immiscibility and lack of homogeneity between the cement slurry and the extra gypsum addition. This is because increased gypsum addition (above a specified quantity) is unlikely to forms a coherent mass with a specified and fixed liquid volume, resulting to delayed and differential setting. The slag cement setting time per unit gypsum addition are as obtained from experiment and derived model are 4.75 and 4.996 mins./% respectively. Statistical analysis of the experimental and derived model-predicted results for each value of the gypsum input concentration considered shows standard errors of 1.8974 and 1.5485% respectively. Deviational analysis indicates 10.46% as the maximum deviation of the model-predicted slag cement setting time from the corresponding experimental value. The validity of the model was rooted on the expression $0.0226 \, \gamma + 0.2101 \, \alpha = 0.0162 \, \alpha^2 + 1$ where both sides of the expression are correspondingly approximately equal. The validity of the derived model-predicted results also was ascertained using SPSS 17.0. The results indicate check variance = 0.001, standard deviation = 0, and model operational confidence of 95.0% at a significant level: 0.05.

# References

[1]  Helmuth, R. (1987). Fly ash in cement and concrete, Portland Cement Association. Journal of Cement and Concrete Research, 30: 201-204

[2]  Chatterji, A. K. (1990). Adsorption of lime and pozzolanic activity. Journal of Scientific Industrial Research, 19B: 493-494.

[3]  Kessler, B., Rollet, M., and Sorrentino, F. (1992). Microstructure of cement pastes as incinerators ash host. Proceedings of 1st International Symposium on cement industry solution to waste Management, Calgary. Pp 235-251.

[4]  Pimienta, P., Remond, S., Rodrigues, N and Bournazel, J. P. (1999). Assessing the Properties of Mortars Containing Musical Solid Waste Incineration Fly Ash. International Congress, Creating with Concrete, University of Dundee. Pp 319-326.

[5]  Remond, S., Pimienta, P., and Bentz, D. P. (2002). Effect of the incorporation of municipal solid waste fly ash in cement paste and Mortars. Journal of cement and concrete research, 10:12-14

[6]  Egunlae, O. O. (2011). The Influence of Gypsum on Foundry Slag for Making Cement. Inter. Res. J. Eng. Sc. & Tech., 8 (1): 33-42

[7]  Nwoye, C. I. (2008). Data Analytical Memory; C-NIKBRAN

[8]  Nwoye, C. I. and J. T. Nwabanne. (2013). Empirical Analysis of Methane Gas Yield Dependence on Organic Loading Rate during Microbial Treatment of Fruit Wastes in Digester. Advances in Applied Science Research, 4 (1): 308-318.

# A Comparative Study of Various Empirical Methods to Estimate the Factor of Safety of Coal Pillars

**A. K. Verma**[*]

Department of Mining Engineering, Indian School of Mines – Dhanbad-04, Jharkhand, India
*Corresponding author: neurogeneticamit@gmail.com

**Abstract** Design of coal pillars in a coal mine remains a challenge inspite of several theories proposed by several researchers over a period of time. India is heavily dependent on coal availability for supply of electricity to it's billion of citizens. This has burdened the coal industry to increase the coal production which ultimately has lead to extraction of coal pillars also. Coal pillars are used to support the overlying roof rock to prevent it from falling. The dilemma with coal pillar stability is that On one hand, the size of the pillar should be as small as possible to enable maximum recovery of coal, while on the other hand, the pillar should be large enough to support the load of overlying strata. The stability of coal pillars has fascinated several researchers and hence many empirical equations have been proposed over the decades. In this paper, parameters like height of pillar, depth of pillar, compressive strength of coal, depth of the coal seam have been taken as input to estimate factor of safety of coal pillar from two mines i.e., Begonia and Bellampalli. It is found that Greenwald (1941), Salamon and Munro (1967), Sheorey (1992) and Maleki (1992) method has estimated failed cases correctly while Sheorey (1992) & Maleki (1992) have not predicted stable case of pillar correctly which is an interesting finding as empirical relations proposed by Sheorey (1992) is assumed to have good prediction in Indian condition.

*Keywords: Pillar strength, factor of safety, Begonia coal mines and Bellampalli coal mines, percentage deviation*

## 1. Introduction

The empirical coal pillar factor of safety approach is considered to represent the most reliable methodology available for analyzing the long-term stability of regular arrays of pillars that are wide with respect to cover depth. Alternative numerical approaches are hampered by our inability to accurately define rock mass properties and develop constitutive laws that fully define rock mass behavior. Board and Pillar mining method is one of the underground mining methods that commonly used for the extraction of coal seam in India. It involves driving two sets of parallel inseam headings, one set being orthogonal to another, thereby forming square or rectangular pillars. If the strength of a pillar in a room-and-pillar mine is exceeded, it will fail, and the load that it carried will be transferred to neighboring pillars. The additional load on these pillars may lead to their failure. Pillar strength can be defined as the maximum resistance of a pillar to axial compression. In flat lying deposits, pillar compression is caused by the weight of the overlying rock mass (Bieniawski, Z., 1968, 1984 & 1992). Empirical evidence suggests that pillar strength is related to both its volume and its shape. Numerous equations have been developed that can be used to estimate the strength of pillars in coal and hard rock mines, and have been reviewed and summarized in the literature. These equations are generally empirically developed and are only applicable for conditions similar to those under which they were developed (Jaiswal, 2009). More recently, numerical model analyses combined with laboratory testing and field monitoring have contributed to the understanding of failure mechanisms and pillar strength.

## 2. Parameter Influencing Stability of Pillars

Pillar strength can be defined as the maximum resistance of a pillar to axial compression. Pillars are the key load bearing elements of an underground coal mine. The stress on the pillars is a function of both the width (w) and the height (h) of the pillars. Several equations for calculating the pillar strength have been proposed by different researchers on the basis of laboratory test of rock.

Strength of coal pillars and their behavior can vary dramatically depending on their shape (Martin, 2000). Three broad categories of pillar behavior and failure mode have been identified, each defined by an approximate range of width-to-height ratios (Mark, 1999):

**Slender pillars:** - Slender pillars are those with w/h ratios less than about 3 or 4. When these pillars are loaded to their maximum capacity, they fail completely, shedding nearly their entire load. When large numbers of slender pillars are used over a large area, the failure of a single pillar can set off a chain reaction, resulting in a sudden, massive collapse accompanied by a powerful air blast.

**Intermediate pillars:** - Intermediate pillars are those whose w/h ratios fall between 4 and 8. These pillars do not shed their entire load when they fail, but neither can they accept any more loads. Instead, they deform until overburden transfers some weight away from it. The result is typically a non-violent pillar "squeeze," which may take place over hours, days, or even weeks. The large roof-to-floor closures that can accompany squeezes can cause hazardous ground conditions and entrap equipment.

**Squat pillars:** - Squat pillars are those with w/h ratios that exceed 10. These pillars can carry very large loads, and may even be strain-hardening (meaning that they may never actually shed load, but just may become more deformable once they "fail."). None the less, the pillar *design* may fail because excessive stress is applied to the roof, rib, or floor, or because the coal bumps. Moreover, the strength of squat pillars can vary considerably depending upon the presence of soft partings, weak roof or floor interfaces, and other geologic factors.

## 2.1. Empirical Pillar Strength Formulas

The design of coal pillar was started in 1773 by Coulomb as traced by Mark, 2006. However, Bunting (1991) designed first pillar for coal mines based on scientific approach. He carried out extensive laboratory test on different sizes of cube (2 to 6 inches size) & prism

(2.25 to 12.25 inches height) of anthracite. He demonstrated that strength of coal prism has relation to their height and width.

Empirical pillar strength formulas were developed as part of major coal brook disaster whose main objective was to establish in situ strength of coal pillars. Salamon & Munro (1967) analyzed 125 case histories of Coal pillar collapse and proposed that the coal pillar strength could be determined using power formula

$$\sigma_p = K_{SM} h_p^\alpha w_p^\beta \qquad (i)$$

Where $\sigma_p$ (MPa) is the pillar strength, $K_{SM}$ (MPa/m$^2$) is strength of a unit volume of coal and w & h are the pillar width & height in meters respectively. Hudson et al. (1972) after extensive laboratory study confirm that strength of a rock mass is too large part controlled by geometry by the geometry of the specimen i.e. w/h ratio. Equation (i) (Mohan, 2001) has been developed for Room & Pillar mining of horizontal coal seams. Following Bunting (1991), a number of pillar design equations were developed around the world as shown Table 1 a,b,c.

Das (1986) who observed that when w/h ratio increases beyond 4-6, the post failure characteristic starts ascending indicating a gain in strength (Figure 1). He also observed that at w/h ratio of 13.5, if pillars of such flatness are left in underground mines for support purpose, they can retain high strength even after failure.

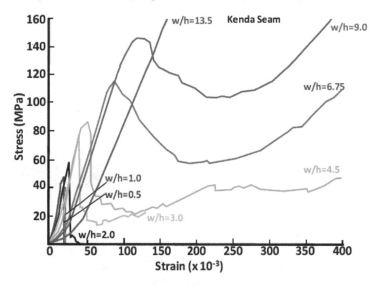

**Figure 1.** Influence of w/h ratio on the post-failure stress-strain behavior of coal (Das, 1986)

**Table 1a. Proposed empirical relations for coal pillar strength determination**

|  | Zern Edward Nathan (1928) | Greenwald (1941) | Salamon and Munro (1967) | Obert and Duvall (1967) |
|---|---|---|---|---|
| **Empirical Relation** | $C_p = C_1 \sqrt{\left(\dfrac{w_p}{h_p}\right)}$ | $C_p = 0.67k \dfrac{\sqrt{w_p}}{h_p^{0.83}}$ | $C_p = k_{SM} h_p^\alpha w_p^\beta$ | $C_p = C_{10}\left[0.778 + 0.222\left(\dfrac{w_p}{h_p}\right)\right]$ |
| **Remarks** | Cp is pillar strength, $C_1$ is the coal strength, $w_p$ is the pillar width and $h_p$ is the pillar width. | Cp is the pillar strength, k is the strength of coal sample, $w_p$ is the pillar width and $h_p$ is the pillar height | Cp is the pillar strength, $k_{SM}$=7.176 kPa, $h_p$ is pillar height, $w_p$ is pillar width, α=-0.66, β=0.46 | Cp is the compressive strength , $C_{10}$ is the compressive strength of specimens having ratio of $\dfrac{d}{h}=1$ , d is the diameter of the specimen, h is the height of the specimen |
| **Limitations** |  | Strength is calculated by unit cube of coal sample | K has to be evaluated by testing of specimen size 30 cm. | $\dfrac{d}{h}=1$ |

**Table 1b. Proposed empirical relations for coal pillar strength determination**

| | Bieniawski (1975) | Logie and Matheson (1982) | Maleki (1992) |
|---|---|---|---|
| Empirical Relation | $C_p = k_B \left[ 0.64 + 0.34 \left( \dfrac{w_p}{h_p} \right) \right]$ | $C_p = k_B \left[ 0.64 + 0.34 \left( \dfrac{w_p}{h_p} \right) \right]^{1.4}$ | $C_p = 3836 \left[ 1 - e^{-0.260 \frac{w_p}{h_p}} \right]$ |
| Remarks | Cp is the pillar strength, $k_B$ is the compressive strength of a 30 cm cube pillar specimen(MPa), $w_p$ is the width of the pillar, $h_p$ height of the pillar | Cp is the compressive strength, $k_B$ is the compressive strength, $w_p$ is the width of the pillar, $h_p$ is the height of the pillar | Cp pillar strength, $w_p$ pillar width, $h_p$ pillar height |
| Limitations | Specimen should be 30 cm cube pillar | | |

**Table 1c. Proposed empirical relations for coal pillar strength determination**

| | Mark-Bieniawski (1997) | Sheorey (1992) |
|---|---|---|
| Empirical Relation | $C_p = S_1 \left[ 0.64 + \left( 0.54 \dfrac{w_p}{h_p} - 0.18 \left( \dfrac{w_p^2}{h_p L_p} \right) \right) \right]$ | $C_p = 0.27 \sigma_{c25} h_p^{-0.36} + \left( \dfrac{H}{250} + 1 \right) \left( \dfrac{w_p}{h_p} - 1 \right)$ |
| Remarks | Cp is pillar strength, $S_1$ is in situ coal strength, $w_p$ pillar width, $h_p$ pillar height, $L_p$ pillar length | Cp is pillar strength, H coal seam depth, $\sigma_c$ strength of coal, $w_p$ pillar width, $h_p$ pillar height |
| Limitations | | |

## 2.2. Case Study

For comparison of suitability of various empirical relations, coal pillars at two mine namely Begonia in Raniganj coal field as failed pillar case and Bellampalli in Andhra Pradesh as stable pillar case have been chosen.

Any method for estimating pillar strength should satisfy the following two statistical conditions:

**(a)** All failed pillar cases should have a safety factor of 1.0. In statistical terms, this merely means that the line of safety factor=1.0 should be the best fit for a plot of pillar strength vs. pillar load.

**(b)** All stable cases must have a safety factor>1.0.

Parameters used for calculation of factor of safety for coal pillars at both Begonia & Bellampalli mine are shown in Table 2. Figure 2 shows the location of both Begonia & Bellampalli mines. Begonia mine is located in Asansol area of West Bengal while Bellampalli mine is located in Andhra Pradesh. Coal at Begonia mine has its origin from Raniganj sub group of upper Permian age while coal at Bellampalli has its origin from Barakar sub group of lower Permian age (Figure 3).

**Table 2. Different parameter value for Begonia (failed) and Bellampalli (stable) case**

| S.No. | Mines | Depth of coal seam | Pillar height (h in m) | Pillar width (w in m) | w/h ratio | Compressive strength of coal sample |
|---|---|---|---|---|---|---|
| 1. | Begonia | 36 | 3 | 3.9 | 1.3 | 26 |
| 2. | Bellampalli | 36 | 3 | 5.4 | 1.8 | 48 |

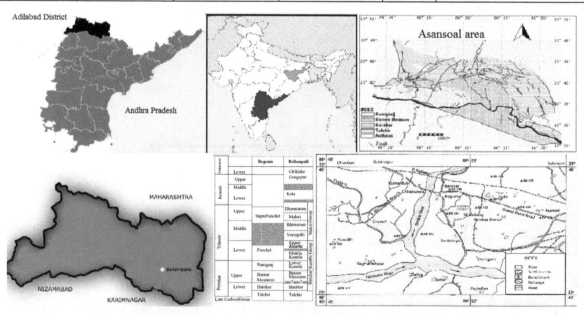

**Figure 2.** Location map and stratigraphy (Mukhopadhyay, 2010) of Begonia and Bellampalli coal mines

Parameters used for calculation of factor of safety for coal pillars at both Begonia & bellampalli mines is shown in Table 2. Figure 2 shows the location both Begonia and Bellampalli mines. Begonia mines is located in Asansol area of West Bengal while Bellampalli mines is located in Andhra Pradesh. Coal at Begonia mines has its origin from Raniganj formation of upper Permian age while coal at Bellampalli has its origin from Barakar formation of lower Permian age (Figure 2).

**Table 3. FoS estimated by different empirical relations and their % deviation from critical factor of safety 1.0**

| Researchers | Failed Case | | Stable Case | |
|---|---|---|---|---|
| | FOS | % deviation from FOS of 1.0 | FOS | %deviation from FOS of 1.0 |
| Zern Edward Nathan (1928) | 1.48 | 48 | 2.9 | 190 |
| Greenwald (1941) | 0.69 | -31 | 1.3 | 30 |
| Salamon and Munro (1967) | 1.17 | 17 | 2.3 | 130 |
| Obert and Duvall (1967) | 1.38 | 38 | 2.6 | 160 |
| Bieniawski (1975) | 1.40 | 40 | 2.7 | 170 |
| Logie and Matheson (1982) | 1.45 | 45 | 3.0 | 200 |
| Mark-Bieniawski (1997) | 1.44 | 44 | 2.8 | 180 |
| Sheorey (1992) | 0.40 | -60 | 0.8 | -20 |
| Maleki (1992) | 0.37 | -63 | 0.45 | -55 |

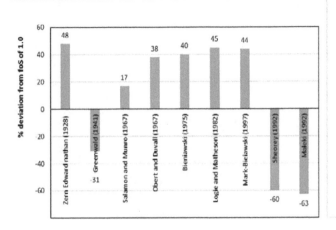

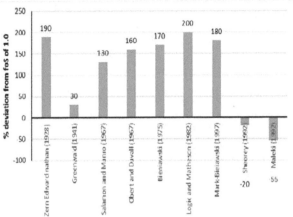

Begonia(failed) Case

Bellampalli(stable) Case

**Figure 3.** percentage deviation from FOS of 1.0 for Begonia (failed case) and Bellampalli (stable case) by different empirical relation

Figure 3 and Table 3 shows the percentage error deviation of factor of safety from 1.0 which is assumed to be the boundary line between the stable and failed cases. The empirical equation which gives us the negative percentage of deviation from the boundary line of FOS of 1.0 is correct for failed case. It can be observed that Greenwald (1941), Salamon and Munro (1967), Sheorey (1992) and Maleki (1992) method has estimated failed cases correctly.

Similarly the empirical equation which gives the positive percentage error deviation from the assumed

boundary line of critical FOS of 1.0 will be correct estimation of FOS for stable coal pillar case. All empirical equation has predicted correct trend of FOS for stable coal pillar case except Sheorey (1992 &1986) & Maleki (1992).

In the stable case only two methods have given the wrong estimation in negative percentage deviation from the boundary line 1.0. For the second case, all the empirical relation given by different researchers has given the positive percentage deviation only 'Sheorey' and 'Maleki' has given the negative parentage deviation.

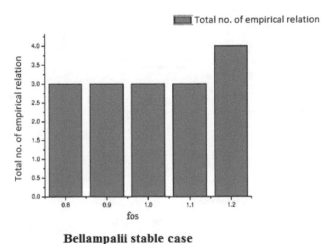

Bellampalii stable case

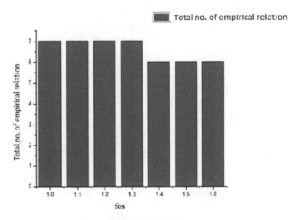

Begonia failed case

**Figure 4.** Number of empirical relation predicting stable cases and failed cases at different critical FOS

Figure 4 show the number of empirical relation predicting the coal pillar stable & failed cases at different value of FOS. Figure 6 show that if we assumed critical FOS of 1.2 , the four empirical can predicted the failed Begonia coal pillar case while FOS range of 0.8 to 1.1, number of empirical equation predicting failed Begonia

coal pillar critical FOS is constant and have value of 3. Similarly for critical FOS range of 1.0 to 1.3, the number empirical relation predicting Bellampalli coal pillar stable case was seven while there are six empirical relations between FOS ranges of 1.4 to 1.6.

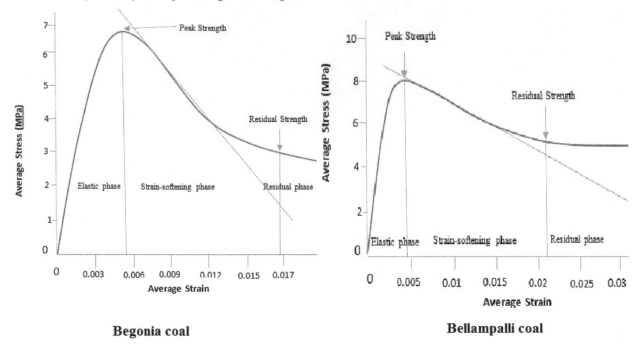

**Begonia coal**                                                    **Bellampalli coal**

**Figure 5.** Average strain and average stress curve for Begonia coal and Bellampalli coal

Figure 5 shows the stress-strain curve and different zone of stable of stress for both Begonia and bellampalli coal.

There are three main regimes: pre- failure, the strain-softening zone, and the residual zone. Peak strength is characterized by a peak failure criterion. It is dependent on the state of stress conditions. Strain-softening behavior is

characterized by both the failure criterion and the plastic potential. The last phases of perfectly plastic are characterized by the residual failure criterion along with a plastic potential. Bellampalli coal has higher strength as compared to Begonia coal.

Strength of Indian pillar coal increases after w/h ratio of 5 which is suggested by observation of Das (1986).

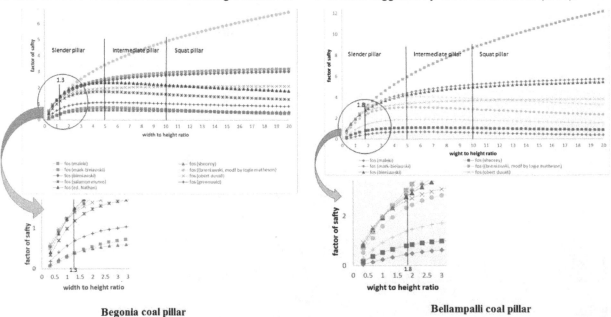

**Begonia coal pillar**                                          **Bellampalli coal pillar**

**Figure 6.** Variation of Begonia coal pillar and Bellampalli coal pillar FOS with w/h ratio

Figure 6 shows increase in FOS with increase in width to height ratio of Bellampalli coal pillar. In the case of

slender pillar and also the case of Bellampalli coal pillar (w/h = 1.8) highest FOS of pillar is predicted by Logie and Matheson (1982), Bieniawski (1975) and Mark-Bieniawski (1997) while least strength is predicted by Maleki(1992) and Sheorey(1992). Similar trend is shown in case of Bellampalli coal pillar (Figure 6).

## 3. Conclusion

In this paper, empirical and analytical pillar design relations proposed by several researchers have been discussed. It is found from the above study that in order to use these relations successfully, it is necessary to obtain the different input parameters realistically which is difficult due to uncertainties in determination of coal properties. Since pillar cases are quite old, the input parameters are generally based on experience. Even though, this empirical technique has a good potential for realistic prediction.

In the present study, the effect of width to height ratio of coal pillar of two mines i.e., Begonia and Bellampalli on its stability has been considered. It has been found that pillar strength is almost linearly dependent on w/h ratio less than 5 and non-linearly dependent on uniaxial compressive strength of the coal specimen. It is concluded that Greenwald (1941), Salamon and Munro (1967), Sheorey (1992) and Maleki (1992) method has estimated failed cases correctly while Sheorey (1992) & Maleki (1992) have not predicted stable case of pillar correctly which is an interesting finding as empirical relations proposed by Sheorey (1992) is assumed to have good prediction in Indian condition. In the case of Begonia coal pillar (w/h = 1.38), highest strength and FOS is predicted by Logie and Matheson (1982), Bieniawski (1975) and Mark- Bieniawski (1997) while the least strength and FOS is predicted by Maleki (1992) and Sheorey (1992). Similarly, in the case of Bellampalli coal pillar (w/h = 1.8) highest strength and FOS of pillar is predicted by Logie and Matheson (1982), Bieniawski (1975) and Mark-Bieniawski (1997) while least strength and FOS is predicted by Maleki (1992) and Sheorey (1992).

## References

[1]   Bieniawski, Z. T, Balkema (1984): A. A. Rock Mechanics Design in Mining and Tunneling; 272 pp.

[2]   Bieniawski, Z. T. (1992): A method revisited: coal pillar strength formula based on field investigations, proceedings, workshop on coal pillar mechanics and design, Bu mines; p 158-165.

[3]   Bieniawski ZT. (1968): In situ strength and deformation characteristics of coal; Volume 2, 325-40.

[4]   Bunting, D. (1991): Chamber Pillars in Deep Anthracite Mines. Trans. AIME; vol. 42, pp. 236-245.

[5]   Christopher Mark. (1981-2006): Chief, Rock Mechanics Section NIOSH-Pittsburgh Research Laboratory Pittsburgh, PA, USA, The evolution of intelligent coal pillar design.

[6]   Das, M.N. and Sheorey, P. R. (1986): *Triaxial Strength behavior* of some Indian coals. Journal of Mines, Metals & Fuels; 34 (3). pp. 118-122.

[7]   Hudson, J.A., Brown, E.T. and Fairhurst, C. (1972): Shape of the complete stress-strain curve for rock. In Proc. 131h Symposium on Rock Mechanics, Urbana, Illinois, Edited by E.J. Cording; pp: 773-795.

[8]   Jaiswal, Ashok, Shrivastva, B.K. (2009): Numerical simulation of coal pillar strength, International Journal of Rock Mechanics & Mining Sciences; Volume 46, 779-788.

[9]   Jawed, M., Sinha, R. K. and Sengupta, S. (2013): Chronological development in coal pillar design for board and pillar workings: A critical appraisal, Journal of Geology and Mining Research; Vol. 5 (1); pp. 1-11.

[10]  Mohan, G. Murali, Sheorey, P.R., Kushwaha. A. (2001): Numerical estimation of pillar strength in coal mines, International Journal of Rock Mechanics & Mining Sciences; Volume 38, 1185-1192.

[11]  Mark, C. (1999): Empirical Methods for Coal Pillar Design. Proceedings of the Second International Workshop on Coal Pillar Mechanics and Design. U.S. Department of Health and Human Services, National Institute for Occupational Safety and Health (NIOSH); IC 9448, pp. 145-154.

[12]  Martin,C.D., Maybee, W.G. (2000): The strength of hard-rock pillars, International Journal of Rock Mechanics & Mining Sciences; Volume 37; 1239-1246.

[13]  Maleki H. (1992): In situ pillar strength and failure mechanisms for US coal seams. In: Proceedings of the workshop on coal pillar mechanics and design, Pittsburgh, US States Bureau of Mines; p. 73-8.

[14]  Mukhopadhyay, G., Mukhopadhyay, S. K., Roychowdhury, M and Parui, P. K. (2010): Stratigraphic Correlation between Different Gondwana Basins of India, journal geological society of india; Vol. 76: pp. 251-266.

[15]  Salamon MDG, Munro. A.H. (1967): A study of the strength of coal pillars. J. South Afr. Inst. Min. Metallurgy; 68:55-67.

[16]  Sheorey, P. R., Das, M. N., Bordia, S. K., Singh B. (1986): Pillar strength approaches based on a new failure criterion for coal seams, International Journal of Mining and Geological Engineering; Volume 4: pp 273-290.

[17]  Sheorey PR. (1992): Design of coal pillar arrays, chain pillars. In: Hudson JA, et al., editors. Comprehensive rock engineering, vol. 2. Oxford: Pergamon; p. 631-70.

# Assessment of the Geotechnical Properties of Lateritic Soils in Minna, North Central Nigeria for Road design and Construction

**Amadi A.N.***, **Akande W. G., Okunlola I. A., Jimoh M.O., Francis Deborah G.**

Department of Geology, Federal University of Technology, Minna, Nigeria
*Corresponding author: geoama76@gmail.com

**Abstract**  Laterite is a highly weathered material, rich in secondary oxides of iron, aluminum, or both. Geotechnical investigation is one of the effective means of detecting and solving pre, syn and post constructional problems. The geotechnical properties of lateritic soils and their suitability for road construction have been evaluated for selected sites in Minna, North-central Nigeria. All analyses were carried out in accordance with the British Standard Institution. The liquid limit ranged from 22.5% to 49.6% with an average value of 34.9%, plastic limits varied from 13.8% to 28% with a mean value of 21.38% while plastic index is of the order of 8.7% to 21.6% with an average value of 13.5%. The maximum dry density ranged from 1.78 g/cm³ to 2.33 g/cm³ with a mean value of 1.858 g/cm³ while the optimum moisture content varied from 6.30% to 14.3% with an average value of 9.74%. The evaluation reveals that the lateritic soils have higher plastic limits, Maximum Dry Densities (MDD) and California Bearing Ratios (CBR) while their liquid limits, plasticity indices and Optimum Moisture Contents (OMC) are lower. The lateritic soils were classified as A-3, A-2-4 and A-2-6 and are adjudged suitable for sub-grade, good fill and sub-base and base materials. This geotechnical information obtained will serve as base-line information for future road foundation design and construction in the study area.

***Keywords:*** *Geotechnical Assessment, Lateritic soil, road construction, Sauka-kahuta, Minna, North-central Nigeria*

## 1. Introduction

The understanding of soil behavior in solving engineering and environmental issues as swelling soil especially expansive lateritic soils that can cause significant damage to road construction and other engineering application is the sole aim of geotechnical engineering (Abubakar, 2006; Oke and Amadi, 2008).One of the major causes of road accident is bad road which is usually caused by wrong application of constructional materials especially laterite as base and sub-base material by construction companies (Oke *et al.,* 2009a; Nwankwoala *et al.*, 2014). For a material to be used as either a base course or sub-base course depends on its strength in transmitting the axle-load to the sub-soil and or sub-grade (the mechanical interlock). The characteristics and durability of any constructional material is a function of its efficiency in response to the load applied on it (Oke *et al.,* 2009b; Nwankwoala and Amadi, 2013). The mineralogical composition of the lateritic soil has an influence on the geotechnical parameters such as specific gravity, shear strength, swelling potential, Atterberg limits,

bearing capacity and petrograpic properties (Amadi *et al.*, 2012). The rate at which newly constructed roads in Minna and environs developed cracks and later damage is worrisome. Hence this study will ensure that the right lateritic soils with the right criteria are used as a base or sub-base material for road construction in Minna and environs, North-Central Nigeria, thereby ensuring their durability.

## 2. Location and Physiography of the study Area

The study area is Sauka-Kahuta industrial layout, behind the Minna building material market. It lies within longitude 06°28'11"E to 06°32'13"E of the Greenwich Meridian and between latitude 09°35'22"N to 09°30'36"N of the Equator (Figure 1). The study area has an undulating topography drained by river chanchaga and its tributaries. The area is within the Guinea Savannah with an annual rainfall of about 1100 mm in the northern part and 1600 mm in the southern part. The rainy season spans between the month of April- October and an optimum temperature of 41°C in dry season and minimum of 22°C during the rainy season (Sheriff, 2012).The study area is

assessed through Talba farm road, Mandella road, other minor untarred road.

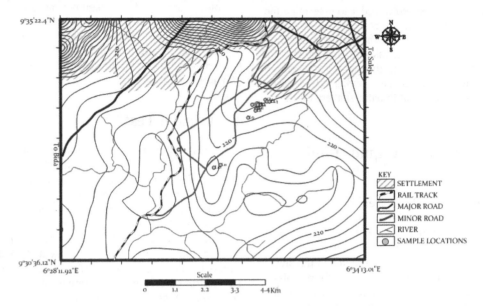

**Figure 1.** Topographical map of the Study Area

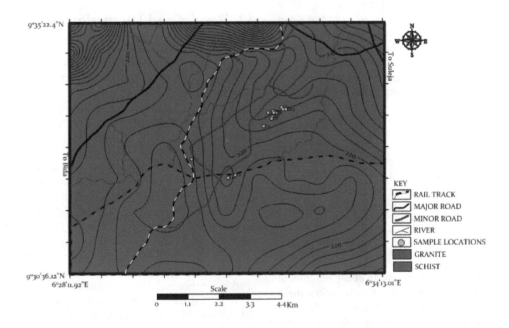

**Figure 2.** Geological map of the Study Area

## 3. Geological Mapping

The mapping exercise was carried out to identify and examine the rock type in the study area. The rock type in the area comprises of granite and schist (Figure 2).

## 4. Methodology of Investigation

The sub-soil conditions was investigated by excavating five trial pits from existing ground level to a maximum of 4.5 m according to British standard code of practice for site investigation (1981), depending on topography and

overburden. Disturbed samples soil samples were collected from the trial pits and analyzed at civil engineering laboratory, Federal University of Technology, Minna, Nigeria for relevant geotechnical analysis.

## 5. Laboratory Analysis

The laboratory analysis was performed according British standard methods of test for soil for civil engineering purposes (BS 1377: Part 1-9, 1990). The laboratory test carried out to determine the suitability of the lateritic soils for use as base and sub-base material using the AASHTO standard method in relation to the generation specification for roads and bridges.

## 6. Sieve Analysis

Sieve analysis was performed in order to determine the soil particle size distribution. Representative sample of approximately 500 g was used for the test after washing and oven-dried. The sample was washed using the BS 200 sieve and the fraction retained on the sieve was air dried and used for the sieve analysis. The sieving was done by mechanical method using an automatic shakers and a set of sieves.

## 7. Atterberg Limits (Liquid Limit and Plastic Limit)

This test determines the clay content in terms of liquid limit, plastic limit, plasticity index and shrinkage potential in order to estimate plasticity, strength and settlement characteristics of the soil sample. For the determination of liguid limit, the soil sample passing through 425 µm sieve, weighing 200 g was mixed with water to form a thick homogeneous paste. The paste was collected inside the Casangrade's apparatus cup with a grove created and the number of blows to close it was recorded. Similarly, for plastic limit determination, the soil sample weighing 200 g was taken from the material passing the 425 µm test sieve and then mixed with water till it became homogenous and plastic to be shaped to ball. The ball of soil was rolled on a glass plate until the thread cracks at approximately 3 mm diameter. The 3 mm diameter sample was placed in the oven at 105°C to determine the plastic limit.

## 8. Moisture content

Moisture content is defined as the ratio of the weight of the water in a soil specimen to the dry weight of the specimen. The moisture content of lateritic soil can be influenced by the mineralogy and formation environment.

## 9. Compaction Test

The densification of soil with mechanical equipment thereby rearranging the soil particles which makes them more closely packed resulting in an increase of the ratio horizontal effective size to the vertical effective stress. The degree of compaction is measured in term of its dry weight and it increasing the bearing capacity of road foundation, stability slopes, controls undesirable volume changes and curb undesirable settlement of structures. The mould is filled and compacted with soil in five layer via 25 blows of a 4.5 rammer.

## 10. California Bearing Ratio

The California bearing ratio (CBR) test is a penetration test carried out to evaluate the mechanical strength of a sub-base or base course material. It measures the shearing resistance, controlled density and moisture content. Both the soaked and unsoaked method of CBR was conducted to characterize the lateritic soil for use as a base or sub-base material. A portion of air-dried soil sample was mixed with about 5% of its weight of water. This was put in CBR mould in 3 layers with each layer compacted with 55 blows using 2.5 kg hammer at drop of 450 mm (standard proctor test). The compacted soil and the mould was weighed and placed under CBR machine and a seating load of approximately 4.5 kg was applied. Load was recorded at penetration of 0.625, 1.9, 2.25, 6.25, 7.5, 10 and 12.5 mm.

## 11. Triaxial Test

In this test, horizontal load was applied as soon as vertical load has been imposed and shearing continued at the rate of 0.25 mm/min until the shear force goes beyond its maximum value and becomes constant or decreases, representing failure condition. Normal stresses of 188.0 kPa, 324.3 kPa, 460.5 kPa and 596.8 kPa were employed in all the direct shear tests. The results of the direct shear tests for the lateritic soils are presented in the form of stress-strain curves and plots of shear stress versus normal stress. From these, the shear strength parameters (angle of cohesion (c) and angle of internal friction ($\Phi$)) were obtained.

## 12. Results and Discussion

The result of the laboratory analyses are summarized in Table 1 while the revised AASHTO system of soil classification is contained in Table 2. The description of the lateritic soils is shown in Table 3 and Figure 3 while the physical properties of soil samples are presented in Table 4. Federal Ministry of Works and Housing general specification for roads and bridges is shown in Table 5. According to Federal Ministry of Works and Housing (1997) specification, the lateritic soil samples are suitable for subgrade, subbase, and base materials as the percentage by weight finer than No. 200 BS test sieve is less than 35% except locations 8 and 10 (Table 1). The liquid limits value, ranged from 15.8% to 49.6%, the plastic limits varied from 12.0% and 28.0% while the plastic index is of the order of 3.8 to 19.4 (Table 1). Federal Ministry of Works and Housing (1997) for road works recommend liquid limits of 50% maximum for subbase and base materials. All the studies soil samples fall within this specification, thus making them suitable for subgrade, subbase and base materials. The plot of plasticity index versus liquid limit is shown in Figure 4. The unsoaked California bearing ratio value for the lateritic soil sample range from 0.0% to 83.8%. Federal Ministry of Works and Housing recommendation for soils for use as: subgrade, subbase and base materials are: $\leq$ 10%, $\leq$ 30% and $\leq$ 80% respectively for unsoaked soil. This implies that locations [2,8,9] with values less than 10% are excellent subgrade materials, locations [1,2,3,8,9,10] having values less than 30% are good materials for subbase.

All the locations except location 5 have their unsoaked CBR value less than 80% which is the maximum value recommended for soils to be used as base materials (Federal Ministry of Works and Housing, 1997). By interpretation the lateritic soils from other locations except location 5 are suitable materials for subgrade, subbase and base materials. Location 5 failed the geotechnical characteristics for use as subgrade, subbase or base material. The maximum dry density for the soil samples varied between 1.81 mg/m³ and 2.35 mg/m³ while that of optimum moisture content ranged between 7.81% and 14.4%. According to O"Flaherty (1988) the range of values that may be anticipated when using the standard proctor test methods are: for clay, maximum dry density (MDD) may fall before 1.44 mg/m³ and 1.685 mg/m³ and optimum moisture content (OMC) may fall between 20-30%. For silty clay MDD is usually between 1.6 mg/m³ and 1.845 mg/m³ and OMC ranged between 15-25%. For sandy clay, MDD usually ranged between 1.76 mg/m³ and 2.165 mg/m³ and OMC between 8 and 15%. Thus, looking at the results of the soil samples, it could be noticed that they are sandyclay. The cohesion (c) of the quick undrainedtriaxial compression test (Figure 5) ranged from 130 KN/m² to 165 KN/m² and the angle of internal friction (Ø) was found to be 8° and this implies low plasticity, high permeability, shear strength and bearing capacity.

The moisture content from the compaction test in ranged from 6.30% to 14.4% with an average value of 10.39 (Table 1 and Table 3) indicating that the soil is generally poorly graded and sandyclay with plastic fines (material passing sieve No. 200) and this finding is in agreement with other determined geotechnical parameters. Federal Ministry of Works and Housing (1972) for road works recommend liquid limits of 50% maximum for subbase and base materials. All the studies soil samples fall within this specification, thus making them suitable for subbase and base materials.

### Table 1. Summary of Laboratory Results

| Trial Pit No | Depth of Sampling | Sieve Analysis % passing | Compaction Test MDD(g/cm³) | OMC (%) | Atterberg Limit LL | PL | PI | California Bearing Ratio Soaked | Unsoaked |
|---|---|---|---|---|---|---|---|---|---|
| L1 | 2.0 | 1.41 | 2.05 | 12.10 | 34.5 | 21.0 | 13.5 | 11.0 | 28.0 |
| L2 | 2.0 | 31.9 | 2.06 | 13.20 | 36.6 | 24.5 | 12.1 | 40.0 | 3.0 |
| L3 | 2.0 | 0.8 | 2.18 | 7.81 | 33.5 | 22.0 | 11.5 | 18.0 | 30.0 |
| L4 | 2.5 | 2.1 | 2.33 | 6.30 | 23.5 | 14.3 | 9.2 | 64.9 | 77.5 |
| L5 | 3.5 | 2.5 | 2.35 | 6.50 | 22.5 | 13.8 | 8.7 | 72.7 | 83.8 |
| L6 | 3.0 | 0.0 | 2.11 | 9.05 | 35.50 | 23.0 | 12.5 | 36.0 | 48.0 |
| L7 | 2.5 | 34.4 | 2.08 | 10.80 | 37.5 | 24.2 | 13.3 | 54.0 | 0.0 |
| L8 | 2.0 | 56.8 | 1.79 | 14.30 | 49.6 | 28.0 | 21.6 | 5.1 | 1.0 |
| L9 | 2.5 | 2.2 | 2.19 | 9.50 | 15.8 | 12.0 | 3.8 | 9.0 | 40.0 |
| L10 | 2.0 | 55.9 | 1.81 | 14.40 | 41.4 | 22.0 | 19.4 | 7.7 | 0.0 |

LL: liquid limit; PL: plastic limit; PI: plastic index
MDD: maximum dry density; OMC: optimum moisture content

### Table 2. Revised AASHTO system of soil classification

| General Classification | General Materials (35% or less passing 0.075 mm) | | | | | | | Silt-clay materials (more than 35% passing 0.075 mm) | | | |
|---|---|---|---|---|---|---|---|---|---|---|---|
| Group Classification | A-1 | | A-3 | A-2 | | | | A-4 | A-5 | A-6 | A-7 A-7-5 A-7-6 |
| | A-1-a | A-1-b | | A-2-4 | A-2-5 | A-2-6 | A-2-7 | | | | |
| Sieve Analysis % passing 2.00 mm (No10) 0.425 mm (No40) 0.725 mm (No200) | 50max 30max 15max | 50max 25max | 51min 10max | 35max | 35max | 35max | 35max | 36min | 36min | 36min | 36min |
| Characteristics of fraction passing Liquid limit Plastic Index | 6max | | N.P | 40max 10max | 41min 10max | 40max 11min | 41min 11min | 40max 10max | 41min 10max | 40max 11min | 40min 11min |
| Usual types of significant Constituent material | Stone fragment Gravel and sand | | Fine Sand | Silty or clayey Gravel and sand | | | | Silty soils | | Clayey soils | |
| General rating | Excellent to Good | | | | | | | Fair to poor | | | |

### Table 3. Description of lateritic soils

| Sample No | M.D.D(g/CM3) | O.M.C(%) | Group symbol | Description |
|---|---|---|---|---|
| L1 | 2.05 | 12.11 | SM-SC | Sandysilt clay mix with slightly plastic fines |
| L2 | 2.059 | 13.2 | SM-SC | Sandysilt clay mix with slightly plastic fines |
| L3 | 2.18 | 7.81 | SC | Siltyclay mix with slightly plastic fines |
| L4 | 2.33 | 6.30 | SC | Siltyclay mix with slightly plastic fines |
| L5 | 2.35 | 6.50 | SC | Siltyclay mix with slightly plastic fines |
| L6 | 2.11 | 9.05 | SC | Siltyclay mix with slightly plastic fines |
| L7 | 2.076 | 10.8 | SM-SC | Sandysilt clay mix with slightly plastic fines |
| L8 | 1.785 | 14.3 | SM | Silty soils, poorly graded sandysilt mix |
| L9 | 2.19 | 9.5 | SM-SC | Sandysilt clay mix with slightly plastic fines |
| L10 | 1.805 | 14.4 | SM | Silty soils, poorly graded sandysilt mix |

### Table 4. Correlation of the Atterberg limit versus plasticity of the lateritic soils

| Location | L1 | L2 | L3 | L4 | L5 | L6 | L7 | L8 | L9 | L10 |
|---|---|---|---|---|---|---|---|---|---|---|
| Sample Nos | 1.0 | 2.0 | 3.0 | 4.0 | 5.0 | 6.0 | 7.0 | 8.0 | 9.0 | 10.0 |
| Liquid limit | 34.5 | 36.6 | 33.5 | 23.5 | 22.5 | 35.5 | 37.5 | 49.6 | 15.8 | 41.4 |
| Plastic limit | 21.0 | 24.5 | 22.0 | 14.3 | 13.8 | 23.0 | 24.2 | 28.0 | 12.0 | 22.0 |
| Plasticity | LP | IP | LP | LP | LP | IP | IP | IP | LP | IP |
| Description | Silty | Intermediate | Silty | Silty | Silty | Intermediate | Intermediate | Intermediate | Silty | Intermediate |

LP: low plasticity; IP: intermediate plasticity

**Table 5. Nigerian Standard of soil classification for roads and bridges**

| Location | 1 | 2 | 3 | 4 | 5 | 6 | 7 | 8 | 9 | 10 |
|---|---|---|---|---|---|---|---|---|---|---|
| Sample | P1 | P2 | P3 | P4 | P5 | P6 | P7 | P8 | P9 | P10 |
| L.L (≤35%) | 34.5 Pass | 36.6 Fail | 33.5 Pass | 23.5 Pass | 22.5 Pass | 35.5 Fail | 34.5 Pass | 49.6 Fail | 34.5 Pass | 41.4 Fail |
| P.I(≤12%) comment | 13.5 Fail | 12.1 Pass | 11.5 Pass | 9.5 Pass | 8.7 Pass | 12.5 Fail | 11.3 Pass | 21.6 Fail | 13.5 Fail | 19.43 Fail |
| C.B.R soaked for subbase (≥30%) | 11.0 Fail | 40.0 Pass | 18.0 Fail | 64.90 Pass | 72.70 Pass | 36.0 Pass | 54 Pass | 5.1 Fail | 9.0 Fail | 7.7 Fail |
| C.B.R unsoaked Base course (≥80%) | 32.0 Fail | ND | 30.0 Fail | 77.5 Pass | 83.8 Pass | 48.0 Fail | ND | ND | 40.0 Fail | ND |
| Overall Rating | Sub-base | Sub-base | Sub-base | Base | Base | Sub-base | Sub-base | Poor | Poor | Poor |

ND: not determined

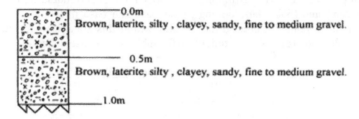

**Figure 3.** Soil profile of trial pit in the study area

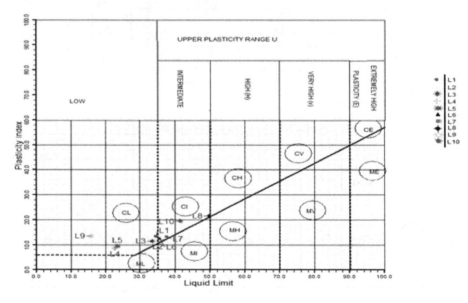

**Figure 4.** Plasticity chart of the Atterberglimits result

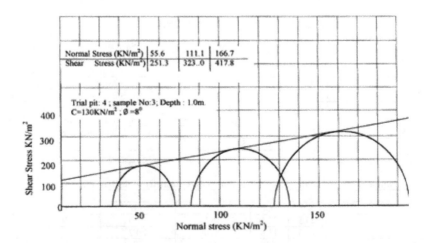

**Figure 5.** Quick undrainedtriaxial compression test indicating the stress Mohr's circles

# 13. Conclusion and Recommendation

The geotechnical properties of Minna, North-central Nigeria has been carried out in compliance with BS 1377 (1997) and head of (1990) methods of soil testing for Civil engineers. The result showed that the studied soil samples are classified as sandyclay, incompressible, easily compactable with good drainage. The soil samples tested from the study area indicate a general cohesive nature with low moisture content, high granular material which is suitable for road construction except location 5. These valuable data obtained from the geotechnical analysis can be useful for civil engineers in the design and construction of roads in Minna and environs for maximum durability and efficiency. It is recommended that engineering confirmatory test be carried out before embarking on any construction such as road. Location 5 which failed lateritic soil should be stabilized with either cement, sand; crushed stone (gravels) of ½ and 3/4 inch size in order to meet the sub-base or base course requirement.

# References

[1]    Abubakar, J.B., (2006). Geotechnical study of lateritic soil in Tipper garage, Katampe Area, Abuja Federal capital territory, pp 4-34.

[2]    Ajibade, A.C., (1972).The Geology of ZungeruSheet. Unpublished M.Sc Thesis, University of Ibadan, Nigeria.

[3]    Ajibade, A.C. and Woakes, M., (1976) Proterozoiccrustal development in the Pan African Regime. In: Geology of Nigeria, Kogbe, C (Ed.). Published by Rock View (Nigeria) Ltd, 57-63.

[4]    Amadi, A. N., Eze, C. J., Igwe, C. O., Okunlola, I. A. and Okoye, N. O., (2012). Architect's and Geologist's view on the causes of building failures in Nigeria. *Modern Applied Science,* 6 (6), 31-38.

[5]    ASTM International (2006). Standard Practice for Classification of Soils for Engineering Purposes (Unified Soil Classification System), Pp 12.

[6]    Bell, D.E., (1983). Risk premiums for decision regret management science, 29, 1156-1166.

[7]    Casagrande, A., (1952). Classification and identification of soils, transactions of the American Society of civil engineering, *Vol. 113,* pp 901.

[8]    Federal Ministry of Works and Housing (1997). General Specification for Roads and Bridges, Volume II, Federal Highway Department, FMWH: Lagos, Nigeria, 317 p.

[9]    Lundgren, R., (1969). "Field Performance of Laterite Soils," *Proceding of the 7th International Conference of Soil Mechanics and Foundation Engineering Vol. 2,* 45-57.

[10]   Nwankwoala, H. O. and Amadi, A. N., (2013). Geotechnical Investigation of Sub-soil and Rock Characteristics in parts of Shiroro-Muya-Chanchaga Area of Niger State, Nigeria. *International Journal of Earth Sciences and Engineering,* 6 (1), 8-17.

[11]   Nwankwoala, H.O., Amadi, A.N., Ushie, F.A. and Warmate, T., (2014). Determination of Subsurface Geotechnical Properties for Foundation Design and Construction in Akenfa Community, Bayelsa State, Nigeria. *American Journal of Civil Engineering and Architecture,* 2 (4), 130-135.

[12]   O'Flaherty, C. A., (1988). Highway Engineering. Vol. 2, Edward Amold Publishers, London UK.

[13]   Oke, S. A. and Amadi, A. N., (2008). An Assessment of the Geotechnical Properties of the Subsoil of parts of Federal University of Technology, Minna, Gidan Kwano Campus, for Foundation Design and Construction. *Journal of Science, Education and Technology,* 1 (2), 87-102.

[14]   Oke, S. A., Amadi, A. N., Abalaka, A. E., Nwosu, J. E. and Ajibade, S. A., (2009). Index and Compaction Properties of Laterite Deposits for Road Construction in Minna Area, Nigeria. *Nigerian Journal of Construction Technology and Management,* 10 (1&2), 28-35.

[15]   Oke, S. A., Okeke, O. E., Amadi, A. N. and Onoduku, U. S., (2009). Geotechnical Properties of the Subsoil for Designing Shallow Foundation in some selected parts of Chanchaga area, Minna, Nigeria. *Journal of Environmental Science,* 1 (1), 45-54.

[16]   Truswell, J.E and Cope, R.N., (1963). The geology of parts of Niger and Zaria province, Northern Nigeria. Bulletin No. 29. Published by the Geological Survey of Nigeria. 29 pp.

[17]   S. Army corps of Engineers (2005): Geotechnical Engineering procedures for foundation Design of Buildings and Structures. *Unified Facilities Criteria (UFC),* Pp 48.

[18]   Sheriff, O.S., (2012). Assesment of the geotechnical properties of soil in Barikin sale-saukahuta Minna, Niger state. Pp 4-50.

# Simulative Analysis of Emitted Carbon during Gas Flaring Based on Quantified Magnitudes of Produced and Flared Gases

**C. I. Nwoye[1,*], I. E. Nwosu[2], N. I. Amalu[3], S. O. Nwakpa[1], M. A. Allen[4], W. C. Onyia[5]**

[1]Department of Metallurgical and Materials Engineering, Nnamdi Azikiwe University Awka, Anambra State, Nigeria
[2]Department of Environmental Technology, Federal University of Technology, Owerri, Nigeria
[3]Project Development Institute Enugu, Nigeria
[4]Department of Mechanical Engineering, Micheal Okpara University, Umuahia, Abia State, Nigeria
[5]Department of Metallurgical and Materials Engineering, Enugu State University of Science &Technology, Enugu, Enugu State, Nigeria
*Corresponding author: nwoyennike@gmail.com

**Abstract** This paper presents a simulative analysis of emitted carbon during gas flaring based on quantified magnitudes of produced and flared gases. Results from both experiment and model prediction show that the quantity of emitted gas increases with increase in both total gas produced (TGP) and total gas flared (TGF). A two-factorial model was derived, validated and used for the empirical analysis. The derived model showed that emitted carbon is a linear function of TGP and TGF. The validity of the derived model expressed as: $\xi = 0.0513\,\vartheta + 0.0776\,\vartheta + 30.7738$ was rooted in the model core expression $\xi - 30.7738 = 0.0513\,\vartheta + 0.0776\,\vartheta$ where both sides of the expression are correspondingly approximately equal. Results from evaluations indicated that the standard error incurred in predicting emitted carbon for each value of the TGP & TGF considered, as obtained from experiment, derived model and regression model were 14.2963, 7.4141 and 14.823 & 1.3657, 7.4084 and 0.0039 % respectively. Further evaluation indicates that emitted carbon per unit TGF as obtained from experiment; derived model and regression model were 0.155, 0.154 and 0.155 Tonnes/Mscfd[-1] respectively. Comparative analysis of the correlations between emitted carbon and TGP & TGF as obtained from experiment; derived model and regression model indicated that they were all > 0.99. The maximum deviation of the model-predicted emitted carbon (from experimental results) was less than 3%. This translated into over 97% operational confidence for the derived model as well as over 0.97 reliability response coefficients of emitted carbon to TGP and TGF.

*Keywords:* emitted carbon prediction, total gas produced, total gas flared, flaring process

## 1. Introduction

The inherent negative impact of air pollution on the environment and ecosystem has raised the need to understudy the various ways in which air could be polluted, the air pollutants as well as possible control measures.

These air pollutants could be primary or secondary. The primary pollutants are emitted directly from processes ie carbon monoxide gas from motor vehicle exhaust or sulphur dioxide released from factories as well as ash from a volcanic eruption. Secondary pollutants are not emitted directly. Rather, they form in the air when primary pollutants react or interact. An important example of a secondary pollutant is ground level ozone - one of the many secondary pollutants that make up photochemical smog.

Gas flaring is the burning of natural gas and other petroleum in flare stacks of upstream oil companies in oil fields during operations. Research [1] has shown that gas flaring is the singular and most common source of global warming; contributing to emissions of carbon monoxide, nitrogen (ii) oxide and methane gas.

Reports [2,3,4] have shown that the presence of pollutants in the atmosphere causes adverse effects on human health and damages to structures. Similar studies [5] have also shown that incomplete combustion do not only contribute to global warming and climate change, as green house gases (GHG), but have major adverse health impacts including acute respiratory infections, chronic obstructive pulmonary disease, asthma, nasopharyngeal and laryngeal cancer.

Studies [2,3] have revealed that primary pollutants undergo chemical reactions to produce a wide variety of secondary pollutants when they are exposed to sunshine, production of Ozone ($O_3$) being most typical. The researchers further typified ozone as an indicator of air quality in urban atmospheres besides the health problems its molecules could cause.

Gas flaring also affects the soil pH. This results to heterogeneous levels of acidity and alkalinity within the area of the flaring process. The low soil pH values at the flare points is suspected to have been as a result of the presence of flare which produced acidic oxides of carbon and nitrogen. These oxides invariably forms carbonic and nitric acids on dissolving in rain water in accordance with research with past findings [6,7].

It has been shown [8] that emission of a variety of compounds such as volatile organic compounds (VOC's), polycyclic aromatic hydrocarbon (PAH's) as well as sooth occur when the flaring process ensues with incomplete combustion.

Following investigation [9] on the Alberta Research Council (ARC) report regarding the compounds predominantly present during incomplete combustion of flared gases, it was discovered that Volatile Organic Compounds (VOC's) and Polycyclic Aromatic Hydrocarbon (PAH's) are the predominant compounds.

It is disheartening that since 1988, neither the Federal Environmental Protection Agency (FEPA) nor the Department of Petroleum Resources (DPR) has implemented anti-flaring policies for excess associated gases nor have they monitored the emissions to ensure compliance with standards [10].

The challenges associated with the control of gas flaring has been specifically reported [8,9] to involve achievement of zero-carbon emissions (through application of Carbon Capture and Storage (CCS)) since carbon is the primary constituent of the flared gas in the form VOC's, PAH's and sooth. These reports indicate that there are no difficulties encountered in removing the other components. There is a great challenge over human health and global warming risk posed by small level leakages over long periods during storage of captured $CO_2$, as well as transportation of the $CO_2$ without any form of leakage to locations were it is needed for other operations. The high cost of this technology is the major factor militating against global conformity to zero-carbon emissions since developing countries cannot withstand the financial implication of the project. The adaptability of the technology to easily maintainable conditions should also be considered. This is because the capture technology is difficult to maintain though it is economically feasible under specific conditions. Based on the foregoing, collaborative efforts are required between chemists and chemical engineers to overcome these challenges by considering a cheaper way of achieving CCS technology, a better mode of storing and transporting captured $CO_2$.

Successful attempts [11,12,13] have been made to remove some these gases during flaring to ensure cleaner atmosphere at flaring sites.

The present work aims at carrying out a simulative analysis of emitted carbon during gas flaring based on quantified magnitudes of gas produced and flared

## 2. Materials and Method

A comprehensive survey of selected communities affected by gas flaring was carried out and their environments carefully monitored for observation. In these communities selected gas locations were considered for the study. Details of the experimental procedures are as stated in past report [1].

## 3. Results and Discussions

Table 1 indicates that the quantity of emitted carbon increases with increase in both total produced and total flared gases.

**Table 1. Variation of emitted carbon with total flared gas and total gas produced [1]**

| $(\delta)$ | $(\vartheta)$ | $(\xi)$ |
|---|---|---|
| 1062.0 | 1062.00 | 164.61 |
| 1612.9 | 1540.58 | 238.70 |
| 2633.5 | 2275.50 | 350.00 |
| 3745.0 | 3045.00 | 471.98 |
| 9009.3 | 6389.02 | 990.30 |

## 3.1. Model Formulation

Computational analysis (using C-NIKBRAN: [14]) of results in Table 1 indicates that

$$\xi - K = N\delta + S\vartheta \qquad (1)$$

Substituting the values of K, N and S into equation (1) reduces it to;

$$\xi - 30.7738 = 0.0513\,\delta + 0.0776\,\vartheta \qquad (2)$$

$$\xi = 0.0513\,\delta + 0.0776\,\vartheta + 30.7738 \qquad (3)$$

Where
K, N and S are equalizing constants
$(\xi)$ = Carbon emitted (Tonnes)
$(\delta)$ = Total gas produced (Mscf/d)
$(\vartheta)$ = Total flared gas (Mscf/d)

### 3.1.1. Boundary and Initial Conditions

The ranges of carbon emitted, total gas produced and flared are 164.61-990.3 tonnes, 1062-9009.35 and 1062-6389.02 (Mscf/d) respectively. The flow rate of the gas was assumed constant.

### 3.1.2. Model Validation

**Table 2. Variation of $\xi - 30.7738$ with $0.0513\,\delta + 0.0776\,\vartheta$**

| $\Xi - 30.7738$ | $0.0513\,\delta + 0.0776\,\vartheta$ |
|---|---|
| 133.8362 | 136.8918 |
| 207.9262 | 202.2908 |
| 319.2262 | 311.6774 |
| 441.2062 | 428.4105 |
| 959.5262 | 957.9651 |

The validity of the derived model was rooted in equation (2) where both sides of the equation are correspondingly approximately almost equal. Furthermore, equation (2) agrees with Table 2 following the values of $\xi - 30.7738$ and $0.0513\,\delta + 0.0776\,\vartheta$ evaluated from Table 1.

Furthermore, the derived model was validated by comparing the model-predicted emitted carbon and that obtained from the experiment. This was done using the 4th Degree Model Validity Test Techniques (4th DMVTT); statistical graphical, computational and deviational analysis.

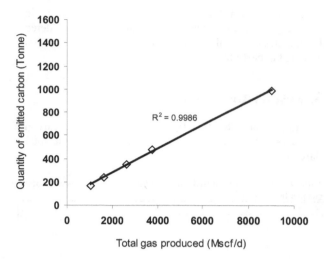

**Figure 1.** Variation of emitted carbon with total gas produced as obtained from experiment [1]

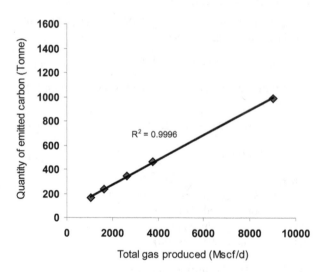

**Figure 2.** Variation of emitted carbon with total gas produced as obtained from derived model

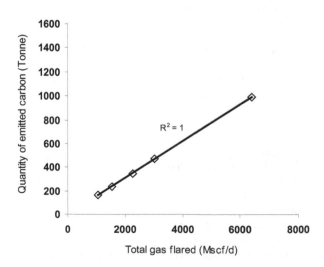

**Figure 3.** Variation of emitted carbon with total gas flared as obtained from experiment [1]

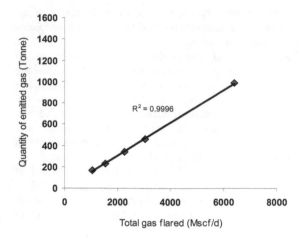

**Figure 4.** Variation of emitted carbon with total gas flared as obtained from derived model

## 3.2. Statistical Analysis

### 3.2.1. Standard Error (STEYX)

The standard errors incurred in predicting emitted carbon for each value of the TGP & TGF considered as obtained from experiment and derived model were 14.2963 and 7.4141 & 1.3657 and 7.4084 % respectively. The standard error was evaluated using Microsoft Excel version 2003.

### 3.2.2. Correlation (CORREL)

The correlation coefficient between emitted carbon and TGP & TGF were evaluated (using Microsoft Excel Version 2003) from results of the experiment and derived model. These evaluations were based on the coefficients of determination $R^2$ shown in Figure 1 - Figure 4.

$$R = \sqrt{R^2} \qquad (4)$$

The evaluated correlations are shown in Table 3 and Table 4. These evaluated results indicate that the derived model predictions are significantly reliable and hence valid considering its proximate agreement with results from actual experiment.

**Table 3. Comparison of the correlations evaluated from derived model predicted and experimental results based on TGP**

| Analysis | Based on TGP | |
|---|---|---|
| | ExD | D-Model |
| CORREL | 0.9993 | 0.9998 |

Comparative analysis of the correlations between emitted carbon and TGP & TGF as obtained from experiment; derived model and regression model indicated that they were all > 0.99.

**Table 4. Comparison of the correlations evaluated from derived model predicted and experimental results based on TGF**

| Analysis | Based on TGF | |
|---|---|---|
| | ExD | D-Model |
| CORREL | 1.0000 | 0.9998 |

## 3.3. Graphical Analysis

Comparative graphical analysis of Figure 5 and Figure 6 show very close alignment of the curves from the experimental (ExD) and model-predicted (MoD) emitted carbon.

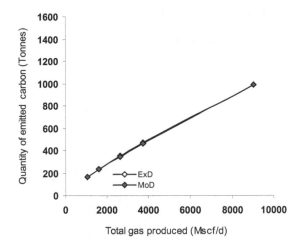

**Figure 5.** Comparison of quantities of emitted carbon (relative to TGP) as obtained from experiment and derived model

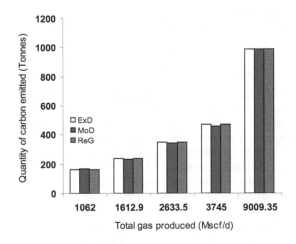

**Figure 7.** 3-D Comparison of emitted carbon (relative to TGP) as obtained from ExD, MoD and ReG

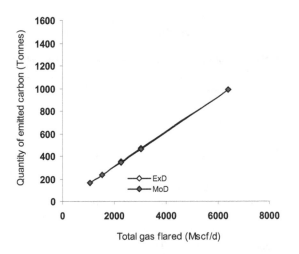

**Figure 6.** Comparison of quantities of emitted carbon (relative to TGF) as obtained from experiment and derived model

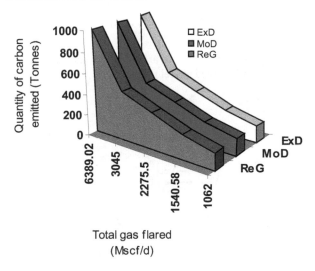

**Figure 8.** 3-D Comparison of emitted carbon (relative to TGF) as obtained from ExD, MoD and ReG

It is strongly believed that the degree of alignment of these curves is indicative of the proximate agreement between both experimental and model-predicted values of the emitted carbon.

### 3.3.1. Comparison of Derived Model with Standard Model

The validity of the derived model was further verified through application of the Least Square Method (LSM) in predicting the trend of the experimental results. Comparative analysis of Figure 7 and Figure 8 shows very close alignment of curves and areas covered by emitted carbon, which precisely translated into significantly similar trend of data point's distribution for experimental (ExD), derived model (MoD) and regression model-predicted (ReG) results of emitted carbon.

Also, the calculated correlations (from Figure 7 and Figure 8) between emitted carbon and TGP & TGF for results obtained from regression model gave 0.9992 & 1.0000 respectively. These values are in proximate agreement with both experimental and derived model-predicted results. The standard errors incurred in predicting emitted carbon for each value of the TGP & TGF considered as obtained from regression model were 14.8230 and 0.0039% respectively.

## 3.4. Computational Analysis

Computational analysis of the experimental and model-predicted emitted carbon was carried out to ascertain the degree of validity of the derived model. This was done by comparing the emitted carbon per unit TGF obtained by calculation from experimental result and model-prediction.

### 3.4.1. Emitted Carbon per unit TGF

The emitted quantity of carbon per unit TGF was calculated from the expression;

$$C_E = \Delta\xi / \Delta\vartheta \qquad (5)$$

Equation (5) is detailed as

$$C_E = \xi_2 - \xi_1 / \vartheta_2 - \vartheta_1 \qquad (6)$$

Where $\Delta\xi$ = Change in emitted carbon at two values of TGF $\vartheta_2, \vartheta_1$.

Considering the points (1062, 164.61) & (6389.02, 990.3), (1062, 167.67) & (6389.02, 988.74) and (1062,163.88) & (6389.02, 990.08) as shown in Figure 8, then designating them as $(\zeta_1, \vartheta_1)$ & $(\zeta_2, \vartheta_2)$ for experimental, derived model and regression model predicted results respectively, and also substituting them into equation (6), gives the slopes: 0.155, 0.154 and 0.155 Tonnes/ Mscfd$^{-1}$ as their respective emitted carbon per unit TGF.

## 3.5. Deviational Analysis

Critical analysis of the emitted carbon obtained from experiment and derived model show low deviations on the part of the model-predicted values relative to values obtained from the experiment. This was attributed to the fact that the flow properties of the flared gases and the physico-chemical interactions between the flared gases and the emitted carbon which played vital roles during the flaring process were not considered during the model formulation. This necessitated the introduction of correction factor, to bring the model-predicted emitted carbon to those of the corresponding experimental values.

The deviation Dv, of model-predicted emitted carbon from the corresponding experimental result is given by

$$D_v = \frac{\vartheta_{MoD} - \vartheta_{ExD}}{\vartheta_{ExD}} \times 100 \qquad (7)$$

Where $\vartheta_{ExD}$ and $\vartheta_{MoD}$ are TGF obtained from experiment and derived model respectively.

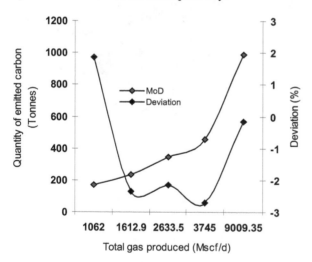

**Figure 9.** Variation of model-predicted quantity of emitted carbon (relative to TGP) with associated deviation from experiment

Deviational analysis of Figure 9 and Figure 10 indicate that the precise maximum deviation of model-predicted emitted carbon from the experimental results is 2.71%. This invariably translated into over 97% operational confidence for the derived model as well as over 0.97 reliability response coefficients of emitted carbon to TGP and TGF.

Consideration of equation (7) and critical analysis of Figure 9 and Figure 10 show that the least and highest magnitudes of deviation of the model-predicted emitted carbon (from the corresponding experimental values) are − 0.16 and - 2.71%. Figure 9 and Figure 10 indicate that these deviations correspond to emitted carbon: 988.74 and 459.18 Tonnes, TGP: 9009.3 and 3745 Mscf/d, as well as TGF: 6389.02 and 3045 Mscf/d respectively.

Correction factor, Cf to the model-predicted results is given by

$$Cf = \frac{\vartheta_{MoD} - \vartheta_{ExD}}{\vartheta_{ExD}} \times 100 \qquad (8)$$

Critical analysis of Figure 9, Figure 10 and Table 5 indicates that the evaluated correction factors are negative of the deviation as shown in equations (7) and (8).

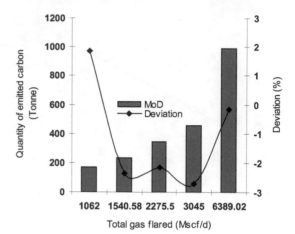

**Figure 10.** Variation of model-predicted quantity of emitted carbon (relative to TGF) with associated deviation from experiment

**Table 5. Variation of correction factor (to model-predicted emitted carbon) with TGP and TGF**

| ($\lambda$) | ($\vartheta$) | Correction factor (%) |
|---|---|---|
| 1062.0 | 1062.00 | + 1.86 |
| 1612.9 | 1540.58 | - 2.36 |
| 2633.5 | 2275.50 | - 2.14 |
| 3745.0 | 3045.00 | - 2.71 |
| 9009.3 | 6389.02 | - 0.16 |

Table 5 shows that the least and highest correction factor (to the model-predicted emitted carbon) are + 0.16 and + 2.71%. Since correction factor is the negative of deviation as shown in equations (7) and (8), Table 5, Figure 9 and Figure 10 indicate that these highlighted correction factors corresponds to to emitted carbon: 988.74 and 459.18 Tonnes, TGP: 9009.3 and 3745 Mscf/d, as well as TGF: 6389.02 and 3045 Mscf/d respectively.

The correction factor took care of the negligence of operational contributions of the flow properties of the flared gases and the physico-chemical interactions between the flared gases and the emitted carbon which actually played vital role during the flaring process. The model predicted results deviated from those of the experiment because these contributions were not considered during the model formulation. Introduction of the corresponding values of Cf from equation (8) into the model gives exactly the corresponding experimental values of emitted carbon.

It is very pertinent to state that the deviation of model predicted results from that of the experiment is just the magnitude of the value. The associated sign preceding the value signifies that the deviation is a deficit (negative sign) or surplus (positive sign).

## 4. Conclusion

Following a simulative analysis of emitted carbon (during flaring process) based on total gas produced and total gas flared, quantity of emitted gas increases with increase in both total gas produced (TGP) and total gas flared (TGF). A two-factorial model, derived and validated was used for the empirical analysis. The emitted carbon is a linear function of TGP and TGF. The validity of the derived model expressed as: was rooted in the model core expression $\xi$ - 30.7738 = 0.0513 $\vartheta$ + 0.0776 $\vartheta$ where both sides of the expression are correspondingly approximately equal. Results from evaluations indicated that the standard error incurred in predicting emitted carbon for each value of the TGP & TGF considered, as obtained from experiment, derived model and regression model were 14.2963, 7.4141 and 14.823 & 1.3657, 7.4084 and 0.0039 % respectively. Further evaluation indicates that emitted carbon per unit TGF as obtained from experiment, derived model and regression model were 0.155, 0.154 and 0.155 Tonnes/Mscfd$^{-1}$ respectively. Comparative analysis of the correlations between emitted carbon and TGP & TGF as obtained from experiment, derived model and regression model were all > 0.99. The maximum deviation of the model-predicted emitted carbon (from experimental results) was less than 3%. This translated into over 97% operational confidence for the derived model as well as over 0.97 reliability response coefficients of emitted carbon to TGP and TGF.

# References

[1] Onyejekwe, I. M. (2012). Health Impact Analyses of Gas Flaring in the Niger Delta. IREJEST, 9 (1): 94-100.

[2] Abraham, A, and Baikunth N., (2000). "Hybrid Intelligent Systems: A Review of a decade of Research", School of Computing and Information Technology, Faculty of Information Technology, Monash University, Australia, Technical Report Series, 5 (2000): 1-55.

[3] Berenji, H. R. and Khedkar, P. (1992). "Learning and Tuning Fuzzy Logic Controllers through Reinforcements", IEEE Transactions on Neural Networks, 3: 724-740.

[4] Czogala, E. and Leski, J. (2000). "Neuro-Fuzzy Intelligent Systems, Studies in Fuzziness and Soft Computing", Springer Verlag, Germany.

[5] Ezzati, M. and Kammen, D. M. (2002). Household Energy, Indoor Air Pollution and Health in Developing Countries. Knowledge Base for Effective Interventions. pp 12.

[6] Hewitt, D. N., Sturges, W. T., and Noa, A. (1995).Global Atmospheric Chemical Changes, New York. Chapman and Hall. pp 56.

[7] Botkin, D. B. and Kella, E. A. (1998): Environmental Science, Earth as living Planet. Second Edition. John Wiley and Sons, Canada. pp 123.

[8] Kindzierski, W. D. (2000). Human Environment Exposure to Hazardous Air Pollutants from Gas Flare. Environmental Reviews. 8: 41-62.

[9] Strosher, M. (1996).Investigation of Flare Gas Emissions in Alberta, Canada: Alberta Research Council.

[10] Manby, B., (1999). The Price of Oil: Corporate Responsibilty and Human Rights Violations in Nigerians Oil Producing Communities. Human Right Watch New York. pp. 202.

[11] Selective catalytic reduction, (SCR) De-NOx Technologies, http://www.de-nox-com/SCR.html

[12] Selective non-catalytic reduction (SNCR), De-NOx Technologies, http://www.de-nox.com/technology.htm

[13] Sulfur Dioxide Scrubbers, Duke Energy, http://www.duke-energy.com/environment/air-quality/sulphur-dioxide-scrubbers.asp

[14] Nwoye, C. I. (2008). Data Analytical Memory; C-NIKBRAN.

# Pd-based Catalysts for Ethanol Oxidation in Alkaline Electrolyte

**A. M. Sheikh[1,*], E. L. Silva[2], L. Moares[2], L. M. Antonini[2], Mohammed Y. Abellah[1], C. F. Malfatti[2]**

[1]Department of Mechanical Engineering, South Valley University, Qena 83521, Egypt
[2]DEMET, Federal University of Rio Grande do Sul, Porto Alegre 91501-970, RS, Brazil
*Corresponding author: ahmed.elshekh@ufrgs.br

**Abstract** Direct ethanol fuel cells (DEFCs) have received tremendous attention from academics recently since they can oxidize the liquid ethanol to produce electrical energy efficiently. In this work, five catalysts, which are single, binary, and ternary ones of Pd, Sn, and Ni respectively, are synthesized by the impregnation-reduction method to speed up the reaction kinetics of ethanol oxidation (EOR) in alkaline medium. The crystal structure is investigated by the X-ray diffraction (XRD) technique. The electrochemical performance is evaluated by the cyclic voltammetry (CV) and chronomperometry (CA) analyses. The best catalytic performance has been achieved by the bi-catalyst $Pd_{40}Ni_{60}/C$ and the tri-catalyst $Pd_{40}Ni_{50}Sn_{10}/C$. The reason, for that high performance, could be exemplified that Ni could generate $OH^-$ species at lower potentials.

*Keywords:* *ethanol oxidation, Pd catalyst, alkaline electrolyte*

## 1. Introduction

Fuel cells are electrochemical devices that could generate electrical energy by oxidation and reduction of the externally supplied fuel and oxidant respectively [1]. One recent system of fuel cells is the direct ethanol fuel cell (DEFCs) in which the liquid ethanol is supplied into the anode. There are other liquid alcohols that could be applied directly into the anode such as the traditionally used methanol, formic acid, ethylene glycol and glycerol. DEFCs have received tremendous attention from the academics and researchers for the high potential advantages that ethanol could offer as a fuel for the anodic oxidation [2]. For instance, ethanol is liquid; thus easy of transport and storage unlike the hydrogen fuel and has satisfying energy density [3]. Moreover, ethanol is produced from biomass products (corn and sugar cane), so that is renewable. And compared to methanol, ethanol is not toxic and has lower membrane crossover rate [4]. Unfortunately, however, ethanol oxidation has sluggish reaction kinetics due to the difficult-to-be-cleaved C-C bond in every molecule added to $CH_3$, and C-O bonds. Thus, there are many intermediate reaction products which are adsorbed strongly on the catalyst surface [5]. Moreover, the complete ethanol oxidation requires 12-electron release [6].

There are two DEFC systems according to the used solution; alkaline and acidic. The acidic ones have been applied extensively between the fuel cell anode and cathode, but the alkaline-electrolyte DEFC has proved intrinsic advantages as compared to the acidic counterparts: faster reaction kinetics, the non-Pt metal catalyst possibility, better water management, higher degree of freedom in selecting fuels, and no need for fluorinated polymers are some of them [1,7,8].

Especially, the use of non-Pt metal catalyst (like Pd) gives a big advantage for the alkaline-medium DEFC due to the fact that Pt is a noble metal and would increase the cell construction cost highly [1]. Pd in alkaline electrolyte has shown higher activity for ethanol oxidation reaction (EOR) compared to Pt in acidic one as result of high reaction speed in which, the $OH^-$ are the active species [5,6]. Moreover, Pd in alkaline electrolyte has been proven to have higher CO tolerance than Pt. CO-poisoning - the attachment of CO molecules into the catalyst surface - reduces the available surface for further fuel oxidation [7]. Thus, Pd-catalysts are currently the best metal alternative to Pt-ones for alcohol oxidation in alkaline solution [9,10].

The surface properties of the catalyst are important consideration for studying catalysis because the reaction takes place on the catalyst surface. Thus there is a vital relation between the activity and surface of the catalyst [9]. There are many research efforts that confirm the good effect of alloying Pd with other elements like Ni, and Sn to increase the catalyst activity. The bi-functional mechanism assumes the second element will form oxygenated species at lower applied potentials, and thus will save $O_2$ to oxidize CO species to $CO_2$ and therefore would facilitate the removal of other intermediate products from the catalyst surface. Furthermore, it supposes that, after CO oxidation, there will be more free

Pd-active sites for ethanol adsorption/oxidation. Z. Zhang *et al.* [10] have prepared PdNi catalysts supported on carbon and have found that Ni addition refreshes the Pd active sites and increases the activity. Like Zhang, Yu-Chen Wei *et al.* [9] have shown that PdNi/C has achieved better performance than Pd/C and the further $CeO_2$ addition has improved the PdNi activity. Moreover, T. Maiyalagan *et al* [11] have supported PdNi on carbon nanofibers (PdNi/CNF) and concluded that the onset EOR potential for the CNF supported catalyst is 200 mV lower than the carbon-supported one. Moreover, the EOR exchange current density for PdNi/CNF was 4 times higher than that for Pd/C [11].

Preparing bi-PdNi catalyst is still in need for further research, so that the endeavors continue to find better results for developing DEFC performance by using PdNi alloy. Zhen Qi *et al.* have fabricated nanocrystalline $Pd_{40}Ni_{60}$ by dealloying the ternary $Al_{75}Pd_{10}Ni_{15}$ alloy [12]. Although they have noticed enhanced amorphous zones and lattice distortions for $Pd_{40}Ni_{60}$, the $Pd_{40}Ni_{60}$ catalytic performance was higher than nano-porous Pd for ethanol and methanol oxidation in alkaline solution [12]. C. Qiu *et al.* have found that the electrocatalytic activity of PdNi thin films depends on the composition, structure and surface morphology obtained under different deposition conditions [13]. In addition to Ni, Sn is known as having improving catalytic effect for Pd for ethanol oxidation in alkaline medium because it could from a *ligand* modifying the electronic structure of Pd and by recovering the active sites, it could change the geometric structure [14]. Wenxin Du *et al.* have formed different-composition PdSn/C catalysts and deduced that $Pd_{86}Sn_{14}/C$ achieved the best performance [15]. Furthermore, PdSn/C has been proved more active and stable for EOR than PdRuSn/C in alkaline solution [2]. In this study, R. Modibedi et al. have used different ethanol concentrations and found that increasing the concentration up to 3M results in higher current density [2]. A similar conclusion has been drawn by Qinggang He *et al.* [7], although some times the Tafel slopes in this study shows the charge transfer for Pt faster than for Pd, yet the Pd has higher poison tolerance making it more stable [7].

In this work, single, bi-, and tri-catalysts of Pd, Ni, and Sn are synthesized by the impregnation-reduction method to improve the catalytic activity of Pd for ethanol oxidation reaction EOR in alkaline medium. Based on the literature review, it should be obvious the improving effects that Sn and Ni would introduce if alloyed with Pd for the catalyst structure and activity. Besides, adding both Ni and Sn, in a ternary alloy, would have further better results on the catalytic activity.

# 2. Experimental

## 2.1. Catalyst Preparation

The impregnation-reduction method was used to synthesize the five catalysts reported in this work. The carbon support is functionalized with nitric acid. The metal loading was 40% in every catalyst. The weights of respective three element precursors ($PdCl_2$, $NiCl_2$, and $SnCl_2$) were first calculated according to their respective weight percentage (Table 1), and then they were mixed in

a solution of ethylene glycol, which is a reducing agent, and deionized water; the percentage of ethylene glycol to deionized water was 75/25 (v/v). Then, the complete solution was dispersed in ultrasonic bath followed by adding the carbon support (0.14 mg) for 30 minutes. The total solution was then agitated in ultrasonic bath for the salts are impregnated. Following, the pH was increased to 12 using a NaOH solution. After that, it was stirred for 3 hours at 130°C followed by washing and centrifuging the catalyst materials. The final step was to dry the catalyst at 80 C.

**Table 1. The impregnated weights of metal precursors for respective catalysts**

| Catalyst | $PdCl_2$ (mg) | $SnCl_2$ (mg) | $NiCl_2$ (mg) |
|---|---|---|---|
| Pd/C | 0.156 | 0.000 | 0 |
| $Pd_{86}Sn_{14}/C$ | 0.131 | 0.028 | 0 |
| $Pd_{60}Ni_{40}/C$ | 0.100 | 0 | 0.073 |
| $Pd_{40}Ni_{60}/C$ | 0.085 | 0 | 0.093 |
| $Pd_{40}Ni_{50}Sn_{10}/C$ | 0.079 | 0.025 | 0.072 |

## 2.2. Structure Characterization

To characterize the surface structure and the chemical composition of the catalyst, the techniques of X-ray diffraction (XRD) is used. XRD is carried out by a Philips, Brucker-Axe-Simens diffractometer; model D5000, year 1992 with wave length 0.154 nm of a Cu-K$\alpha$ radiation generated at 40 kV and 40 mA and the objective is to find the chemical composition and the chacterize the crystal structure of the catalyst.

## 2.3. Electrochemical Evaluation

The electrochemical activity of the five prepared catalysts was evaluated using the techniques of cyclic voltammetry (CV), and chronomperometry (CA). A three-electrode half-cell was prepared used for both the CV and CA experiments. This cell is constituted of the reference electrode (RE), which is a saturated calomelan electrode (SCE), the counter electrode (CE), which was a platinum wire, and the working electrode (WE). WE is a graphite disc with an area of 0.29 cm$^2$. It was prepared as it follows: 5 mg of the catalyst powder were mixed with 2 mL of ethanol (Merck) and 25 μL of Nafion (5%, Aldrich). The mixture was ultrasonically suspended to obtain ink slurry. Then, 136 mL of the slurry was spread on the working electrode to form a thin layer. The three-electrode half-cell was used both in the CV and CA annalyses and the scan rate was 50 mV/s. The solution concentration, applied in both CV and CA, was 1.0 M ethanol + 1.0 M NaOH.

# 3. Results and Discussion

## 3.1. XRD Analysis

Figure1 shows the XRD diffraction patterns for the five synthesized catalysts by the impregnation-reduction method in this work. It is shown the clear diffraction peaks of *fcc* facets of Pd (111), Pd (200), Pd (220), and Pd (311) are at 40.5°, 46.9°, 68.47°, and 82.43° respectively. From the same figure, it should be clear the largest peak shift noticed from the Pd/C is for $Pd_{68}Sn_{14}/C$ catalysts for which Pd peaks (111), (200), (220), and (311) have been shifted about of 0.6° (Figure1, *right*), which means Pd and

Sn have been alloyed together. Also, it is clear from the XRD pattern of Pd$_{86}$Sn$_{14}$/C the formation of SnO$_2$ (101), and (211) peaks at 33.4° and 51.3° respectively. This result agrees with the findings of [2,15]. On the contrary, Q. He *et al.* [7] has noticed intense peaks of Sn (311) and (211) at nearly 42.5° and 75.5° on their catalyst diffractogram Pd$_{2.5}$Sn/C. About the Ni addition effect on the Pd structure, after adding the smallest Ni amount (40% in Pd$_{60}$Ni$_{40}$/C), the new PdNi peaks were closer to the original Pd peaks with a smaller shift (about 0.3° to the left) than that of Pd$_{86}$Sn$_{14}$/C ones. Furthermore, with increasing the Ni amount (to 60% in Pd$_{40}$Ni$_{60}$/C), the new PdNi peaks get even closer to the original Pd peaks and the shift in this case was even smaller about 0.2° to the left. It is clear also for Pd$_{60}$Ni$_{40}$/C and Pd$_{40}$Ni$_{60}$/C from XRD (Figure1, *left*), there are two separate Ni(OH)$_2$ peaks of (100) and (110) at 34.7° and 60.5°. These two Ni(OH)$_2$ have been noticed in the study of S. Y. Shen et al. [6] and also Z. Zhang et al. [10] have noticed them for Pd$_1$Ni$_1$/C they synthesized using NaBH$_4$ reduction. Furthermore, NiO (220) at peak at 60° had been reported [11]. The smaller peak shift and the clear formation of Ni(OH)$_2$ could be revised to the less homogeneous dispersion among the PdNi nanoparticles, although Pd and Ni form solid solutions in all elemental proportions as it is clear from their phase diagram (Figure 3). This can be declared by the study of Pai-Cheng Su et al. [8], who have argued that the degree of Ni and Pd alloying (36) is lower than that of Pd and Au revising this effect to the lower atomic radius difference between Pd and Au (7 pm) than that between Pd and Ni (13 pm). Other studies argue that the weak alloying between Pd and Ni is the result of low-temperature reduction synthesis [5,10] although adding Ni in high proportion to Pt produces very good alloys using similar methods [6]. In case of the ternary catalyst Pd$_{40}$Ni$_{50}$Sn$_{10}$/C, the Pd peaks are positioned nearly in the same diffraction angles of the binary Pd$_{40}$Ni$_{60}$/C ones.

Again in this one, the peaks of Ni(OH)$_2$ (110) and SnO$_2$(211) are noticed. There is one third peak at nearly 34 that may be revised to Ni(OH)$_2$ or SnO$_2$ formation or both of them. Conversely, adding a much greater quantity of Ni does not produce very homogeneous nano-alloy with Pd [6]. However, Pai-Cheng Su *et al.* have alloyed Pd with both Ni and Au with ratio (3:1) and found that for either the binary or the ternary Pd catalyst, the particle size will decrease after adding Ni or/and Au to Pd [8]. In this study they highlighted that the Ni and Au addition to the catalyst may suppress the grain growth of the Pd, and thus the particle size is decreased.

The Scherrer's equation (eq.1) is used to calculate the crystallite sizes for all catalysts. The palladium peak (220) is applied to deduce the crystallite sizes.

$$\tau = \frac{(k\lambda)}{\beta \cos(\theta)} \tag{1}$$

Where $\tau$ is the crystallite size nm, $\lambda$ is the wave length (=0.154 nm for Cu-K$_\alpha$), $\beta$ is the full width at half-maximum (FWHM) (in radians), k is a constant (0.94 to spherical crystallites) and $\theta$ is the diffraction angle. Depending on the XRD results, the interplanar distance $d_{hkl}$ with Miller Indices (*hkl*) for the diverse catalysts could be estimated according to Bragg's equation (2):

$$n\lambda = 2d_{hkl}.\sin\theta \tag{2}$$

Where, $n$ is the reflection order (1 for first order), $\lambda$ is the wave length (in nanometers), and $\theta$ is the half of $2\theta$ diffraction angle for the different Pd (111), and (220) peaks. Thus, it is possible to calculate the lattice parameter $a$ by equation (3).

$$d_{hkl} = \frac{a}{\sqrt{h^2 + k^2 + l^2}} \tag{3}$$

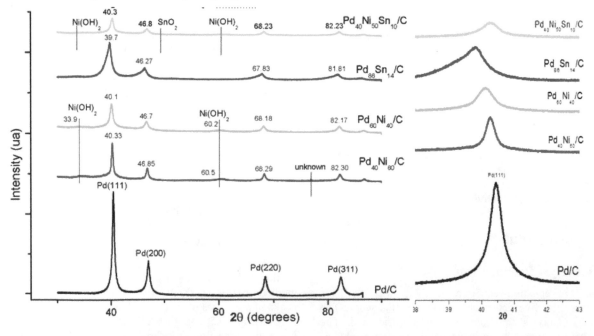

**Figure 1.** (*left*) XRD patterns of the prepared catalysts, (*right*) Enlarged Pd (111) peak

Table (2) shows the obtained crystallite sizes and lattice parameters for the different catalyst. According to this table, the Sn addition (Pd$_{86}$Sn$_{14}$/C and Pd$_{40}$Ni$_{50}$Sn$_{10}$/C)

resluts in increasing the lattice parameter although the inverse conclusion was drawn in ref [2], but more importantly the Sn presence clearly has reduced the Pd crystallite size from about 13 to 5 nm (Table 2). The study of Q. He *et al.* [7] has ended out, after adding 25% of Sn to Pd/C, the Pd crystal lattice has become orthorhombic system having *Pnma* space group with different lattice constants (a = 5.65 Å, b = 4.31 Å, and c = 8.12 Å). This effect could be explained that the maximum Sn amount, to be added to Pd/C and yet produces solid solution, is 17% as it could be declared from the PdSn phase diagram (Figure 3). The same conclusion about the effect of higher Sn proportion on the Pd crystal lattice (orthorhombic one) was confirmed by the study of W. Du et al. [15] after adding 53% of Sn to Pd/C.

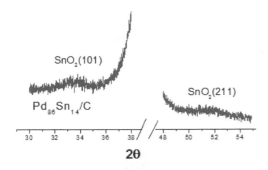

**Figure 2.** The XRD distinctive peaks for $Pd_{86}Sn_{14}/C$

**Table 2. The XRD data results for the 5 prepared catalysts**

| Catalyst | τ (nm) | a (Å) | β° |
|---|---|---|---|
| Pd/C | 13,43 | 3,87 | 0,77 |
| $Pd_{86}Sn_{14}/C$ | 5,38 | 3,92 | 1,75 |
| $Pd_{60}Ni_{40}/C$ | 11,57 | 3,89 | 0,84 |
| $Pd_{40}Ni_{60}/C$ | 15,25 | 3,88 | 0,57 |
| $Pd_{40}Ni_{50}Sn_{10}/C$ | 10,40 | 3,88 | 0,94 |

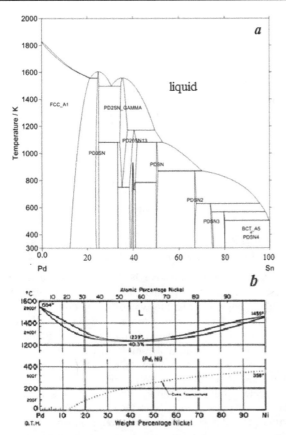

**Figure 3.** Phase diagrams of (*a*) Pd-Sn [16], (*b*) Pd-Ni [17]

On the other hand, the Ni addition (40%) and further Ni (60%) addition have caused the crystallite size to increase which may be explained by the poor alloying degree between Pd and Ni which is the same conclusion of [8]. From table (1) also, it can observed for both Pd (200) and Pd (220) that the crystallite size of the ternary $Pd_{40}Ni_{50}Sn_{10}/C$ catalyst has decreased less than those of Pd/C and $Pd_{86}Sn_{14}/C$ to go suggest that the 10% Sn addition still plays a key role in formalizing the catalyst structure even though there are 50% of nickel which is expected to increase the crystallite size as what happened with the two binary PdNi/C catalysts. Since there is only peak noticed for $Ni(OH)_2$, but not NiO, NiOOH, metallic Ni, it may be assumed that Ni, in the major part, exists in amorphous or little crystalline nature. This conclusion goes with [8].

## 3.2. Electrochemical Evaluation

The cyclic voltammetry (CV) technique is used to evaluate the electrochemical activity of the variable Pd catalysts. Figure 4 shows the CV performance for the prepared Pd catalysts supported on XC-72 Vulcan carbon. The CV voltamograms show that alloying Pd with other elements can enhance the electrochemical activity for ethanol oxidation in alkaline medium clearly. For instance, in case of Pd/C, the ethanol oxidation reaction (EOR) starts at 500 mV ($E_{onset}$) and the oxidation current peak is 48 mA, while using any other catalyst, the EOR starts at lower potential, and achieves higher current peak.

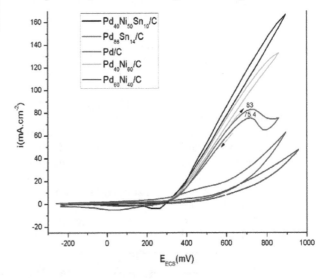

**Figure 4.** CV voltamograms for EOR for Pd catalysts in 1.0M Ethanol+1.0M NaOH (50 mV/s)

Thus, beneficial effects of adding Ni and Sn could be investigated. For example, adding Sn ($Pd_{86}Sn_{14}/C$) has decreased the $E_{onset}$ and increased both the $i_{peak}$, and ethanol diffusion in the pores of Pd/C. It is useful to consider the conclusion of Z. Liang *et al.* [5], who explained the CV analysis of Pd/C in 1M KOH + 1M ethanol. The presence of ethanol has led the $H_2$ peak suppression by the dissociative adsorption of ethanol. It results in the formation of $(CH_3CO)_{ads}$ species at the catalyst surface. They explained that the $(CH_3CO)_{ads}$ have blocked the Pd active sites, but due to the alkaline electrolyte, the OH⁻ species could be generated in lower potentials and facilitate the oxidation of them and thus increasing the current density. By the further OH⁻ formation (in the forward scan), more Pd active sites are liberated for further ethanol oxidation and thus increasing the current density until it comes to the oxidation current peak at which the oxidation of Pd surface starts. By the further oxidation of Pd surface, the active sites of Pd are blocked and thus reducing the activity until the forward scan ends. Starting the inverse scan, the PdO is reduced, thus the active sites are recovered again and the current density increases.

**Table 3. The catalytic activity data of the various catalysts**

| Catalyst | $E_{onset}$ (mV) | $i_{peak}$ (mA/cm$^2$) | D (cm$^2$.s$^{-1}$) |
|---|---|---|---|
| Pd/C | 500 | 47,5 | $3,53*10^{-10}$ |
| Pd$_{86}$Sn$_{14}$/C | 440 | 62,7 | $6,15*10^{-10}$ |
| Pd$_{60}$Ni$_{40}$/C | 330 | 83 | $1,1*10^{-9}$ |
| Pd$_{40}$Ni$_{60}$/C | 320 | 133,1 | $2,83*10^{-9}$ |
| Pd$_{40}$Ni$_{50}$Sn$_{10}$/C | 300 | 166,8 | $4,41*10^{-9}$ |

On the contrary of Liang findings [5], in this work there is no oxidation peak noticed in the forward scan, but directly after the maximum current value, the inverse scan starts. It means that the present catalyst surfaces are not oxidized during EOR. This is the case of all the catalysts in this study except Pd$_{60}$Ni$_{40}$/C which is oxidized partially and then reduced in the inverse scan. Although adding 14% of Sn has improved the Pd/C activity, adding Ni has a more prominent effect in increasing the activity. For example, adding 40% of Ni (Pd$_{60}$Ni$_{40}$/C) achieved even lower onset potential (330 mV) than Pd$_{68}$Sn$_{14}$/C for ethanol oxidation together with a noticed oxidation current peak, 83 mA/cm$^2$. This effect could be explained by the bi-functional mechanism that Ni can perform toward the catalytic Pd performance. According to [8], The Ni is capable to generate OH$_{ads}$ at lower potentials to oxidize CO-like species into CO$_2$. Furthermore, with increasing the Ni content to 60% (Pd$_{40}$Ni$_{60}$/C), the catalytic performance was even developed with E$_{onset}$ of 320 mV and the maximum current density was 133 mA.cm$^{-2}$. Thus, according to Figure 4 and Table 3, also the coefficient of diffusion has been sharpened more than Pd/C and Pd$_{86}$Sn$_{14}$/C. These promotional effects of Ni could be revised to its capacity to generate OH⁻ in lower potentials and thus help recover the Pd active sites by facilitating the oxidation of intermediate products [8,6,10,11,13].

Equation (4) is applied to evaluate the ethanol diffusion cofficient D (cm$^2$.s$^{-1}$).

$$D = \sqrt{\frac{I_{peak}}{\left(2.69\times10^5\right)n^{3/2}A\mu^{1/2}c}} \qquad (4)\ [18]$$

Where $D$ is diffusion coefficient (cm$^2$.s$^{-1}$), $I_{peak}$ is the peak current from CV graphs (A), $n$ is number of electrons, $A$=surface area = 0.29 cm$^2$ in this study, $\mu$= scan rate mV/s, and $C$= solution concentration (mol.cm$^{-3}$). Thus, according to Figure 4 and Table 3, a catalytic comparison could be made among the five catalysts through the peak current density (i$_{peak}$), oxidation onset potential (E$_{onset}$) and the coefficient of diffusion (D cm$^2$.s$^{-1}$), which was calculated according to eq. 4 [18].

P.-Cheng Su *et al.* [8] have argued that the abundant surface Ni may block the Pd sites but is beneficial for catalyst stability through the bi-functional mechanism. In this work, being used the ethylene glycol as a reducing

agent has the same role noticed for NaBH$_4$ which can reduce PdO, and thus OH adsorption on Pd surface leads to the strip away the carbonaceous species and increases the current [8].

Having seen the developing effects of Ni and Sn if added into Pd/C to increase its activity for EOR in alkaline medium, it was sought to sharpen the activity by combining Pd, Ni, and Sn in one ternary catalyst. It was necessary for the Sn proportion to be small (less than 14%), but the Ni one is better to be high (50% or more). So that the nominal proportions of tri-catalyst were 40% Pd, 50% Ni, and Sn 10%. It is clear also for this ternary alloy catalyst that it has achieved the best catalytic activity proven by the least E$_{onset}$ and the highest current peak (Figure 4). Furthermore, while using this catalyst, it was maintained the highest contact between ethanol and the Pd active sites which is based on the highest diffusion coefficient D (Table 3).

The chronomperometry test results are shown in Figure 5. It is obvious that Pd has zero activity for EOR if applied in acidic medium (red) while if the electrolyte is alkaline (black) the EOR activity is raised considerably. Furthermore, the 14% Sn addition (green) has improved the Pd stability sharply than Pd/C. The catalytic activity improvement continues with the addition of Ni (both 40 and 60%). Yet, while the current density achieved by the Ni–containing catalyst is higher than the rest of catalysts, there is a continuous fluctuation of the values up and down which may be revised the weak alloying degree between Pd and Ni on the catalyst surface.

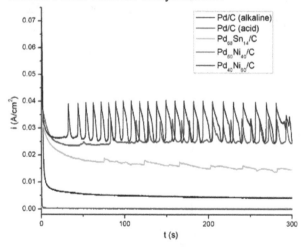

**Figure 5.** The CA results for the single and binary prepared catalysts in 1.0M EtOH + 1.0M NaOH (300 mV)

Unlike the binary catalysts, the time implemented for CA test in case of the tri catalysts was 1 hr (Figure 6). This time elongation was motivated by the conclusion of [8] who have argued that the ternary catalyst has shown low catalytic activity compared to the binary counterparts. However, when they applied long-term CA analysis they noted an obvious enhancement of the activity and they have revised it to the liberation of Pd active sites after intermediate product oxidation. On the other side, this conclusion is reinforced in this work since the catalytic activity of tri-catalyst is lower than the binary ones both in the onset of the analysis and it continues to be lower after long time.

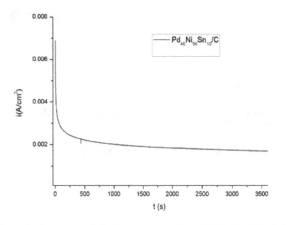

**Figure 6.** CA test result for the ternary catalyst $Pd_{40}Ni_{50}Sn_{10}$/C in 1.0M NaOH + 1.0M EtOH (300 mV)

It should be obvious from the CV and CA results that the best catalytic performance would result from the binary catalyst $Pd_{40}Ni_{60}$/C, which achieved the highest current density in the CA experiments. On the side, the ternary alloy, regardless of having the highest current densities in the CV experiments have very low current in CA analysis.

## 4. Conclusion

In this work, the impregnation-reduction technique was used to synthesize single, bi-, and tri-catalysts from Pd, Ni, and Sn for ethanol oxidation reaction EOR in alkaline-medium DEFC. Ni has enhanced the catalytic activity highly of Pd for EOR as it could generate $OH^-$ species at lower potentials. Yet, the Ni does not have good alloying degree with Pd. On the other side, adding a small amount of Sn will result in a good alloy formation between Pd and Sn, and moreover, will develop the catalytic activity for EOR though less than Ni. Combining the three elements in one tri-catalyst, however, did not result in producing a well-dispersed alloy structure though there are not strange peaks noticed from XRD. Furthermore, the tri-catalyst has shown a slightly less onset potential for EOR than the $Pd_{40}Ni_{60}$/C which means the catalytic performance is improved by alloying the three elements.

## Acknowledgements

The authors are grateful for the financial support of the agency Brazilian Government entity focused in human resources formation, (CAPES) for this work.

## Reference

[1] E. Antolini, E.R. Gonzalez, "Alkaline direct alcohol fuel cells," Journal of Power Sources 195 (2010) 3431-3450.

[2] R. M. Modibedi, T. Masombuka, M. K. Mathe, ''Carbon supported Pd–Sn and Pd–Ru–Sn nanocatalysts for ethanol electro-oxidation in alkaline medium'', International Journal of Hydrogen Energy, Vol 36, 8, April 2011, 4664-4672.

[3] Centi; Paola Lanzafame.Catalysis for alternative energy generation. Siglinda Perathoner. Springer Science+Business Media New York, Chapter 1, 2012.

[4] M.Z.F. Kamarudin; S.K. Kamarudin; M.S. Masdar; W.R.W. Daud. Review: Direct ethanol fuel cells. *International Journal of Hydrogen Energy*, V. 38, n. 22, pp. *9438-9453*, 2013.

[5] Z.X. Liang, T.S. Zhao, J.B. Xu, L.D. Zhu, "Mechanism study of the ethanol oxidation reaction on palladium in alkaline media", Electroch. Acta, Vol 54, 8, March 2009, 2203-2208.

[6] Y. Shen, T.S. Zhao, J.B. Xu, Y.S. Li, "Synthesis of PdNi catalysts for the oxidation of ethanol in alkaline direct ethanol fuel cells", J. Power Sources, Vol 195, 4, February 2010, 1001-1006.

[7] Q. He, W. Chen, S. Mukerjee, S. Chen, F. Laufek, "Carbon-supported PdM (M= Au and Sn) nanocatalysts for the electrooxidation of ethanol in high pH media, Journal of Power Sources, Volume 187, 2, February 2009, 298-304.

[8] P.-Cheng Su, H.-Shuo Chen, T.-Yao Chen, C.-Wei Liu, C.-Hao Lee, J.-Fu Lee, T.-Shan Chan, K.-Wen Wang, "Enhancement of electrochemical properties of Pd/C catalysts toward ethanol oxidation reaction in alkaline solution through Ni and Au alloying", International Journal of Hydrogen Energy, Vol 38, 11, April 2013, 4474-4482.

[9] Y.-C. Wei, C.-W. Liu, W.-D. Kang, C.-M. Lai, L.-D. Tsai, K.-W. Wang, "Electro-catalytic activity enhancement of Pd–Ni electrocatalysts for the ethanol electro-oxidation in alkaline medium: The promotional effect of $CeO_2$ addition", J. Electroanal. Chem., Vol 660, 1, September 2011, 64-70.

[10] Z. Zhang, L. Xin, K. Sun, W. Li, "Pd–Ni electrocatalysts for efficient ethanol oxidation reaction in alkaline electrolyte", Inter. J. Hydrogen Energy, Vol 36, 20, 2011, 12686-12697.

[11] T. Maiyalagan, K. Scott, "Performance of carbon nanofiber supported Pd–Ni catalysts for electro-oxidation of ethanol in alkaline medium", J. Power Sources, Vol 195, 16, 15 August 2010, 5246-5251.

[12] Z. Qi, H. Geng, X. Wang, C. Zhao, H. Ji, C. Zhang, J. Xu, Z. Zhang, "Novel nanocrystalline PdNi alloy catalyst for methanol and ethanol electro-oxidation in alkaline media", J. Power Sources, Vol 196, 14, July 2011, 5823-5828.

[13] C. Qiu, R. Shang, Y. Xie, Y. Bu, C. Li, H. Ma, "Electrocatalytic activity of bimetallic Pd–Ni thin films towards the oxidation of methanol and ethanol", Materials Chemistry and Physics, Vol 120, Issues 2-3, April 2010, 323-330.

[14] Laurent Piccolo. Nanoalloys-synthesis, structure, and properties. Springer-Verlag London 2012, pp. 369-304

[15] W. Du, K. Mackenzie, D. Milano, N. Deskins, D. Su, X. Teng," Palladium–Tin Alloyed Catalysts for the Ethanol Oxidation Reaction in an Alkaline Medium," ACS Catalysis **2012** *2* (2), 287-297.

[16] Available at http://resource.npl.co.uk/mtdata/phdiagrams/pdsn.htm, accessed on 5th Jul 2014

[17] Available at http://infomet.com.br/diagramas-fases-ver.php?e=mostrar&id_diagrama=342&btn_filtrar=Ok, accessed on 5th Jul 2014

[18] Z. Qi. PEM Fuel cells electrocatalysts and catalytic layers. *Springer Verlag*, Chapter 11, London, 2008.

# Reliability Level of Al-Mn Alloy Corrosion Rate Dependence on Its As-Cast Manganese Content and Pre-Installed Weight in Sea Water Environment

**C. I. Nwoye[1,*], P. C. Nwosu[2], E. C. Chinwuko[3], S. O. Nwakpa[1], I. E. Nwosu[4], N. E. Idenyi[5]**

[1]Department of Metallurgical and Materials Engineering, Nnamdi Azikiwe University, Awka, Nigeria
[2]Department of Mechanical Engineering, Federal Polytechnic Nekede, Nigeria
[3]Department of Industrial and Production Engineering, Nnamdi Azikiwe University, Awka, Nigeria
[4]Department of Environmental Technology, Federal University of Technology, Owerri, Nigeria
[5]Department of Industrial Physics Ebonyi State University, Abakiliki, Nigeria
*Corresponding author: nwoyennike@gmail.com

**Abstract** The corrosion rate of Al-Mn alloy in sea water environment was studied, considering the catastrophic effect of salty water on aluminum alloy. The reliability level of the alloy corrosion rate dependence on its as-cast manganese content and pre-installed weight under service condition; in sea water environment was evaluation. Analysis of the surface structures of corroded Al-Mn alloys were carried out to evaluate the phase distribution morphology. The reliability response coefficient of the alloy corrosion rate to the combined influence of as-cast manganese content $\vartheta$ and pre-installed alloy weight $\gamma$ were evaluated to ascertain the reliability of the highlighted dependence. Analysis of the surface structure of the corroded alloy revealed in all cases widely distributed oxide film of the alloy in whitish form. Increased oxidation of the as-cast Al-Mn matix (up to 3%Mn input) was due to increased permeability of the formed oxide film (to corrosive species in the sea water) as result of increased as-cast manganese content of the alloy (up to 3% Mn). Above 3% Mn input, the corrosion rate dropped indicating oxide film coherency and resistance to inflow of oxygen into the alloy. A two-factorial polynomial model was derived, validated and used for the predictive evaluation of the Al-Mn alloy corrosion rate. The validity of the model; $\zeta = -0.0347\vartheta^3 + 0.2431\vartheta^2 - 0.4848\vartheta - \gamma^2 + 0.002\gamma + 0.2863$ was rooted on the core model expression $\zeta + \gamma^2 - 0.002\gamma - 0.2863 = -0.0347\vartheta^3 + 0.2431\vartheta^2 - 0.4848\vartheta$ where both sides of the expression are correspondingly approximately equal. The corrosion rate per unit as-cast manganese content as obtained from experiment and derived model were 0.0714 and 0.0714 mm/yr/ % respectively. Standard errors incurred in predicting the corrosion rate for each value of the as-cast manganese content & pre-installed alloy weight considered as obtained from experiment and derived model were 0.0333 and 0.03346 % & 0.0350 and 0.0363 % respectively. Deviational analysis indicates that the maximum deviation of model-predicted corrosion rate from the experimental results was less than 9%. This translated into over 91% operational confidence and response level for the derived model as well as over 0.91 reliability response coefficient of corrosion rate to the collective operational contributions of as-cast manganese content and pre-installed alloy weight in the sea environment.

*Keywords:* reliability level, Al-Mn Corrosion Rate Dependence, as-cast manganese content, pre-installed weight, sea water environment

## 1. Introduction

The susceptibility of aluminum and its alloys to corrosion attack and its attendant abrupt in-service failures have raised an urgent need for intensive and extensive research & development geared towards controlling and inhibiting the corrosion attack. Achievement of a high level of quality and safety assurance in the selection and usage of metals and alloys requires a comprehensive and intensive study of the corrosion type, its mode of attack, its testing, evaluation as well as monitoring.

Metallic corrosion has been reported [1] to be basically electrochemical in nature, involving both oxidation and reduction reactions. During oxidation, the resulting metal ions may either go into the corroding solution or form an insoluble compound since there is loss of the metal atom's valence electrons. The researcher concluded that these electrons are transferred to at least one other chemical species during reduction. The scientist further reveal that the character of the corrosion environment dictates which of several possible reduction reactions will occur.

A number of metals and alloys passivate, or lose their chemical reactivity significantly, under some environmental circumstances called to play [2,3]. This phenomenon basically involve the formation of a thin protective oxide film which varies with wet-dry cycle, atmospheric conditions, type and amount of pollutants, the chemical composition and metallurgical history of the metals or alloys as well as physico-chemical properties of coating. Research [1] carried out on stainless steels and aluminum alloys revealed that they both exhibit this type of behavior. Results of the research showed that the alloy's S-shaped electrochemical potential-versus-log current density curve ideally explains the phenomenon: active-to-passive behavior. The research also showed that intersections with reduction polarization curves in active and passive regions correspond, respectively, to high and low corrosion rates.

The high level of resistance of aluminum and its alloys to corrosion in many environments stems significantly on their ability to passivates. The protective films rapidly reforms if damaged. However, it was reported [1] that a change in the character of the environment (e.g., alteration in the concentration of the active corrosive species) may cause a passivated material to revert to an active state. The report revealed that a sharp increase in corrosion rate, by as much as 100,000 times could result from subsequent damage to a preexisting passive film. The behavior is linear as it is for normal metals at relatively low potential values, within the "active" region. Furthermore, the current density suddenly decreases to a very low value that remains independent of potential with increasing potential. This is referred to as the "passive" region. Finally, the current density again increases with potential in the "transpassive" region at even higher potential values.

Al-Mn alloys has been reported [4] to be susceptible to corrosion attack if exposed in the atmosphere because of the presence of moisture. The corrosion of this alloy stems from the strong affinity aluminium has for oxygen which results to its oxidation and subsequent formation of oxide film. Similar research [5] revealed that with time, this film becomes passive to further oxidation and stable in aqueous media when the pH is between 4.0 and 8.5. It is important to state that the passive films can break and fall of, hence exposing the surface of the alloy to further corrosion.

Studies [1,6,7,8] have shown series of methods for calculating the corrosion rate. Even though there is an electric current associated with electrochemical corrosion reactions, corrosion rate can be expressed in terms of this current, or, more specifically, current density- that is, the current per unit surface area of material corroding, which is designated i [1]. The corrosion rate r, in units of mol/m$^2$s, is therefore determined using the expression;

$$\left(\frac{r}{nF}\right) = i \qquad (1)$$

where, n is the number of electrons associated with the ionization of each metal atom, and is 96,500 C/mol.

The possible exposure time for aluminium-manganese alloy (having particular corrosion rates and as-cast weights) in sea water environment has been evaluated [6] using a validated empirical model. The validity of the derived model;

$$\alpha = 26.67\,\gamma + 0.55\,\beta - 0.29 \qquad (2)$$

was rooted on the core expression: $0.0375\,\alpha = \gamma + 0.0206\,\beta - 0.0109$ where both sides of the expression are correspondingly approximately equal. The depth of corrosion penetration (at increasing corrosion rate: 0.0104-0.0157 mm/yr) as predicted by derived model and obtained from experiment are $0.7208 \times 10^{-4}$ & $1.0123 \times 10^{-4}$ mm and $2.5460 \times 10^{-4}$ & $1.8240 \times 10^{-4}$ mm (at decreasing corrosion rate: 0.0157-0.0062 mm/yr) respectively. Statistical analysis of model-predicted and experimentally evaluated exposure time for each value of as-cast weight and alloy corrosion rate considered shows a standard error of 0.0017 & 0.0044 % and 0.0140 & 0.0150 % respectively. Deviational analysis indicates that the maximum deviation of the model-predicted alloy exposure time from the corresponding experimental value is less than 10%.

Assessment of the open system corrosion rate of aluminium-manganese alloy in sea water environment was carried out [7] based on the alloy weight loss and exposure time. A model was derived and used as a tool for the assessment. It is made up of a quadratic and natural logarithmic function. The validity of the model;

$$C_R = 98.76\,\alpha^2 - 11.8051\alpha + 0.0445\ln\gamma + 0.612 \qquad (3)$$

was rooted on the core expression: $1.0126 \times 10^{-2}\,C_R = \alpha^2 - 11.9538 \times 10^{-2}\,\alpha + 4.5059 \times 10^{-4}\ln\gamma + 6.1968 \times 10^{-3}$ where both sides of the expression are correspondingly approximately equal. The resultant depth of corrosion penetration as predicted by derived model, regression model and obtained from experiment are 0.0102, 0.01 and 0.0112 mm respectively, while the corrosion rate per unit weight loss of the alloy as predicted by derived model, regression model and obtained from experiment are 7.7830, 7.6774 and 8.5777 mm/yr/g respectively. The maximum deviation of the model-predicted alloy corrosion rates from the corresponding experimental values is less than 27%. Statistical analysis of model-predicted, regression-predicted and experimentally evaluated corrosion rates for each value of exposure time and alloy weight loss considered shows a standard error of 0.0657, 0.0709 & 0.0715 % and 0.0190 & $2.83 \times 10^{-5}$ & 0.0068 % respectively.

The predictability of Al-Mn alloy corrosion rate in atmospheric environment was assessed [8] based on direct input of initial weights of the alloy and its exposure times. The validity of the two-factorial polynomial derived model;

$$\beta = -3.4674\alpha^2 + 0.3655\alpha - 0.0013\gamma^2 \\ + 0.007\gamma - 0.0031 \qquad (4)$$

is rooted on the core expression $0.2884\,\beta = -\alpha^2 + 0.1054\alpha - 3.7489 \times 10^{-4}\,\gamma^2 + 2.0186 \times 10^{-3}\gamma - 8.9396 \times 10^{-4}$ where both sides of the expression are correspondingly approximately equal. Corrosion rate per unit initial weight of exposed alloy as predicted by derived model and obtained from experiment are 1.8421 and 1.6316 (mm/yr) kg$^{-1}$ respectively. Similarly, between exposure time: 0.0192 - 0.0628 yr, the depth of corrosion penetration on the exposed alloy as predicted by derived model and obtained from experiment are $1.5260 \times 10^{-4}$ and $1.3516 \times 10^{-4}$ mm respectively. Deviational analysis indicates that the maximum deviation of the model-predicted corrosion penetration rate from the corresponding experimental value is less than 11%.

Statistical analysis of model-predicted and experimentally evaluated corrosion rates as well as depth of corrosion penetration for each value of alloy initial weight and exposure time considered show standard errors of 0.0014 and 0.0015 % as well as 9.48 x10$^{-4}$ and 8.64 x10$^{-4}$ %, respectively.

The aim of this research is to ascertain the reliability level of Al-Mn alloy corrosion rate dependence on its as-cast manganese content and its pre-installed weight in sea water environment. The difference between the present and past researches [6] and [7] is just that the possible exposure time of Al-Mn in the sea water environment was evaluated based on the alloy as-cast weight and its corrosion rate [6] while in [7], the corrosion rate of the same alloy was predicted based on its exposure time and weight loss in the sea water environment.

## 2. Materials and Methods

Materials used for the experiment are virgin aluminium of 99% purity and pure granulated manganese. The other materials used were acetone, sodium chloride, distilled water, beakers and measuring cylinders. The equipment used were lathe machine, drilling machine, crucible furnace, analytical digital weighing machine, reagents for etching and metallurgical microscope.

## 2.1. Specimen Preparation and Experimentation

**Figure 1.** As-cast Al-Mn alloy

**Figure 2.** Corroded pieces of Al-Mn alloy cut and exposed to sea water environment

Computation for each of the Al-Mn alloy compositions was carefully worked out, and the alloying materials charged into the surface crucible furnace. The molten alloy was cast into cylindrical shapes and allowed to cool in air (at room temperature). The cooled cylinders were machined to specific dimensions, cut into test samples and weighed. Each sample coupon was drilled with 5mm drill bit to provide hole for the suspension of the strings and submersion of the sample in sea water environment.

The surface of each of the test coupons was thoroughly polished (before and after corrosion) with emery cloth according to ASTM standards.

The method adopted for this phase of the research is the weight loss technique. The test coupons were exposed to the sea water environment and withdrawn after a known period of time. The withdrawn coupons were washed with distilled water, cleaned with acetone and dried in open air before weighing to determine the final weight.

## 3. Results and Discussion

### 3.1. Surface Structural Analysis of Corroded Al-Mn Alloy

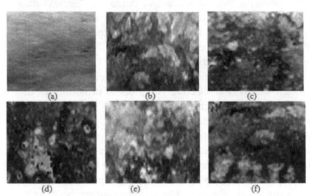

**Figure 3.** Surface structure of Al-Mn alloy (a) control (b), (c) (d), (e) and (f) for as-cast manganese contents: 2, 2.5, 2.8, 3 and 4 wt % respectively. (x200)

Figure 3(a) depicts a slightly corroded Al-Mn alloy which serves as control alloy, since it was not exposed to the sea water environment. The surface structure of this alloy shows an ash coloured background; a little away from the normal colour of aluminium, due to absence of manganese. Figure 3(a) also shows inconspicuous presence of oxide film in the control alloy exposed to room atmosphere.

Figure 3(b - f) show different levels of intensive corrosion attacks at different as-cast manganese contents. The white patches observed in Figure 3(b – f) are strongly believed to be oxide films produced during the initial corrosion attack on the Al-Mn alloy. Also, the greenish patches also observed in Figure 3(b – f) were attributed to increased oxidation of the alloy matrix as a result of increased as-cast manganese content in alloy exposed to the sea water environment.

Table 1 and Table 2 show that the corrosion rate of Al-Mn alloy increases with increase in the alloy as-cast manganese content (up to 3%). Increased alloy as-cast manganese content ensures prolonged physico-chemical interactions between the Al-Mn alloy and corrosion-induced agents resident in the sea water. Increased pre-installed weight (2-3% Mn input) increased the alloy corrosion rate through greater metal/alloy removal per unit area following activities of the corrosion-induced aggressive species (resident in the sea water) on the alloy. Table 1 and Table 2 present similar results except the conversion of alloy pre-installed weight from gramme (g) to kilogramme (Kg).

**Table 1. Variation of corrosion rate $\zeta$ of Al-Mn alloy with its as-cast manganese content $\vartheta$ and pre-installed weight $\gamma$**

| ($\zeta$) (mm/yr) | ($\vartheta$) (%) | ($\gamma$) (g) |
|---|---|---|
| 0.0099 | 1.0 | 11.812 |
| 0.0115 | 2.0 | 12.513 |
| 0.0472 | 2.5 | 12.421 |
| 0.0687 | 2.8 | 12.221 |
| 0.0829 | 3.0 | 12.112 |
| 0.0159 | 4.0 | 12.498 |

**Table 2. Variation of corrosion rate $\zeta$ of Al-Mn alloy with its as-cast manganese content $\vartheta$ and pre-installed weight $\gamma$ (in kg)**

| ($\zeta$) (mm/yr) | ($\vartheta$) (%) | ($\gamma$) (kg) |
|---|---|---|
| 0.0099 | 1.0 | 0.0118 |
| 0.0115 | 2.0 | 0.0125 |
| 0.0472 | 2.5 | 0.0124 |
| 0.0687 | 2.8 | 0.0122 |
| 0.0829 | 3.0 | 0.0121 |
| 0.0159 | 4.0 | 0.0125 |

Aluminum and its alloys are highly corrosion resistant in many environments due to its ability to passivate, and the protection offered by the initially formed film. Table 2 suggests that increased oxidation of the as-cast Al-Mn matix (up to 3%Mn input) was due to increased permeability of the formed oxide film (to corrosive species in the sea water) as result of increased as-cast manganese content of the alloy (up to 3% Mn). Above 3% Mn input, the corrosion rate dropped (Table 2), indicating oxide film coherency and resistance to inflow of oxygen into the alloy. This implied that the increased corrosion attack (as input Mn increased up to 3%) on the alloy occurred because the passivated material initially formed on the alloy were increasingly permeable to the active corrosive species in the sea water within this as-cast Mn content.

### 3.1.1. Boundary and Initial Conditions

Consider solid Al-Mn alloy exposed to atmosphere environment and interacting with some corrosion-induced agents. The atmosphere is assumed to be affected by unwanted dissolved gases. Range of input manganese concentration considered: 1.0 – 4.0%. Alloy pre-installed weight range considered: 0.0118-0.0125 kg. The quantity and purity of aluminium used were 99 wt % and 99% respectively.

The boundary conditions are: aerobic environment to enhance Al-Mn alloy oxidation (since the atmosphere contains oxygen. At the bottom of the exposed alloy, a zero gradient for the gas scalar are assumed. The exposed alloy is stationary. The sides of the solid are taken to be symmetries.

Computational analysis of generated experimental data shown in Table 2, gave rise to Table 3 which indicate that;

$$\zeta + \gamma^2 - K\gamma - S = -N\vartheta^3 + S_e \vartheta^2 - N_e \vartheta \qquad (5)$$

Introducing the values of K, S, N, $S_e$, and $N_e$ into equation (5) reduces it to;

$$\zeta + \gamma^2 - 0.002\,\gamma - 0.2863 = -0.0347\,\vartheta^3 + 0.2431\vartheta^2 - 0.4848\,\vartheta \qquad (6)$$

$$\zeta = -0.0347\,\vartheta^3 + 0.2431\,\vartheta^2 - 0.4848\,\vartheta - \gamma^2 + 0.002\,\gamma + 0.2863 \qquad (7)$$

Where K = 0.002, S = 0.2863, N = 0.0347, $S_e$= 0.2431 and $N_e$ = 0.4848 are empirical constants (determined using C-NIKBRAN [9]

($\vartheta$) = As-cast manganese content (%)

($\gamma$) =Alloy pre-installed weight (kg)

($\zeta$) = Corrosion rate (mm/yr)

The derived model is equation (7). Computational analysis of Table 2 gave rise to Table 3. The derived model is two-factorial polynomial in nature since it is composed of two input process factors: as-cast manganese content and alloy pre-installed weight. This implies that the predicted corrosion rate of Al-Mn alloy in the sea water environment is dependent on just two factors: as-cast manganese content and alloy pre-installed weight.

**Table 3. Variation of $\zeta + \gamma^2 - 0.002\,\gamma - 0.2863$ with $- 0.0347\,\vartheta^3 + 0.2431\vartheta^2 - 0.4848\,\vartheta$**

| $\zeta + \gamma^2 - 0.002\,\gamma - 0.2863$ | $- 0.0347\,\vartheta^3 + 0.2431\vartheta^2 - 0.4848\,\vartheta$ |
|---|---|
| - 0.2763 | - 0.2764 |
| - 0.2749 | - 0.2748 |
| - 0.2390 | - 0.2348 |
| - 0.2175 | - 0.2132 |
| - 0.2033 | - 0.2034 |
| - 0.2703 | - 0.2704 |

### 3.2. Model Validity

The validity of the model is strongly rooted on equation (6) (core model equation) where both sides of the equation are correspondingly approximately equal. Table 3 also agrees with equation (6) following the values of $\zeta + \gamma^2 - 0.002\,\gamma - 0.2863$ and $- 0.0347\,\vartheta^3 + 0.2431\,\vartheta^2 - 0.4848\,\vartheta$ evaluated from the experimental results in Table 1. Furthermore, the derived model was validated by comparing the corrosion rate predicted by the model and that obtained from the experiment. This was done using various analytical techniques.

### 3.2.1. Statistical Analysis

#### 3.2.1.1. Standard Error (STEYX)

The standard errors incurred in predicting Al-Mn corrosion rate for each value of as-cast manganese content & pre-installed weight considered as obtained from experiment and derived model were 0.0333 and 0.03346% & 0.0350 and 0.0363 % respectively. The standard error was evaluated using Microsoft Excel version 2003.

#### 3.2.1.2. Correlation

The correlation coefficient between Al-Mn corrosion rate and as-cast manganese content and alloy pre-installed weight were evaluated (using Microsoft Excel Version 2003) from results of the experiment and derived model. These evaluations were based on the coefficients of determination $R^2$ shown in Figure 4- Figure 7.

$$R = \sqrt{R^2} \qquad (8)$$

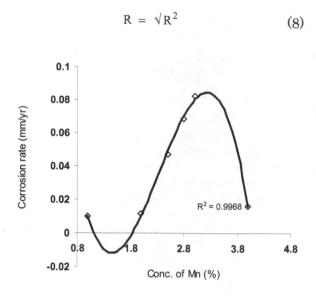

**Figure 4.** Coefficient of determination between Al-Mn alloy corrosion rate and as-cast manganese content as obtained from the experiment

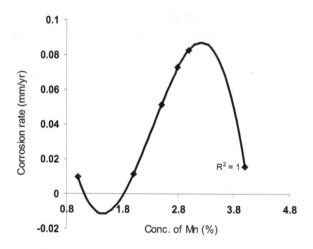

**Figure 5.** Coefficient of determination between Al-alloy Mn corrosion rate and as-cast manganese content as predicted by derived model

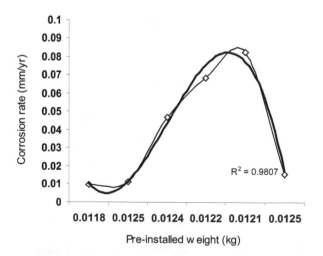

**Figure 6.** Coefficient of determination between Al-Mn alloy corrosion rate and its pre-installed weight as obtained from the experiment

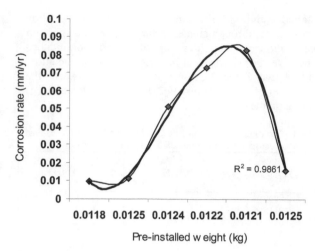

**Figure 7.** Coefficient of determination between Al-Mn alloy corrosion rate and its pre-installed weight as obtained from the experiment

**Table 4. Comparison of the correlations between corrosion rate and as-cast manganese content as evaluated from experimental (ExD) and derived model (MoD) predicted results**

| Analysis | Based on as-cast Mn content | |
|---|---|---|
| | ExD | D-Model |
| CORREL | 0.9984 | 1.0000 |

The evaluated correlations are shown in Tables 4 and 5. These evaluated results indicate that the derived model predictions are significantly reliable and hence valid considering its proximate agreement with results from actual experiment.

**Table 5. Comparison of the correlations between corrosion rate and pre-installed weight and as evaluated from experimental and derived model predicted results**

| Analysis | Based on pre-installed weight | |
|---|---|---|
| | ExD | D-Model |
| CORREL | 0.9903 | 0.9930 |

## 3.3. Graphical Analysis

Graphical analysis of Figure 8 and Figure 9 shows very close alignment of the curves from derived model and experiment. It is strongly believed that the degree of alignment of these curves is indicative of the proximate agreement between ExD and MoD predicted results.

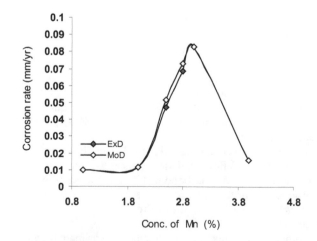

**Figure 8.** Comparison of the Al-Mn alloy corrosion rate (relative to as-cast manganese content) as obtained from experiment and derived model

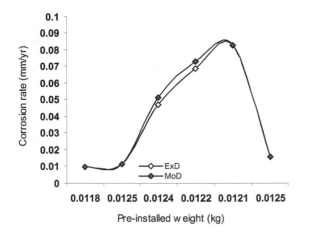

**Figure 9.** Comparison of the Al-Mn corrosion rate (relative to alloy pre-installed weight) as obtained from experiment and derived model

## 3.4. Computational Analysis

Computational analysis of the experimental and model-predicted corrosion rate per unit as-cast manganese content was carried out to ascertain the degree of validity of the derived model. This was done by comparing the corrosion rate per unit as-cast manganese content obtained by calculation, using experimental and model-predicted results.

*Corrosion rate of Al-Mn alloy per unit as-cast manganese content during the period of exposure in sea water environment $\zeta_S$ (mm/yr /%) was calculated from the equation;*

$$\zeta_S = \zeta / \vartheta \qquad (9)$$

Rewritten as

$$\zeta_S = \Delta\zeta / \Delta\vartheta \qquad (10)$$

Equation (10) is detailed as

$$\zeta_S = \zeta_2 - \zeta_1 / \vartheta_2 - \vartheta_1 \qquad (11)$$

Where

$\Delta\zeta$ = Change in the corrosion rates ($\zeta_2 - \zeta_1$) at two values of input manganese concentration $\vartheta_2$, $\vartheta_1$

$\Delta\vartheta$ = Change in the pre-installed weights $\vartheta_2$, $\vartheta_1$

Considering the points (2, 0.0115) & (3, 0.0829), and (2, 0.0114) & (3, 0.0828) as shown in Figure 4, 5 and 8 then designating them as ($\zeta_1$, $\vartheta_1$) & ($\zeta_2$, $\vartheta_2$) for experimental and derived model predicted results respectively, and also substituting them into equation (11), gives positive slopes: 0.0714 and 0.0714 mm/yr /% as their corrosion rate per unit input manganese concentration respectively.

## 3.5 Deviational Analysis

Comparative analysis of the corrosion rates precisely obtained from experiment and derived model shows that the model-predicted values deviated from experimental results. This was attributed to the fact that the effects of the surface properties of the Al-Mn alloy which played vital roles during the corrosion process were not considered during the model formulation. This necessitated the introduction of correction factor, to bring the model-predicted corrosion rate to those of the corresponding experimental values.

The deviation Dv, of model-predicted corrosion rate from the corresponding experimental result was given by

$$Dv = \frac{\zeta_{MoD} - \zeta_{ExD}}{\zeta_{ExD}} \times 100 \qquad (12)$$

Where $\zeta_{ExD}$ and $\zeta_{MoD}$ are corrosion rates evaluated from experiment and derived model respectively.

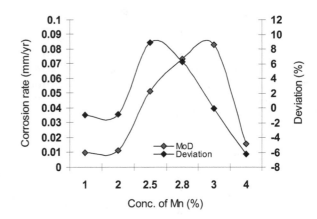

**Figure 10.** Variation of model-predicted corrosion rate with associated deviation from experimental results (relative to as-cast manganese content)

Figure 10 and Figure 11 show that the least and highest magnitudes of deviation of the model-predicted corrosion rate (from the corresponding experimental values) are − 0.12 and + 8.89%.

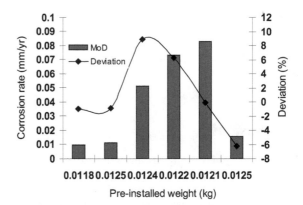

**Figure 11.** Variation of model-predicted corrosion rate with associated deviation from experimental results (relative to pre-installed weight)

It could be seen from Figure 10 and Figure 11 that these deviations correspond to corrosion rates: 0.0828 and 0.0514 mm/yr, as-cast manganese contents: 3 and 2.5% as well as alloy pre-installed weights: 0.0121 and 0.0124 kg respectively.

Comparative analysis of Figure 10 and Figure 11 also shows that the maximum deviation of model-predicted corrosion rate from the experimental results was less than 9%. This translated into over 91% operational confidence and response level for the derived model as well as over 0.91 reliability response coefficient of corrosion rate to the collective operational contributions of the as-cast manganese content and pre-installed alloy weight (under service) in the sea water environment.

Correction factor, Cf to the model-predicted results was given by

$$Cf = \frac{\zeta_{MoD} - \zeta_{ExD}}{\zeta_{ExD}} \times 100 \qquad (13)$$

Analysis of Table 6 in comparison with Figure 10 and 11 indicates that the evaluated correction factors are negative of the deviation as shown in equations (12) and (13).

The correction factor took care of the negligence of operational contributions of the effects of surface properties of the Al-Mn alloy which actually affected the corrosion process. The model predicted results deviated from those of the experiment because these contributions were not considered during the model formulation. Introduction of the corresponding values of Cf from equation (13) into the model gives exactly the corresponding experimental corrosion rate

**Table 6. Variation of correction factor with as-cast manganese content and pre-installed alloy weight**

| Mn $\vartheta$ (%) | Weight $\gamma$ (kg) | Cf (%) |
|---|---|---|
| 1 | 0.0118 | - 1.01 |
| 2 | 0.0125 | - 0.87 |
| 2.5 | 0.0124 | + 8.89 |
| 2.8 | 0.0122 | + 6.26 |
| 3 | 0.0121 | - 0.12 |
| 4 | 0.0125 | - 6.29 |

Table 6 shows that the least and highest magnitudes of correction factor to the model-predicted corrosion rate are + 0.12 and - 8.89%. Since correction factor is the negative of deviation (equations (12) and (13)), Table 6, Figure 10 and 11 indicate that these correction factors correspond to corrosion rates: 0.0828 and 0.0514 mm/yr, as-cast manganese contents: 3 and 2.5% as well as alloy pre-installed weights: 0.0121 and 0.0124 kg respectively.

It is important to state that the deviation of model predicted results from that of the experiment is just the magnitude of the value. The associated sign preceding the value signifies that the deviation is a deficit (negative sign) or surplus (positive sign).

# 4. Conclusions

The reliability level of Al-Mn alloy corrosion rate dependence on its as-cast manganese content and pre-installed weight under service condition; in sea water environment has been evaluation. Increased oxidation of the as-cast Al-Mn matix (up to 3%Mn input) was due to increased permeability of the formed oxide film (to corrosive species in the sea water) as result of increased as-cast manganese content of the alloy (up to 3% Mn). Above 3% Mn input, the corrosion rate dropped indicating oxide film coherency and resistance to inflow of oxygen into the alloy. Oxide films of the corroded Al-Mn alloy were widely distributed in whitish form within the alloy surface structures. The validity of the two-factorial polynomial model derived, validated and used for the predictive evaluation. was rooted in the core model expression $\zeta + \gamma^2 - 0.002 \gamma - 0.2863 = - 0.0347 \vartheta^3 + 0.2431 \vartheta^2 - 0.4848 \vartheta$ where both sides of the expression are correspondingly approximately equal. The corrosion rate per unit as-cast manganese content as obtained from experiment and derived model were 0.0714 and 0.0714 mm/yr/ % respectively. Standard errors incurred in predicting the corrosion rate for each value of the as-cast manganese content & pre-installed alloy weight considered as obtained from experiment and derived model were 0.0333 and 0.03346 % & 0.0350 and 0.0363 % respectively. Deviational analysis indicates that the maximum deviation of model-predicted corrosion rate from the experimental results was less than 9%. This translated into over 91% operational confidence and response level for the derived model as well as over 0.91 reliability response coefficient of corrosion rate to the collective operational contributions of as-cast manganese content and pre-installed alloy weight in the sea environment.

# References

[1] Callister Jr, W. D. (2007). Materials Science and Engineering, 7th Edition, John Wiley & Sons Inc., USA.

[2] Ekuma, C. E., and Idenyi, N. E. (2007). Statistical Analysis of the influence of Environment on Prediction of Corrosion from its Parameters. Res. J. Phy., USA, 1(1):27-34.

[3] Stratmann, S. G., and Strekcel, H. (1990). On the Atmospheric Corrosion of Metals which are Covered with Thin Electrolyte Layers. II. Experimental Results. Corros. Sci.., 30:697-714.

[4] Polmear, I., J. (1981). Light Alloys. Edward Arnold Publishers Ltd.

[5] Ekuma, C. E., Idenyi, N. E., and Umahi, A. E. (2007). The Effects of Zinc Addition on the Corrosion Susceptibility of Aluminium Alloys in Various Tetraoxosulphate (vi) Acid Environments. J. of Appl. Sci., 7(2):237-241.

[6] Nwoye, C. I., Neife, S., Ameh, E. M., Nwobasi, A. and Idenyi, N. E. (2013). Predictability of Al-Mn Alloy Exposure Time Based on Its As-Cast Weight and Corrosion Rate in Sea Water Environment. Journal of Minerals and Materials Characterization and Engineering, 1:307-314.

[7] Nwoye, C. I., Idenyi, N. E., Asuke, A. and Ameh, E. M. (2013). Open System Assessment of Corrosion Rate of Aluminum-Manganese Alloy in Sea Water Environment Based on Exposure Time and Alloy Weight Loss. J. Mater. Environ. Sci., 4(6): 943-952. http://dx.doi.org/10.4236/jmmce.2013.16046

[8] Nwoye, C. I., Idenyi, N. E., and Odo, J. U. (2012). Predictability of Corrosion Rates of Aluminum-Manganese Alloys Based on Initial Weights and Exposure Time in the Atmosphere, Nigerian Journal of Materials Science and Engineering, 3(1):8-14.

[9] Nwoye, C. I. (2008). C-NIKBRAN Data Analytical Memory (Software).

# Statistical Assessment of Groundwater Quality in Ogbomosho, Southwest Nigeria

**Olasehinde P. I.[1], Amadi A. N.[1,*], Dan-Hassan M. A.[2], Jimoh M. O.[1], Okunlola I. A[3]**

[1]Department of Geology, Federal University of Technology, Minna, Nigeria
[2]Rural Water Supply and Sanitation Department, FCT Water Board, Garki, Abuja
[3]Department of Chemical and Geological Sciences, Al-Hikmah University, Ilorin, Nigeria
*Corresponding author: geoama76@gmail.com

**Abstract** Groundwater quality in Ogbomosho area of southwest Nigeria was investigated in this study using multivariate statistical analysis. Factor analysis was applied to the Hydrochemical data in order to extract the principal factors responsible for the different Hydrochemical facies. By using Kaiser Normalization, the principal factors were extracted from the data. The analysis reveals six sources of solutes which correspond to six possible sources of groundwater pollution. Five factors (1, 2, 3, 4 and 6) originate from the natural sources while factor 5 is from anthropogenic source. Based on the calculated water quality index, the groundwater in the area falls under poor water and it was attributed to the enrichment of the groundwater with fluoride, major ions and heavy metals. The water type in the area is calcium-bicarbonate type. The efficacy of factor analysis and water quality index in the characterization of groundwater geochemistry in Ogbomosho, southwest Nigeria has been demonstrated in the present study.

**Keywords:** *statistical assessment, groundwater quality, ogbomosho, southwest Nigeria*

## 1. Introduction

Water supply and good sanitation remains a vital component of urban and rural infrastructure. The insufficiency in quality and quantity of water supply remains a challenge in many developing countries. Water supply in many urban and semi-urban area in Nigeria are grossly inadequate and most villagers trek several kilometers in search for water in rural areas. The absence or seasonal nature of surface water sources has shifted attention in the exploration and development of groundwater resources. According to WHO (2006) more than one billion people lack access to good water supply and sanitation globally.

Groundwater forms a major source of drinking water and a crucial part in the maintenance of plant and animal life (Amadi, 2010). It is a vital resource of life but is increasingly being polluted in the wake of modern civilization, industrialization, urbanization and population growth that has led to the degradation of groundwater resources as well as several other anthropogenic activities that are impacting daily on the area (Yisa and Jimoh, 2010). Water being an important component of the ecosystem, any imbalance either in quantity or quality affects the whole ecosystem negatively (Aminu and Amadi, 2014).

Groundwater contains impurities whose nature and amount vary. Metals are introduced in the groundwater system through weathering of rocks and leaching of soils, dissolution of aerosol particles and other human activities such as mining and metal processing. The increase in the use of metal based fertilizer in agricultural revolution of the government could result in continued rise in the concentration of metal pollution in shallow freshwater aquifers due to surface run off and infiltration mechanism (Amadi *et al.*, 2014). Studies revealed that 85% of all communicable diseases affecting humans are either water borne or water related (WHO, 2006; Amadi *et al.*, 2013). The need to evaluate the quality of groundwater in Ogbomosho and environs for domestic and other purposes gave rise to this study.

## 2. Material and Methods

### 2.1. Location and Accessibility

Ogbomosho is situated at 57 km northwest of Oshogbo, capital of Osun State, Southwestern Nigeria. Ogbomosho town lies between longitudes 4°10′E to 4°20′E of the Greenwich Meridian and between latitudes 8°00′N to 8°15′N of the Equator. The area is accessible through Ogbomosho-Ilorin road. Ogbomosho is relatively rugged with undulating topography with elevation ranging between 330 m and 390 m averaging about 360 m above the sea level (Figure 1). The study area falls within a tropical rain forest and is characterized by several hills. The area is well drained by several rivers (Figure 2).

## 2.2. Geology and Hydrogeology of the Area

The major lithologic units in the area include Migmatite-gneiss, granites and quartzites (Figure 3). The quartzites occur as long elongated ridges trending NW-SE and are mostly massive. The gneisses are the most dominant rock type. They occur as granite-gneiss and banded gneiss with coarsed to medium grained texture. Noticeable minerals include quartz, feldspar and biotite. Structural features exhibited by these rocks are foliation, faults, joints and micro-folds which have implications on groundwater potential. Basement complex rocks are regarded as poor aquifers because of the lack primary porosity and permeability. However, secondary porosity and permeability are imposed on them by fracturing, fissuring, jointing and weathering through which water percolates and migrates.

## 2.3. Sampling

Sampling stations were selected, taking into account the direction of groundwater flow, direction of prevailing winds and the density of the population within the studied area. Glassware and vessels were treated in 10% (v/v) nitric acid solution for 24 h and were washed with distilled and deionized water. The samples for cation determination were collected in polypropylene containers, labeled and immediately few drops of $HNO_3$ (ultra-pure grade) to pH < 2 were added to prevent loss of metals, bacterial and fungal growth and then stored in a refrigerator while samples for anion analysis were collected in glass containers. The physical parameters were determined insitu in the field using appropriate instruments. The samples were stored on ice in cooler boxes and transported to the laboratory. A GPS was used to mark and identify the sampling sites and the values obtained were used to generate the digital terrain model of the area (Figure 4).

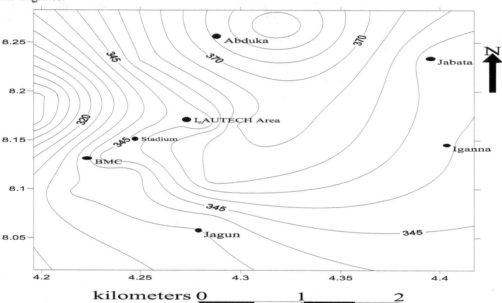

**Figure 1.** Topo map of the study area

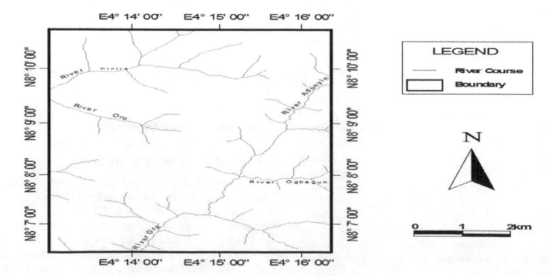

**Figure 2.** Drainage map of the study area

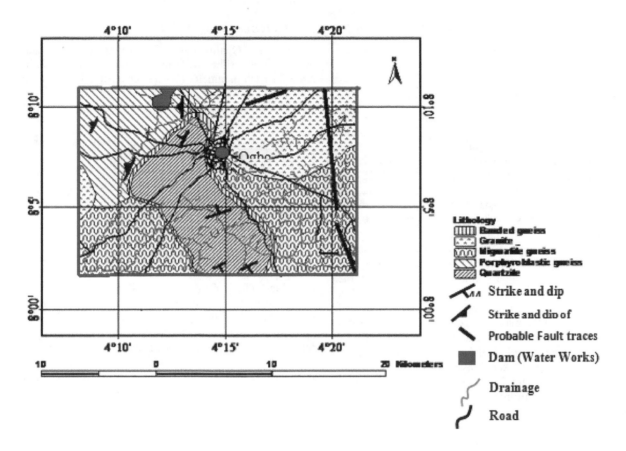

**Figure 3.** Geology map of the area

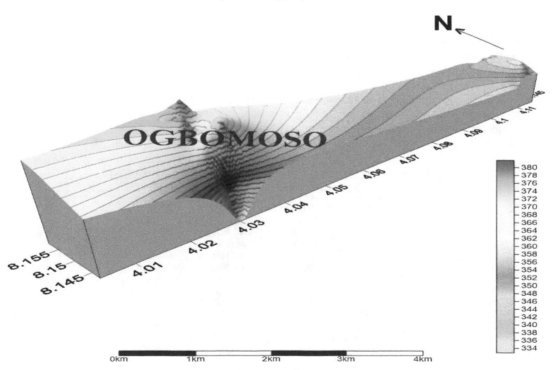

**Figure 4.** Digital terrain model of Ogbomosho, Southwest Nigeria

## 2.4. Laboratory Analysis

The method used for the determination of these physico-chemical parameters was described by APHA, 2005). The chemicals and reagent used for the analysis were of analar grade. The pH and conductivity were determined with a consort digital pH meter and consort digital conductometer respectively. JENWAY 6310 spectrophotometer was used to determine calcium and magnesium while JENWAY PFP-7 flame photometer was used for the determination of potassium and sodium. The heavy metals were analyzed using Atomic Absorption

Spectrophotometer, Model Pye Unicam SP-9 Cambridge, UK.

## 2.5. Statistical Analysis

### 2.5.1. Factor Analysis

Factor analysis (FA) is a statistical technique that focuses on data reduction in order to identify a small number of factors that explain most of the variables observed in a much larger number of manifest variables (Abdullah and Aris, 2007; Amadi, et al., 2010). It attempts to identify new underlying variables or factors that give a better understanding of the pattern of correlation within a set of observed variables (Praus, 2005). Factor analysis is based more on explaining the covariance structure of the variables than with explaining the variances (Lambarkis, et al., 2004). The purpose of factor analysis is to interpret the structure within the variance-covariance matrix of a multivariate data collection. It uses the extraction of the eigenvalues and eigenvectors from the matrix of correlation or covariance. The information gained about the interdependencies between observed variables can be used later to reduce the set of variables in a dataset (Prasad and Narayana, 2004; Olobaniyi and Owoyemi, 2006). SPSS-window-16 version was the statistical software used to perform factor analysis on the datasets.

## 2.6. Water Quality Index

Water quality index (WQI) is one of the most efficient and effective means of communicating information about the quality of water to all stakeholders in the water sector. It is a good platform for the assessment and management of water resources. It is a scale used to estimate an overall quality of water based on the values of the water quality parameters (Amadi, 2011). It is a composite rating that reflects the impact of different water quality parameters on a given water resources. WQI is calculated from the point view of the suitability of groundwater for human consumption (Amadi et al., 2010).

## 2.7. Calculation of WQI

The Water Quality Index (WQI) was calculated using the Weighted Arithmetic Index method. The quality rating scale for each parameter ($qi$) was calculated by using this equation:

$$qi = (Ci / Si) \times 100$$

A quality rating scale $(qi)$ for each parameter is assigned by dividing the mean concentration $(Ci)$ in each water sample by its respective standard $(Si)$ and the result multiplied by 100.

Similarly, relative weight $(Wi)$ was calculated using the equation:

$$Wi = 1 / Si$$

Thus the relative weight is inversely proportional to the recommended standard $(Si)$ of the corresponding parameter. The overall Water Quality Index (WQI) was calculated by aggregating the quality rating $(qi)$ with unit weight $(wi)$ linearly. Where: qi: the quality of the ith parameter, wi: the unit weight of the ith parameter and n: the number of the parameter considered.

Generally, WQI were discussed for a specific and intended use of water. In this study the WQI for drinking purposes is considered and permissible WQI for the drinking water is taken as 100. The WQI value less than 100 implies that the water is of good quality while values greater than 100 are an indication that the water is poor in quality.

# 3. Results and Discussion

The statistical summary of the physico-chemical parameters of groundwater samples in Ogbomosho and environs are contained in Table 1. The computed WQI values of the groundwater are shown in Table 2 while the standard water quality classification is summarized in Tables 3. The results of Varimax rotated factor loading on the data are illustrated in Table 4.

**Table 1. Statistical summary of the Groundwater physico-chemical parameters analyzed**

| Parameters | Min. | Max. | Mean | Range | St.Deviation | Variance | Skewness | Kurtosis |
|---|---|---|---|---|---|---|---|---|
| Alkalinity | 26.00 | 69.00 | 44.4429 | 43.00 | 15.63350 | 244.406 | 0.306 | -0.892 |
| Bicarbonae | 48.00 | 84.00 | 64.0714 | 36.00 | 11.89127 | 141.402 | 0.382 | 0.312 |
| Cadmium | 0.00 | 0.01 | 0.0030 | 0.01 | 0.00216 | 0.000 | 0.000 | -1.200 |
| Calc.Hardness | 135.00 | 604.00 | 3.3757E2 | 469.00 | 177.86124 | 3.163E4 | 0.281 | -1.451 |
| Calcium | 69.00 | 301.00 | 1.8143E2 | 232.00 | 88.62629 | 7.855E3 | 0.046 | -1.828 |
| Carbonate | 15.30 | 35.00 | 22.8857 | 19.70 | 6.63486 | 44.021 | 1.011 | 0.793 |
| Chromium | 0.01 | 0.09 | 0.0457 | 0.08 | 0.02992 | 0.001 | 0.528 | -1.117 |
| Colour | 0.08 | 0.82 | 0.3243 | 0.74 | 0.25967 | 0.067 | 1.294 | 1.473 |
| Conductivity | 6.87 | 1705.00 | 5.5087E2 | 1698.13 | 604.49632 | 3.654E5 | 1.234 | 1.542 |
| Copper | 0.10 | 1.06 | 0.4171 | 0.96 | 0.35617 | 0.127 | 1.166 | 0.433 |
| Fluoride | 1.35 | 2.69 | 2.1686 | 1.34 | 0.46980 | 0.221 | -0.792 | 0.010 |
| Iron | 0.01 | 0.48 | 0.2143 | 0.47 | 0.16861 | 0.028 | 0.496 | -0.655 |
| Magnesium | 0.11 | 0.79 | 0.3586 | 0.68 | 0.23227 | 0.054 | 1.003 | 1.127 |
| Manganese | 0.00 | 0.07 | 0.0256 | 0.07 | 0.02789 | 0.001 | 1.034 | -0.737 |
| Nickel | 0.00 | 0.09 | 0.0414 | 0.09 | 0.03532 | 0.001 | 0.344 | -1.788 |
| Nitrate | 0.00 | 42.00 | 26.6714 | 42.00 | 14.71255 | 216.459 | -1.088 | 0.592 |
| Nitrite | 0.03 | 0.17 | 0.0943 | 0.14 | 0.05442 | 0.003 | 0.229 | -1.672 |
| pH | 6.71 | 7.23 | 6.9357 | 0.52 | 0.20024 | 0.040 | 0.835 | -0.924 |
| Potassium | 0.41 | 18.51 | 8.6257 | 18.10 | 6.87857 | 47.315 | 0.297 | -1.503 |
| Sodium | 0.14 | 9.68 | 2.2086 | 9.54 | 3.41371 | 11.653 | 2.297 | 5.473 |
| Sulphate | 0.00 | 60.60 | 25.4600 | 60.60 | 21.38724 | 457.414 | 0.668 | -0.526 |
| T.Hardness | 205.00 | 920.00 | 5.2929E2 | 715.00 | 273.58345 | 7.485E4 | 0.214 | -1.663 |
| TDS | 81.30 | 1108.00 | 4.6116E2 | 1026.70 | 350.05770 | 1.225E5 | 0.917 | 1.304 |
| Zinc | 0.00 | 0.81 | 0.3214 | 0.81 | 0.36921 | 0.136 | 0.503 | -2.283 |

**Table 2. Calculated Water Quality Index for Groundwater in Ogbomosho area**

| Parameters | Mean ($C_i$) | NSDWQ ($S_i$) | $C_i/S_i$ | $q_i$ | $W_i$ | $q_i.w_i$ |
|---|---|---|---|---|---|---|
| pH | 6.936 | 6.50-8.50 | 0.925 | 92.48 | 0.133 | 12.299 |
| Conductivity | 675.571 | 1000.00 | 0.676 | 67.557 | 0.001 | 0.0676 |
| TDS | 461.157 | 500.00 | 0.922 | 92.231 | 0.002 | 0.1845 |
| Colour | 0.324 | 15.00 | 0.022 | 2.16 | 0.0667 | 0.14407 |
| Alkalinity | 44.44 | 200.00 | 0.222 | 22.22 | 0.005 | 0.1111 |
| Calcium | 181.428 | 200.00 | 0.907 | 90.714 | 0.005 | 0.4536 |
| Chloride | 0.0457 | 0.30 | 0.152 | 15.233 | 3.33 | 50.726 |
| Chromium | 0.0743 | 0.05 | 1.486 | 148.600 | 20.00 | 2972.00 |
| Copper | 0.417 | 1.00 | 0.417 | 41.700 | 1.00 | 41.700 |
| Fluoride | 2.168 | 1.50 | 1.445 | 144.533 | 0.667 | 96.4035 |
| Calcium | 337.57 | 200.00 | 1.688 | 168.785 | 0.005 | 0.8439 |
| Magnesium | 0.358 | 200.00 | 0.0018 | 0.179 | 0.005 | 0.000895 |
| Iron | 0.214 | 0.30 | 0.713 | 71.333 | 3.33 | 237.539 |
| Manganese | 0.0255 | 0.20 | 0.128 | 12.75 | 5.00 | 63.75 |
| Nickel | 0.0414 | 0.02 | 2.070 | 207.00 | 50.00 | 10350.00 |
| Nitrite | 0.0942 | 0.20 | 0.471 | 47.10 | 5.00 | 235.50 |
| Potassium | 8.625 | 100.00 | 0.086 | 8.625 | 0.01 | 0.0863 |
| Sulfate | 25.46 | 100.00 | 0.255 | 25.46 | 0.01 | 0.2546 |
| Zinc | 0.321 | 3.00 | 0.107 | 10.7 | 0.33 | 3.531 |
| Carbonate | 22.885 | 250.00 | 0.0915 | 9.154 | 0.004 | 0.0366 |
| T. Carbonate | 64.071 | 250.00 | 0.256 | 25.628 | 0.004 | 0.1025 |
| T. Hardness | 529.285 | 500.00 | 1.059 | 105.857 | 0.002 | 0.2117 |
| Sodium | 2.20 | 200.00 | 0.0104 | 1.1035 | 0.005 | 0.005578 |
| Cadmium | 0.003 | 0.003 | 1.00 | 100 | 0.01 | 1.00 |
| Nitrate | 26.671 | 50.00 | 0.533 | 53.342 | 0.02 | 1.067 |
|  |  |  |  |  | 88.945 | 14068.018 |

**Table 3. Classification of groundwater from the hand-dug wells based on WQI**

| WQI Value | Category | Water Samples (%) |
|---|---|---|
| <50 | Excellent | 11 |
| 50-100 | Good water | 29 |
| 100-200 | Poor water | 25 |
| 200-300 | Very poor water | 20 |
| >300 | Unsuitable for drinking | 15 |

**Table 4. Factor loading of the Groundwater data after Varimax rotation**

| Parameters | \multicolumn{6}{c}{Factors} | | | | | |
|---|---|---|---|---|---|---|
|  | 1 | 2 | 3 | 4 | 5 | 6 |
| pH | .228 | .050 | .553 | .681 | .127 | .599 |
| Conductivity | .746 | .119 | -.560 | -.022 | .292 | .176 |
| TDS | .799 | -.018 | -.195 | .081 | .523 | .209 |
| Colour | -.093 | -.163 | -.385 | .570 | .767 | -.083 |
| Alkalinity | .822 | .451 | -.004 | -.172 | -.154 | -.260 |
| Calcium | .807 | -.291 | .503 | .023 | .095 | .038 |
| Chromium | -.436 | -.617 | .364 | .521 | .009 | .162 |
| Copper | .418 | .616 | -.328 | .419 | .270 | .299 |
| Fluoride | -.360 | .187 | .574 | .430 | .062 | .650 |
| Calc.Hardness | .864 | -.170 | .425 | -.131 | -.158 | .031 |
| Magnesium | .827 | -.057 | .319 | -.233 | -.388 | .075 |
| Iron | -.729 | .131 | .568 | .075 | .130 | .326 |
| Manganese | -.201 | -.753 | .298 | .583 | .123 | -.236 |
| Nickel | -.432 | .100 | .726 | -.337 | .401 | .056 |
| Nitrate | .456 | .539 | .398 | -.038 | .511 | -.283 |
| Nitrite | -.103 | .369 | .458 | -.251 | .567 | .409 |
| Potassium | .482 | .822 | .203 | .119 | .189 | .030 |
| Sulphate | .545 | -.771 | -.407 | .158 | .127 | .044 |
| Zinc | -.577 | .788 | -.065 | -.097 | .179 | .016 |
| Carbonate | .265 | .518 | .592 | .518 | -.206 | .001 |
| Bicarbonae | .849 | .265 | .288 | -.093 | -.211 | -.270 |
| Cadmium | -.362 | -.285 | .708 | -.535 | .024 | -.018 |
| T.Hardness | .859 | -.197 | .442 | -.119 | -.115 | .012 |
| Sodium | .234 | .758 | .071 | .600 | -.071 | -.029 |

The overall water quality index using the formula:

$$WQI = \frac{\sum qwi}{\sum wi} = 14068.018/88.945 = 158.165$$

corresponding to poor water (Table 3). The following quality parameters (calcium, chromium, copper, fluoride, iron, manganese, nickel and zinc) exceed their respective maximum permissible limit in some locations and may be the reason for the poor quality of the groundwater system in the area. The enrichment of these parameters in the groundwater system led to the observed high conductivity and total dissolved solid.

Factor analysis was applied to dataset and it generated six significant factors (Eigenvalues >1) which explained 87.3% of the variance in datasets and this suggests six different sources of pollution. The first factor consists of calcium, magnesium, sulphate, bicarbonate, alkalinity, total hardness, conductivity and total dissolved solid (Table 4) which accounts for 28.2% of the total variance. The enrichment of these elements in the groundwater system can be attributed to bedrock dissolution, weathering and rock/water interaction processes. Hardness of water is caused by calcium and magnesium ions and can be tied to bedrock geochemistry. The major ions are responsible for the high conductivity and total dissolved solid of the groundwater system.

Factor 2 explains 20.5% of the total variance and it includes potassium, nitrate, copper, zinc, carbonate and sodium and their dominance in groundwater is related to process of aquifer recharge mechanism as well rock/water interaction. Factor 3 has a high loading from pH, calcium, iron, nickel, carbonate, cadmium, fluoride (Table 4) and constitutes 14.6% of the total variance. Iron is one of the most abundant metals in the earth's crust and an essential element in human nutrition. Estimates of minimum daily requirement for iron depend on age, sex, physiological status and iron bioavailability. Excessive iron in the body does not present any health hazard, only the turbidity, taste and appearance of the drinking water will usually be affected (Amadi et al., 2010).

Factor 4 has a moderate loading 10.9% and comprises of pH, colour, chromium, manganese, carbonate and sodium. Factors 3 and 4 consist of mostly pH, carbonate and heavy metals. The dissolution of the host-rock accounts for the abundance of carbonate and heavy metals in the groundwater system in the area. When compared with the Nigerian Standard for Drinking Water Quality and World Health Organization (NSDWQ, 2007; WHO, 2006) the concentration of heavy metals such as iron, nickel, copper, and cadmium were slightly higher than the recommended maximum permissible limit in some location and is purely due to geogenic influence. Slightly

acidic water favours rapid reaction leading to chemical weathering and release of ions into the groundwater system. Factor 5 accounts for 7.8% with TDS nitrate, nitrite and colour. Nitrate pollution of groundwater system is an indication of urban pollution and may be attributed to fertilizer application as well as leachate from dumpsites and soakaways. Factor 6 has a low loading of 5.3% coming from pH and fluoride. The scree plot of the factors is shown in Figure 5.

Fluoride content in groundwater of the area ranged from 1.35 mg/l to 2.69 mg/l with a mean value of 2.16 mg/l (Table 1) and mean value is higher than the maximum permissible limit of 1.5 mg/l (WHO, 2006; NSDWQ, 2007). Fluorite, a hydrothermal mineral in granite and due to its fast dissolution kinetics, is probably the source of fluoride in the groundwater in the area. This implies that fluoride-rich groundwater in the area emanates from the granite aquifers (Figure 3). High concentration of fluoride in ground water causes a disease known fluorosis which affects mainly the teeth and bones (Chidambaram et al., 2003, Amadi et al., 2013). The outcomes of this investigated is targeted to serves as reference points and baseline information for metallic and fluoride contamination of groundwater system in Ogbomosho area of Southwest Nigeria. These findings suggest that the enrichment factors of the ions on the groundwater are of geogenic mean and related to the local geology of the area. The alkaline pH and high bicarbonate are responsible for release of fluoride-bearing minerals into groundwater (Chae et al., 2007).

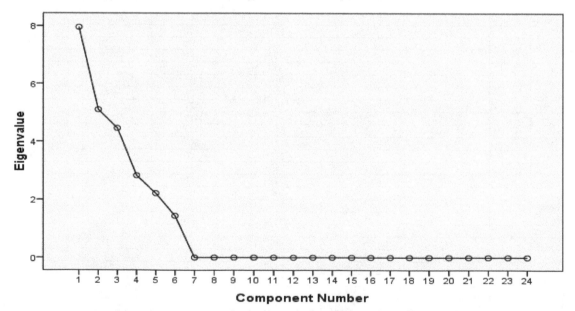

**Figure 5.** Factor Scree Plot

## 3.1. Piper and Stiff Diagram

The concentration of 8 major ions (Na$^+$, K$^+$, Mg$^{2+}$, Ca$^{2+}$, Cl$^-$, CO$_3$, HCO$_3$ and SO$_4$) are represented on the Piper trilinear diagram (Figure 6) by grouping the (K$^+$ with Na$^+$) and the (CO$_3$ with HCO$_3$), thus reducing the number of parameters for plotting to 6. On the piper diagram, the relative concentration of the cations and anions are plotted in the lower triangles, and the resulting two points are extended into the central field to represent the total ion

concentration. The degree of mixing between waters can also be shown on the piper diagram (Figure 5). The Piper diagram is used to classify the hydrochemical facies of the water samples according to their dominant ions. The water in the area is Calcium-Bicarbonate type and it can be attributed to the outcome of the rock/water interaction in the area. This is a reflection of the wide range and high standard deviation and variance observed in the ionic concentration of some parameters in the dataset (Table 1).

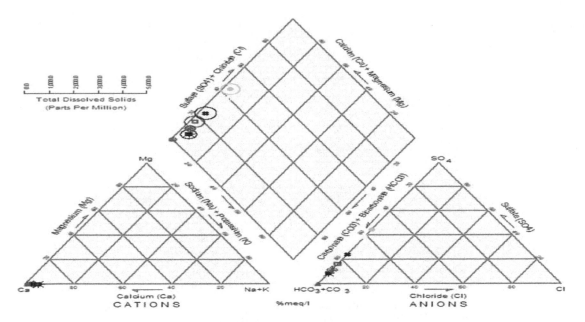

**Figure 6.** Piper diagram of groundwater in the area

Stiff diagrams are plotted for individual samples as a method of graphically comparing the concentration of major anions and cations for several individual samples (Figure 7). The shape formed by the Stiff diagrams will quickly identify samples that have similar compositions and are particularly useful when used as map symbols to show the geographic location of different water facies.

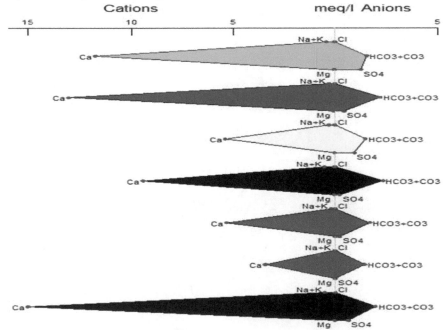

**Figure 7.** Stiff plots of groundwater in the area

## 4. Conclusions

The result of the multivariate statistical analysis, as applied to the hydrochemical data set in Ogbomosho, southwest Nigeria, provides an insight into the underlying factors controlling groundwater hydrogeochemical processes. The observed wide range, high standard deviation and variance in some of the parameters are indications that there are substantial differences in the groundwater quality within the study area. The WQI value was 158.16 which classify the groundwater in the area as poor in quality. The high value of WQI obtained was due to the high concentration of fluoride and trace elements in the groundwater and their presence can be attributed to both natural and anthropogenic sources. The dissolution of these elements in water accounts for the observed high conductivity and total dissolved solid. Factors analysis reduces the dataset into six major components representing the different sources of the contamination. Contributors of factors 1 to 4 and 6 are lithogenic/natural phenomenon while factor 4 is anthropogenic in origin. The water in the area is Calcium-Bicarbonate type from Piper and Stiff diagrams. The effectiveness of multivariate statistical analysis in groundwater quality studies have been demonstrated in this study.

# Reference

[1] Amadi, A. N. (2011). Assessing the Effects of Aladimma Dumpsite on Soil and Groundwater Using Water Quality Index and Factor Analysis. Australian Journal of Basic and Applied Sciences, 2011, 5 (11): 763-770.

[2] Amadi, A. N., Olasehinde, P.I. Yisa, J. (2010). Characterization of Groundwater Chemistry in the Coastal plain-sand Aquifer of Owerri using Factor Analysis. Int. J. Phys. Sci., 5 (8): 1306-1314.

[3] Abdullah, M. H. and Aris, A. Z. (2007). Groundwater quality of Sipadan Island, Sabah: Revisited-2004. Proceedings of the 2nd Regional Symposium on Environment and Natural Resources, pp: 254-257.

[4] Amadi, A. N., (2010). Effects of urbanization on groundwater quality: A case study of Port-Harcourt, Southern Nigeria. Natur. Appl. Sci. J., ISSN: 1119-9296; www.naasjournal-ng.org., 11 (2): 143-152.

[5] Amadi, A. N., Nwankwoala, H. O., Olasehinde, P. I., Okoye, N. O., Okunlola, I. A. and Alkali, Y. B., (2012). Investigation of aquifer quality in Bonny Island, Eastern Niger Delta, Nigeria using geophysical and geochemical techniques. *Journal of Emerging Trends in Engineering and Apllied Sciences, 3* (1), 180-184.

[6] Amadi, A.N., Olasehinde, P.I., Yisa, J., Okosun, E.A., Nwankwoala, H.O.and Alkali, Y. B. (2013). Geostatistical Assessment of Groundwater Quality from Coastal Aquifers of Eastern Niger Delta, Nigeria. *Geosciences, 2 (3); 51-59.*

[7] Amadi, A. N., Olasehinde, P. I., Dan-Hassan, M. A, Okoye, N. O. and Ezeagu, G. G., (2014). Hydrochemical Facies Classification and Groundwater Quality Studies in Eastern Niger Delta, Nigeria.

*International Journal of Engineering Research and Development,* 10 (3), 01-09.

[8] *Aminu, T. and Amadi A. N., (2014). Fluoride Contamination of shallow Groundwater in parts of Zango Local Government Area of Katsina State, Northwest Nigeria.* Journal of Geosciences and Geomatics, 2 (4), 178-184.

[9] APHA, (1995). Standards methods for the examination of water and wastewater. 19th Ed. American water works association, water environmental federation, Washington DC.

[10] Chae, G.T, Seong, T.M., Kim, Bernhard, K., Kyoung-Ho, K. and Seong-Yong, K., (2007). Fluorine geochemistry in bedrock groundwater in the water-rock interaction and hydrologic mixing in Pocheon SPA area, South Korea, Total Environment, 385 (1-3), 272-283.

[11] Lambarkis, N., V. Antonakos and G. Panagopoulos, (2004). The use of multi-component statistical analysis in hydrogeological environmental research. Water Resources, 38 (7), 1862-1872.

[12] NSDWQ, (2007). Nigerian Standard for Drinking Water Quality. Nigerian Industrial Standard, NIS., 554: 13-14.

[13] Olobaniyi, S.B. and. Owoyemi, F.B., (2006). Characterization by factor analysis of the chemical facies of groundwater in the deltaic plain-sands aquifer of Warri, Western Niger Delta, Nigeria. African Jour. Of Science and Tech., 7 (1), 73-81.

[14] Praus, P., (2005). Water quality assessment using SVD-based principal component analysis of hydrogeological data. Water SA., 31: 417-422.

[15] SPSS-16, (2009). Statistical Package for the Social Sciences. SPSS Inc. Chicago, USA

[16] World Health Organisation., (2006). Guideline for Drinking Water Quality, Geneva. (3rd ed.), Geneva, 346-385.

[17] Yisa, J. and T. Jimoh, (2010). Analytical studies on water quality index of River Landzu. Am. J. Applied Sci.

# PERMISSIONS

# LIST OF CONTRIBUTORS

**Dhananjai Verma, Ashutosh Kainthola, S S Gupte and T N Singh**
Department of Earth Sciences, Indian Institute of Technology Bombay, Mumbai, India

**S.P. Pradhan and T. N. Singh**
Department of Earth Sciences, Indian Institute of Technology Bombay, Mumbai, India

**V.K. Singh and V. Vishal**
Mine Fire Division, CSIR-Central Institute of Mine and Fuel Research, Dhanbad, India

**H. H. M. Darweesh**
Refractories, Ceramics and Building Materials Department, Egypt

**M. R. Abo El-Suoud**
Botany Department, National Research Centre, Egypt

**Viktor Boiko**
Department of Geobuilding and Mining Technologies, Institute of Energy Saving and Energy Management, National Technical University of Ukraine "Kiev Polytechnic Institute", Kyiv, Ukraine

**H. H. M. Darweesh**
Refractories, Ceramics and Building Materials Dept. National Research Centre, Cairo, Egypt

**M. G. El-Meligy**
Cellulose and Paper Dept., National Research Centre, Cairo, Egypt

**Kaveh Khaksar**
Institute of Scientific Applied Higher Education of Jihad-e-Agriculture, Education and Extension Organization, Ministry of Agriculture, Department of Soil Science, Karaj, Iran

**Mohammad Masudul Alam**
Department of Petroleum and Mineral Resources Engineering, Bangladesh University of Engineering Technology, Dhaka, Bangladesh

**Mir Raisul Islam and Md. Ashraful Islam Khan**
Department of Petroleum and Mining Engineering, Shahjalal University of Science and Technology, Sylhet, Bangladesh

**Kaveh Khaksar**
Institute of Scientific Applied Higher Education of Jihad-e-Agriculture, Department of Soil Science, Education and Extension Organization, Ministry of Agriculture, Karaj, Iran

**Keyvan Khaksar**
Faculty of Basic Sciences, Qom Branch, Islamic Azad University, Qom, Iran

**Saeid Haghighi**
Department of Soil Sciences, Rudehen Branch, Islamic Azad University, Rudehen, Iran

**Junjie Chen and Deguang Xu**
School of Mechanical and Power Engineering, Henan Polytechnic University, Jiaozuo, China

**C. I. Nwoye**
Department of Metallurgical and Materials Engineering, Nnamdi Azikiwe University, Awka, Nigeria

**E. C. Chinwuko**
Department of Industrial and Production Engineering, Nnamdi Azikiwe University, Awka, Nigeria

**I. E. Nwosu**
Department of Environmental Technology, Federal University of Technology, Owerri, Nigeria

**W. C. Onyia**
Department of Metallurgical and Materials Engineering, Enugu State University of Science & Technology Enugu, Nigeria

**N. I. Amalu**
Project Development Institute Enugu, Nigeria

**P. C. Nwosu**
Department of Mechanical Engineering, Federal Polytechnic, Nekede, Nigeria

**Rajesh Kumar Singh and Rajeev Kumar**
Department of Chemistry, Jagdam College, J P University, Chapra, India

**J.M. Ishaku and A.M. Abbo**
Department of Geology, School of Pure and Applied Sciences, Modibbo Adama University of Technology, PMB 2076, Yola, Nigeria

**B.A. Ankidawa**
Department of Agricultural and Environmental Engineering, School of Engineering and Engineering Technology, Modibbo Adama University of Technology, PMB 2076,Yola, Nigeria

**Devidas S. Nimaje and Debi P. Tripathy**
Department of Mining Engineering, National Institute of Technology, Rourkela, Odisha, India

**Opara A.I., Oparaku O.I., Essien A.G., Echetama H.N. and Muze N.E.**
Department of Geosciences, Federal University of Technology, PMB 1526 Owerri

**Emberga T.T.**
Department of Physics and Industrial Physics, Federal Polytechnic Nekede, Owerri

**Onyewuchi R.A.**
Department of Geology, University of Portharcourt, Choba, Rivers State

**Onwe R.M**
Department of Gelogy/Geophysics, Federal University Ndufu-Alike Ikwo, Abakaliki

**Siwei He**
CAMCE Mining and Tunneling, Lougheed Hwy, Burnaby, Canada

**Xianli Xiang and Gun Huang**
Guizhou University of Engineering Science, Bijie, Guizhou, China

**Zheldak T.A., Slesarev V.V. and Volovenko D.O.**
Department of Systems Analysis and Control, National Mining University, Dnipropetrovs'k, Ukraine

**Soh Tamehe Landry, Ganno Sylvestre and Nzenti Jean Paul**
Laboratory of Petrology and Structural Geology, University of Yaoundé I, Cameroon

**Kouankap Nono Gus Djibril**
Laboratory of Petrology and Structural Geology, University of Yaoundé I, Cameroon Department of Geology, HTTC, University of Bamenda, Cameroon

**Ngnotue Timoleon**
Department of Geology, University of Dschang, Dschang, Cameroon

**Kankeu Boniface**
Institut de Recherches Géologiques et Minières, Yaoundé, Cameroun

**Mohammad Kashem Hossen Chowdhury, Md. Ashraful Islam Khan and Mir Raisul Islam**
Department of Petroleum and Mining Engineering, Shahjalal University of Science and Technology, Sylhet, Bangladesh

**C. I. Nwoye and C. C. Nwangwu**
Department of Metallurgical and Materials Engineering, Nnamdi Azikiwe University, Awka, Nigeria

**I. Obuekwe**
Department of Metallurgical and Materials Engineering, Nnamdi Azikiwe University, Awka, Nigeria
Scientific Equipment Development Institute, Enugu, Nigeria

**C. N. Mbah and S. E. Ede**
Department of Metallurgical and Materials Engineering, Enugu State University of Science & Technology, Enugu, Nigeria

**D. D. Abubakar**
Department of Metallurgical and Materials Engineering, Nnamdi Azikiwe University, Awka, Nigeria
Ajaokuta Steel Company, Kogi State, Nigeria

**A. K. Verma**
Department of Mining Engineering, Indian School of Mines – Dhanbad-04, Jharkhand, India

**Amadi A.N.\*, Akande W. G., Okunlola I. A., Jimoh M.O. and Francis Deborah G. Department of Geology, Federal University of Technology, Minna, Nigeria**

**C. I. Nwoye and S. O. Nwakpa**
Department of Metallurgical and Materials Engineering, Nnamdi Azikiwe University Awka, Anambra State, Nigeria

**I. E. Nwosu**
Department of Environmental Technology, Federal University of Technology, Owerri, Nigeria

**N. I. Amalu**
Project Development Institute Enugu, Nigeria

**M. A. Allen**
Department of Mechanical Engineering, Micheal Okpara University, Umuahia, Abia State, Nigeria

**W. C. Onyia**
Department of Metallurgical and Materials Engineering, Enugu State University of Science &Technology, Enugu, Enugu State, Nigeria

**A. M. Sheikh and Mohammed Y. Abellah**
Department of Mechanical Engineering, South Valley University, Qena 83521, Egypt

**C. F. Malfatti, E. L. Silva, L. Moares and L. M. Antonini**
DEMET, Federal University of Rio Grande do Sul, Porto Alegre 91501-970, RS, Brazil

**C. I. Nwoye and S. O. Nwakpa**
Department of Metallurgical and Materials Engineering, Nnamdi Azikiwe University, Awka, Nigeria

**P. C. Nwosu**
Department of Mechanical Engineering, Federal Polytechnic Nekede, Nigeria

**E. C. Chinwuko**
Department of Industrial and Production Engineering, Nnamdi Azikiwe University, Awka, Nigeria

**I. E. Nwosu**
Department of Environmental Technology, Federal University of Technology, Owerri, Nigeria

**N. E. Idenyi**
Department of Industrial Physics Ebonyi State University, Abakiliki, Nigeria

**Olasehinde P. I., Amadi A. N. and Jimoh M. O.**
Department of Geology, Federal University of Technology, Minna, Nigeria

**Dan-Hassan M. A.**
Rural Water Supply and Sanitation Department, FCT Water Board, Garki, Abuja

**Okunlola I. A**
Department of Chemical and Geological Sciences, Al-Hikmah University, Ilorin, Nigeria

# Index

**A**

Agrowaste Composites, 14
Alkaline Electrolyte, 150, 153, 155
Alloy Corrosion Rate, 156-158, 160, 162
Anomalous Concentrations, 113, 119-120
Apparent Porosity, 14-15, 18, 21, 27-31
Artificial Neural Network, 6, 80, 82, 89-90
Average Grade, 105, 107-108

**B**

Bayesian Networks, 109, 112
Building Foundation, 23
Bulk Density, 14-15, 18-19, 21, 27-31

**C**

Cement Admixture, 27, 31
Cement Industry, 122, 131
Close Blasting, 23
Coal Field, 1, 6
Coal Mines, 1, 12, 49-50, 53, 55, 89, 132, 137
Coal Pillars, 132, 137
Compressive Strength, 15, 19, 21, 27-28, 30-31, 129-130, 132, 137
Construction Sector, 122, 126
Continuous Casting Machines (ccm), 109
Corrosion, 22, 57-65, 67-68, 156-162
Corrosion Protection, 64, 67-68
Corrosion Rate, 57, 59-62, 64-65, 67, 156-162
Cut-off Grade, 105, 107-108
Cyclic Voltammetry (cv), 150, 153

**D**

Decision Trees, 109-112
Dump Slope Geometry, 7

**E**

Economic Feasibility, 122
Empirical Evaluation, 126, 131
Empirical Methods, 132, 137
Environmental Effects, 33, 38, 50
Environmental Pollution, 33-34, 80
Ethanol Oxidation, 150-151, 153-155
Euler Deconvolution, 91-94, 101-104

**F**

Factor Of Safety (fos), 2, 7-8
Feasibility Study, 39-40, 122
Finite Difference Method (fdm), 7
Finite Element Approach, 1, 13
Fire Risk, 80-82, 84-86, 88-90
Flyash Utilisation, 7
Foundry Slag, 126-127, 131
Free Lime, 14, 18, 20-21

**G**

Galvanized Steel, 57-62
Geo Resources, 39
Geological Analysis, 39
Geological Interpretation, 91
Geometric Configuration, 23
Geotechnical Properties, 138, 143
Gouap-nkollo Prospect, 113-115, 117-120
Gravel Mines, 33
Greenhouse Gases, 49
Groundwater Geochemistry, 69, 78, 163
Groundwater Quality, 69, 74, 78-79, 163, 169-170
Gypsum Addition, 126-131

**H**

Hard Rock Extraction, 122
Hardening, 14, 21, 28, 44, 112
Hauling Cost, 105, 107-108
Hydrogeochemistry, 69, 78-79

**I**

Indian Coals, 80-82, 87-89, 137
Internal Dump Slope, 1-6

**L**

Lanthanum, 113, 118, 120
Lateritic Soils, 138-141
Linear Features, 91, 103

**M**

Magnetic Basement Depth, 91, 103
Manganese Content, 156, 158-162
Mechanical Properties, 1-2, 14-15, 20-21, 24, 28, 31-32, 64, 110, 130

Metallic Mineral Resources, 33
Metallogenic, 44, 46, 48
Metamorphism, 43-44, 93, 116, 121
Multilayer Perceptron (mlp), 80, 83

**N**
Numerical Modeling, 1, 56
Numerical Simulation, 7-8, 55, 137

**O**
Open Pit Mining, 105
Operational Dependence, 57, 62
Optimization, 82, 85, 93, 96, 109-110, 126-127, 130-131
Ordinary Portland Cement (opc), 14, 27-28
Ore Tonnage, 105

**P**
Percentage Deviation, 132, 135
Phosphate Fertilizer Industry, 64
Pillar Strength, 132, 137
Potentiostat, 64-65
Precambrian Rocks, 44-48
Precambrian Stratigraphy, 44
Production Process, 50, 109
Pulp White Liquor Waste, 27
Pulp White Liquor Waste (pwl), 27

**Q**
Quarry Mining Method, 122
Quaternary Sediments, 33, 35

**R**
Radial Basis Function (rbf), 80, 85
Rare Earth Elements (ree), 113
Residual Sodium Carbonate (rsc), 69, 75
Reversal Reactor, 49, 52-53, 55
Road Design, 138

**S**
Sea Water Environment, 57-59, 62, 156-158, 161-162
Sedimentary Formations, 45
Sedimentation, 35-36, 40-41, 121
Seismic Effect Prognosis, 23
Seismic Wave, 23
Slag Cement Minimum Setting Time (scmst), 126
Sodium Absorption Ratio (sar), 69
Spectral Inversion, 91-92, 100
Stability Analysis, 1-2, 6, 9, 12
Stainless Steel, 15, 28, 64-67, 116
Statistical Assessment, 79, 163
Steel Deoxidation, 109-110
Stratigraphy, 34-35, 38, 41-42, 44-45, 48
Stream Sediment Geochemical Survey, 113
Structural Weight Loss, 57-58, 62
Sustainable Energy Source, 49, 55

**T**
Total Dissolved Solid (tds), 69

**V**
Ventilation Air Methane, 49-50, 52, 55-56

Printed in the USA
CPSIA information can be obtained
at www.ICGtesting.com
JSHW051446221024
72173JS00006B/1590